Dimensional Metrology

This book provides in-depth coverage of metrology principles for students, practicing engineers, technologists and researchers.

Dimensional Metrology presents and explains mathematical principles and treatments and practical applications of metrology, with numerous chapter exercises that link theory to the solution of practical problems. Computer-based classes of dimensional metrology are covered, such as CMM-technology, areal surface measurement and X-ray computed tomography. Readers are shown how to perform and evaluate dimensional measurements and interpret the results. Measuring instruments and methods are explained so that readers can determine which one to use for specific applications.

This book aims to give both technicians and academic researchers in the field a thorough understanding of both the mathematical principles and uses and their applications. It can well act as the basis for a course series at the bachelor's and master's level for students in mechanical engineering.

Dimensional Metrology

Han Haitjema and Richard Leach

CRC Press
Taylor & Francis Group
Boca Raton London New York

CRC Press is an imprint of the
Taylor & Francis Group, an **informa** business

Designed cover image: Han Haitjema

First edition published 2026
by CRC Press
2385 NW Executive Center Drive, Suite 320, Boca Raton FL 33431

and by CRC Press
4 Park Square, Milton Park, Abingdon, Oxon, OX14 4RN

CRC Press is an imprint of Taylor & Francis Group, LLC

ISBN: 978-0-367-42092-5 (hbk)
ISBN: 978-1-041-07047-4 (pbk)
ISBN: 978-0-367-82174-6 (ebk)

DOI: 10.1201/9780367821746

Typeset in Times
by Newgen Publishing UK

Contents

Preface

This book is the reflection of courses and lectures that the authors have given over the years at different places and occasions and reflects the knowledge they gained to be relevant in the field.

The book aims to give an authoritative and in-depth treatment of the basic principles and their application, as well as an overview of the wide spectrum of the field. It gives both the technician and the academic researcher in the field a thorough treatment of both the mathematical principles and uses and their applications; enough basics both for a direct application and further investigations of the subjects are covered. Exercises are an integral part of the book that further illustrate the relevance of the theory for practical problems and circumstances.

In their journey through the subject, they greatly appreciated and valued international communities such as the European Society for Precision Engineering (Euspen) and the International Academy for Production Engineering (CIRP), which made them colleagues and friends and made them join forces to compile this book.

Acknowledgments

The authors acknowledge their employers for giving them the opportunity to explore the fascinating world of dimensional metrology. For RL: The National Physical Laboratory and University of Nottingham; for HH: the VSL, TU Eindhoven, KU Leuven and the Mitutoyo company. They want to acknowledge their mentors who showed them the way in the subject, especially the late Prof. Piet Schellekens at the Eindhoven University and Prof. Derek Chetwynd at the University of Warwick. The colleagues who helped, assisted, guided and criticized them and who were also close friends are too many to mention. Jerôme Senden and the Trescal Company are acknowledged for their permission to make illustrative photographs.

Authors

Han Haitjema is Emeritus Professor at the KU Leuven, Belgium, and a metrologist at Mitutoyo Europe. Prior to these positions, he was the director of the Mitutoyo Research Center Europe, an assistant professor at the TU-Eindhoven in the Precision Engineering group of Professor Schellekens and a scientist covering dimensional metrology at the VSL, the National Metrology Institute of the Netherlands. He loves dimensional metrology in all its aspects, from fundamental physics to quality control, and the application of mathematical and basic principles to new and existing calibration methods. He is a fellow of the CIRP, a member of the board of experts for the Dutch national standards and the editor-in-chief of the journal *Metrology*.

Richard Leach is Professor of Metrology at the University of Nottingham and grew the Manufacturing Metrology Team. Prior to this position, he was at the National Physical Laboratory from 1990 to 2014. His primary love is instrument building, from concept to final installation, and his current interests are the dimensional measurement of precision and additive manufactured structures. His research themes include the measurement of surface topography, development of methods for measuring 3D structures, uncertainty evaluation and theoretical methods to model optical instruments. During his career, he was a leader of several professional societies and a visiting professor at Loughborough University and the Harbin Institute of Technology. These days he is partially retired and focusing on an ERC advanced grant on using advanced mathematics and machine learning to increase optical instrument data processing.

1 Basic Concepts

1.1 INTRODUCTION

Dimensional metrology is both a classical aspect of manufacturing engineering and a fascinating scientific subject that develops at the forefront of technical and scientific development. It has been formally defined by the International Academy of Production Engineering (CIRP) as: 'Dimensional metrology is the science and application of geometrical measurements' (Leach et al. 2019). Dimensional metrology is the cornerstone of manufacturing technology, where it is imperative for the further development of precision engineering and production of mechanical and optical parts (Leach and Smith 2018). Dimensional metrology comprises essential parts of physics when it comes to primary standards (stabilized lasers, frequency combs) and essential parts of mechanical engineering when it comes to its everyday applications (calipers, micrometers), and it uses elementary mathematical principles (symmetry, geometrical forms, filtering, least-squares principles) in its application. The computer revolution in the recent decennia has enabled data processing and evaluation at new levels, enabling new classes of dimensional metrology, such as automated shape measurement, areal surface measurement and X-ray computed tomography. With all these aspects, dimensional metrology holds and keeps its position as a special discipline that combines physics, mathematics and mechanical engineering principles in its own unique way. In a broader context, metrology is the science and application of measurement. Within this discipline, a distinction can be made between theoretical, practical and legal aspects; see Figure 1.1. 'Measurement theory' deals with theoretical aspects, such as the basic concepts and the indication of quantities and units, matters that ensure that measurements can take place equivocally. Another important aspect of measurement theory is the analysis of errors that occur during measurement, and the quantification thereof that is necessary to evaluate an uncertainty. Other theoretical aspects are the physical and mathematical principles on which measurements are based, such as the physical description of interferometry, the compression/deflection of materials, Fourier analysis of surface topographies, least-squares fitting of a cylinder to

DOI: 10.1201/9780367821746-1

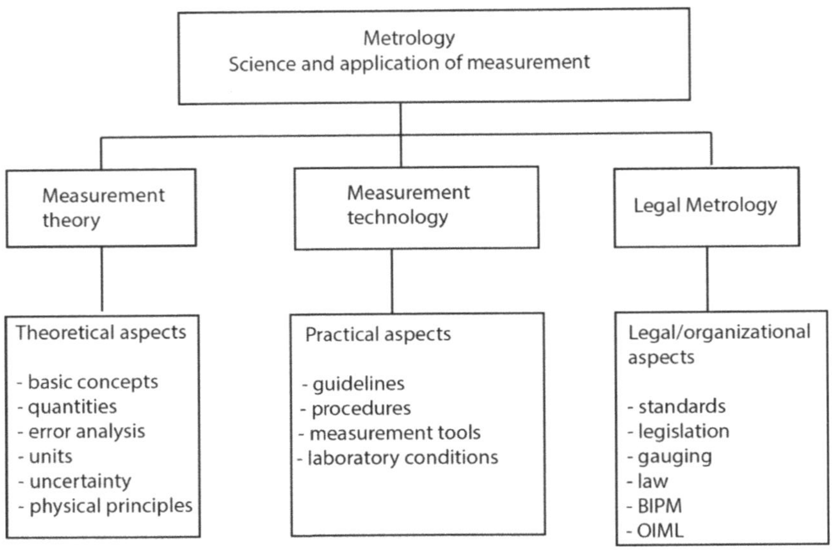

FIGURE 1.1 Various aspects of metrology.

a point cloud of measured coordinates and many other aspects. 'Measurement technology' is concerned with the practical aspects of measurement. This includes the measuring tools, instructions and procedures for the use of the measuring instruments and the execution of the measurements. Finally, 'legal metrology' deals with the organizational and legal aspects of metrology – aspects that are important to be able to make unambiguous statements about measurement results on products where such results are relevant in economic or legal transactions. Measurement technology is extensive and applies to science, industry and to daily life. This is why it is useful to make a distinction between the main areas of application. The classification presented (Figure 1.2) is not overly strict, since the areas overlap. Precision metrology is in particular the metrology that is carried out by standard measurement laboratories and specialized laboratories of universities and companies. For example, national metrology institutes (NMIs) are the National Institute of Standards and Technology (NIST) in the United States, Physikalisch-Technische Bundesanstalt (PTB) in Germany, the National Physical Laboratory (NPL) in the United Kingdom, the National Institute of Measurement (NIM) in China and the National Metrology Institute of Japan (NMIJ) in Japan. These laboratories are mainly concerned with the development of standards, the establishment of measurement methods for the realization of units and the accurate determination of the value of natural constants. 'Laboratory metrology' generally relates to research and development of the measurement technology itself. This is in contrast to 'industrial metrology', which mainly focuses on the

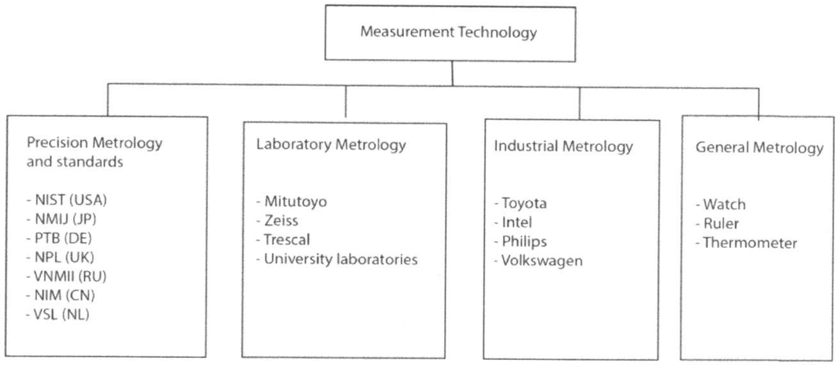

FIGURE 1.2 Application areas of metrology, with examples.

control of products and production processes. However, the distinction between these two application areas is not overly strict. An important feature of laboratory metrology is the fact that single measurements and calibrations of reference standards are often carried out in separate metrology laboratories, whereas in industrial metrology, the same aspects or products are often measured in series, near or in the production process.

1.2 METROLOGY ORGANIZATIONS

1.2.1 NMIs and the Bureau International des Poids et Mésures

Most industrialized countries have a national metrology institute, abbreviated as an NMI. NMIs are engaged in research in the area of (new) standards and units, and also ensure that these standards and units are available for science and industry. Some institutions have been privatized to a greater or lesser extent, but there is almost always a link with a national government. The Bureau International des Poids et Mésures (BIPM) is an independent international organization; its office is located near Paris and is a reminder of the significant French role in the introduction of the SI system. Where originally the BIPM was the place where the standards were maintained, the physical dissemination has stopped with the redefinition of the kilogram in 2019. The role of the BIPM has become the coordination of worldwide research into fundamental constants and comparisons at the highest level (so-called 'key comparisons'). It may seem that the presence of an NMI in almost every single country is somewhat overdone in a time of worldwide scientific cooperation, and there is certainly an element of national pride in the maintenance of such an institute in every country, but on the other hand, the revision of the SI system in 2019 could never have taken place without far-reaching cooperation and joint efforts of these institutions. The European NMIs work together in EURAMET; from there, there is a link with the European Union for the allocation of funds for joint research. An important aspect of these collaborations is the acceptance of each other's results; that means the confidence that the measured value, and its associated uncertainty as stated on reports and certificates, is correct. To this end, measurement (key) comparisons are organized and mutual visits/examinations ('peer review') take place to guarantee a general harmonized approach. In order to guarantee this further, institutions often have accreditation according to ISO 17025:2017 which guarantees not only general quality assurance aspects such as a complaint procedures, storage of original measurement results and other common administrational issues, but also technical issues such as traceability to SI units, participation to comparisons and proper uncertainty evaluations.

1.2.2 Legal Metrology and the International Organization of Legal Metrology

Metrology also has legal aspects regarding 'fair trade': supplying the appropriate amount of petrol at the filling station, the right length of fabric, the right amount of grain, etc. On the other hand, there are law enforcement issues, such as traffic speed or alcohol controls, where traceability, uncertainty and appropriate measurement are important. Usually in a country, the enforcement of the metrological laws is in the hands of the government; often it is affiliated to the NMIs. The coordinating organization worldwide is the International Organization of Legal Metrology (OIML).

1.2.3 Accreditation Organizations, Quality Assurance and the International Laboratory Accreditation Organization

Within a country, laboratories can be accredited for making certain measurements or calibrations that can be officially regarded as traceable (see Section 1.7) and that can be used as references for further measurements or comparisons. Such accreditations are granted by a national institution. Examples of such institutions are the United Kingdom Accreditation Service (UKAS) in the United Kingdom or the National Association of Testing Authorities (NATA) in Australia. At a regional level, there are, for example, the European Accreditation (EA) organization and the Asian Pacific Accreditation Cooperation (APAC). Worldwide, these organizations cooperate in the International Laboratory Accreditation Organization (ILAC). For a company or organization to obtain an accreditation, a number of matters are important. Technically, there is the requirement for competence and the correct execution and reporting of measurements. Also related to the subject of this book is the requirement for traceability and a correct method of uncertainty evaluation. Organizationally, there are requirements, for example, the registration and archiving of raw measurement data, competence of personnel, having a complaints procedure. There is also a system of mutual recognition and inspection visits. Calibration and testing laboratories are evaluated according to the standard ISO 17025:2017.

1.2.4 Normalization/Standardization Organizations

Standardization organizations are not to be confused with NMIs. Standardization institutes produce (documentary/paper) standards in which prevailing practices agreed by countries or industries are laid down. These standards are generally not legally binding, but laws can refer to them and it is good practice to adhere to them where possible. For dimensional metrology, examples are ISO 3650:1998 in which gage blocks are specified; the ISO 10360 series on coordinate measuring systems (see Chapter 8); and, last but not least, actually first, ISO 1:2022 specifying that length measurements must be carried out at, or reduced to, 20 °C. National organizations are, for example, Nederlandse Norm (NEN) in the Netherlands and Deutsche Industrie Normen (DIN) in Germany. For Europe, there is the European Committee for Standardization (CEN) and worldwide there is the International Organization for Standardization (ISO) based in Switzerland. Historically, national committees have made standards that apply within a country; however, ISO standards are increasingly being adopted by all national organizations – often translated. There is a voting system in which countries vote on the acceptance of a standard, but in practice, it is often also representatives of interested industries that determine the content of standards. Also, companies or industries can define standards and demand suppliers to adhere to them. An example of this are the VDA standards issued by the German car industry.

Most standards concerning dimensional metrology are developed in the framework of the GPS-Masterplan. Here GPS means 'Geometrical Product Specification' (not Global Positioning System). According to the GPS scheme, standards are ordered in a matrix of which the columns constitute chains of standards and the rows describe various characteristics of geometrical features. See Table 1.1. This system is known as the 'GPS matrix model', and each standard in this system contains this matrix with an indication of which elements the standard covers. For example, ISO 3650:1998 (gage blocks) covers chain 6 of size and distance.

A chain of standards consists of a set of standards related to a given characteristic. The standards are collected in a few groups – 'links' of the chain. The six chain links are the following:

1. *Product documentation indication – codification.* The standards placed in this link define drawing symbols (specific codes) and designations, establish rules of their application and explain how to read and generally understand specifications.
2. *Definition of tolerances.* The standards contain theoretical definitions of tolerances and their numerical values (as translated from code symbols). By means of these standards, theoretical exact features with associated tolerances can be defined.
3. *Definitions of characteristics of actual (real) feature.* Based on these standards, the geometry of a nonideal, real workpiece can be unambiguously defined in relation to tolerance code symbols on the drawing. The definitions are based on a set of data points of considered features.

TABLE 1.1
General GPS Matrix Layout; with the Aspects Covered by ISO 3650:1998 Indicated as an Example

Chain Link Number		1	2	3	4	5	6
	Geometrical Characteristic of Feature						
1	Size						
2	Distance						
3	Radius						
4	Angle						
5	Form of line independent of datum						
6	Form of line dependent on datum						
7	Form of surface independent of datum						
8	Form of surface dependent on datum						
9	Orientation						
10	Location						
11	Circular run-out						
12	Total run-out						
13	Datums						
14	Roughness profile						
15	Waviness profile						
16	Primary profile						
17	Surface imperfections						
18	Edges						

4. *Assessment of the workpiece deviations – comparison with specified limits.* The main task of these standards is to state how to prove conformance or nonconformance of a real workpiece with specifications, taking into account the uncertainty of the inspection procedures.
5. *Measurement equipment requirements.* The standards in this link describe characteristics (particularly metrological ones) of measurement instruments in a general manner or related to specific types of equipment.
6. *Calibration requirements – measurement standards.* These standards establish the characteristics of calibration standards used in calibration procedures of the equipment described in chain link 5.

It is good practice to refer to ISO standards in publications, measurement reports and elsewhere – where possible, indicating the publication year of the standards, because these standards may change over time. Some examples of correct references:

'The gage block length complies with class 1 according to ISO 3650:1998'.

'The roughness is Ra = (0.15 ± 0.02) µm, where Ra is defined according to ISO 21920-2:2021. The profile is filtered with a Gaussian filter according to ISO 16610-21:2011 with cut-off lengths λs = 2.5 µm and λc = 0.8 mm'.

'This coordinate measuring system has been verified according to ISO 10360-2: 2009'.

1.3 THE MEASUREMENT PROCESS

1.3.1 Definition and Scope

In the previous sections, a brief outline has been given of the area of metrology, and the science of and application of measurement. In this section, the definition, objectives and descriptions of the measurement process will be discussed. The first question that may arise is: what actually is a measurement? The answer can be found by looking at the definition in the internationally agreed Vocabulary of Metrology (BIPM 2012). The definition is:

> Measurement is the process of experimentally obtaining one or more values that can reasonably be attributed to a quantity.

Or, shorter:

> Measurement is the performance of actions for determining the value of a quantity.

By being able to express quantities in numbers, two possible objectives of a measurement can be achieved, namely:

a. finding relationships between different quantities
b. checking of realized products, values, and other numerical quantities.

By measuring certain quantities, one can try to determine relationships between these quantities and express them in laws of physics or engineering rules. As a result, it is also then possible to determine, by measuring certain quantities and using these laws, the values of other quantities that are difficult to determine directly. For example, Hooke's law that describes how the elongation Δl of a wire having a length l is determined by the force F, the cross-section A and the elastic modulus E is given by

$$\frac{\Delta l}{l} = \frac{F}{AE}. \tag{1.1}$$

By measuring Δl and the mechanical load F, E can be calculated from these measurements and the measured or the A and l. Hooke's law was once derived from measurements, now it forms the basis for the direct measurement of a material property. When manufacturing products, measurement is essential for determining whether the sizes and tolerances on a technical drawing are met. In economic transactions, it can be checked whether the proper amount of a product was delivered, for example, a liter of gasoline, a kilogram of cheese or a meter of fabric.

1.3.2 The Ideal Measurement Process

In order to be able to compare a quantity with a unit in practice, one needs a material realization of the unit, in the form of a reference standard. In practice, measuring always means: comparing a quantity to a reference standard. Mathematically, the ideal case can be represented as follows:

$$A = \{A\} \cdot [A]. \tag{1.2}$$

where: A is the quantity to be measured, $[A]$ is the unit in which A is expressed and $\{A\}$ is a ratio.

The number $\{A\}$ is thus the numerical value of the quantity A when it is expressed in the unit $[A]$. For example, $l = 4$ mm. The measurement process for the ideal case is shown in Figure 1.3.

The quantity is also shown in a materialized form as the 'object'. This has to be considered broadly: pressure and temperature are therefore objects. The comparison of this object with the standard is generally carried out by an observer with an instrument. The result is a ratio, so it has a dimensionless number. To this number the unit is added that is associated with the standard. The number associated with the standard is added to this number. The number and unit together form the measurement result. The numerical value can thus be represented as $\{A\} = A / [A]$ (for example, $4 = l$/mm). This is the most appropriate way to display numbers in tables and graphs. For example: along an axis in a graph or in the heading of a column in a table where a number of measurements of a length l are given, it is better to write 'l/mm' instead of 'l (mm)' or 'l [mm]'.

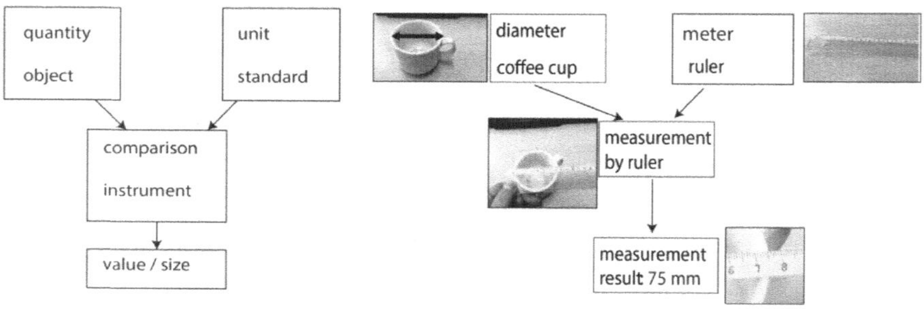

FIGURE 1.3 The ideal measurement process.

1.3.3 The Real Measurement Process

The ideal measurement process, as outlined in Section 1.3.2, does not exist in practice. Internal and external sources of interference always cause certain deviations in the determination of a quantity. These sources of interference can in principle affect all parts of the measuring process, so both the reference standard, the measuring object and the measuring instrument are subject to errors that make their exact value uncertain. A similar real measuring process is shown in Figure 1.4. The errors that occur are traditionally divided into two types:

- systematic errors
- random errors.

Systematic errors have a sign and quantity value that remain exactly the same in exactly repeated circumstances. These errors can sometimes be determined absolutely, but often this is not the case. By adding the systematic errors as a correction value (including the sign) to the measurement result, these errors can sometimes be eliminated (known as 'adjustment'). An uncertainty in a particular systematic error must always be taken into account, even after correction.

Random errors fluctuate with repetition of the measurement process, both with respect to the sign and the quantity value. The absolute value of this deviation is, therefore, not fixed for a particular measurement. A measurement result can, therefore, not be indicated as a single value but must be represented by an interval. The result of a measurement is actually the result plus a measure for the dispersion (the interval). Although it is still common to speak of systematic and random errors, for calculating uncertainties, it is more meaningful to distinguish

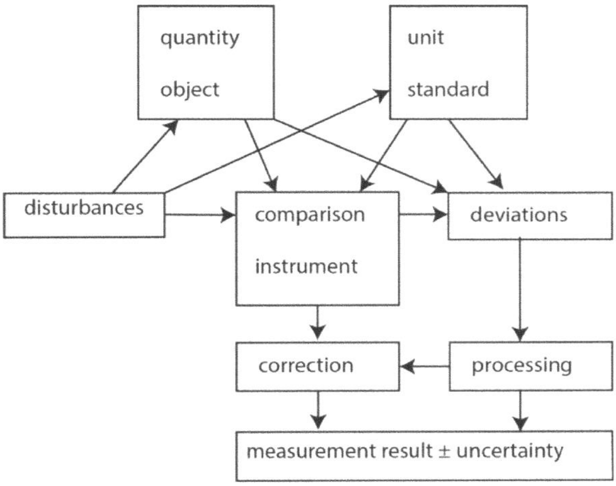

FIGURE 1.4 The real measurement process.

the manner in which the numerical values for a dispersion range are obtained. One distinguishes between a 'type A' and 'type B' approaches to uncertainty, which roughly corresponds to the distinction between random and systematic errors (but, only roughly!). In Chapter 3, evaluation of using both approaches will be discussed in more detail. Measurement is, therefore, always performed directly or indirectly using a reference standard. Since this standard is also subject to deviations, it must be calibrated against a higher standard ('higher' in this context means with less uncertainty), that is, a more precise realization of this standard. The formal definition of calibration is: 'Calibration is an operation that, under specified conditions, in a first step, establishes a relation between the *quantity values* with *measurement uncertainties* provided by *measurement standards* and corresponding *indications* with associated measurement uncertainties and, in a second step, uses this information to establish a relation for obtaining a *measurement result* from an indication'. Calibration, therefore, comes down to identifying the random and systematic errors, where the uncertainty in the systematic deviation is also taken into account. To quantify the deviations from the higher standard, these must also be calibrated against an even higher standard. Eventually, one arrives at the highest standard. If one wants certainty about the size of the deviations from the standard used, for example, a length, then this standard must be related to the highest standard via a closed chain of calibrated standards – in this case, via a reference length and a frequency, to the velocity of light in a vacuum. This is an example of traceability, the formal name for the closed chain of calibrations.

1.4 THE INTERNATIONAL SYSTEM OF UNITS

In order to be able to relate quantities to each other in an unambiguous manner and to be able to compare them, it is important that there is also unambiguity in the use of units. However, this does not come automatically. In the course of time, different systems of units have been established in many countries. This led to a lot of confusion, lack of clarity and errors, especially when it is not clear nor clearly indicated which units are being used. Even today, the use of the Anglo-Saxon 'Imperial' system (including the inch as length unit) causes much confusion; airplanes have crashed because they were refueled in liters, where gallons should have been used. This is why international agreements have been made for the use of units. The starting point is the International System of Units (SI). In the SI, a number of base and derived units are included. In addition, prefixes for the formation of decimal multiples and parts are also included. For more details and the official formulations and definitions, the reader is referred to the SI guide.

1.4.1 SI Constants and Base Units

Since May 2019, the SI has been based on fundamental constants that have a defined value; with zero uncertainty – this is a key point and a significant difference from how the SI was formed prior to May 2019. Table 1.2 gives these defining

TABLE 1.2
The Seven Defining Constants of the SI System and the Units That They Define

Defining Constant	Symbol	Numeric Value	Unit
Hyperfine transition frequency of Cesium 133	$\Delta\nu_{Cs}$	9 192 631 770	Hz
Planck constant	h	$6.626\ 070\ 15 \times 10^{-34}$ J s	J s
Speed of light in vacuum	c	299 792 458	$m\ s^{-1}$
Elementary charge	e	$1.602\ 176\ 634 \times 10^{-19}$	C
Boltzmann constant	k	$1.380\ 649 \times 10^{-23}$	$J\ K^{-1}$
Avogadro constant	N_A	$6.022\ 140\ 76 \times 10^{23}$	mol^{-1}
Luminous efficacy	K_{cd}	683	$lm\ W^{-1}$

TABLE 1.3
The SI Base Units

Base Quantity		Base SI Unit	
Name	**Typical Symbol**	**Name**	**Symbol**
Time	t	second	s
Length	l, x, r	meter	m
Mass	m	kilogram	kg
Electric current	I	ampere	A
Thermodynamic temperature	T	kelvin	K
Amount of substance	n	mole	mol
Luminous intensity	I_v	candela	cd

constants. Table 1.3 gives the base units of the SI system; these are defined based on the defining constants.

The definitions of the seven base units are slightly simplified for readability:

- The *second* is defined as the duration of 9 192 631 770 periods of the unperturbed ground-state hyperfine transition of the cesium 133 atom (since 1967);
- The *meter* is the length traveled by light in vacuum during a time interval of 1/299 792 458 of a second (since 1983);
- The *kilogram* is defined by fixing the Planck constant to be $h = 6.62\ 607\ 015 \cdot 10^{-34}$ kg $m^2\ s^{-1}$ (since 2019);
- The *ampere* is defined by fixing the elementary charge to be $e = 1.602\ 176\ 634 \cdot 10^{-19}$ A s (since 2019);

- The *kelvin* is defined by fixing the Boltzmann constant to be $k = 1.380\,649 \cdot 10^{-23}$ kg m^2 s^{-2} (since 2019);
- The *mole* is defined by fixing the Avogadro constant to be $N_A = 6.022\,140\,76 \cdot 10^{23}$. One mole contains N_A elementary entities (since 2019);
- The *candela* is the luminous intensity, in a given direction, of a source that emits monochromatic radiation of frequency $540 \cdot 10^{12}$ Hz and has a radiant intensity in that direction of $(1/K_{cd}) = (1/683)$ W/Sr (since 1979).

These definitions make little sense for practical, daily measurements. Documents have been published by the BIPM that describe how a definition can be realized; these are known as '*Mise en Practique*'. For length measurement technology, stabilized lasers are described that emit a light frequency defined within a specified uncertainty. Such a light source can then be used for measuring distances or displacements by means of interferometry and is called a 'primary standard'. The most drastic recent change is the abolition in 2019 of the kilogram as an artifact that was stored in a safe at the BIPM near Paris. Over the years, differences of the various spare copies of the kilograms were observed and a redefinition was decided upon. The primary realization of the kilogram now consists of a Watt balance: a mass is balanced with an electromechanical force. Via quantum standards for the volt and the ohm, Planck's constant remains the determining factor. In practice, masses will traditionally be compared in a way that can ultimately be traced back to a standard mass that has been linked to the natural constants. Similarly, the ampere can be realized by the combination of quantum standards for the volt and the ohm, leaving the elementary charge as the variable to be defined. The connection of all units to a fundamental physical constant is illustrated in the logo that was adopted in 2018 for the SI; it is given in Figure 1.5.

1.4.2 Derived SI Units

By combining base units, derived SI units can be constructed. Table 1.4 gives some examples of derived SI units. Derived SI units can have their own name and symbol. Some of these are listed in Table 1.5. By combining derived SI units, new derived SI units can be derived. Table 1.6 gives some examples. An SI unit can be the same for different quantities. An example is the joule per kelvin (J/K); this is both the SI unit for heat capacity and for the quantity entropy. Therefore, both the quantity and the unit must be documented when measurement results are reported in texts and on measuring instruments. There are a number of prefixes for forming decimal multiples and parts of SI units. These are shown in Table 1.7. Note that large numbers are called differently in some languages; 10^9 is commonly called 'milliard' or similar in many languages and 10^{12} is 'billion' or similar. These prefixes may not be used with degrees Celsius, arc degree, arc minute, arc second, minute, hour, day, hectare and kilogram, but they can be used with gram, for example, milligram. The use of the SI units is not voluntary. For example, in the EU, The European directive 2004/22/ EC specifies that the use of SI units

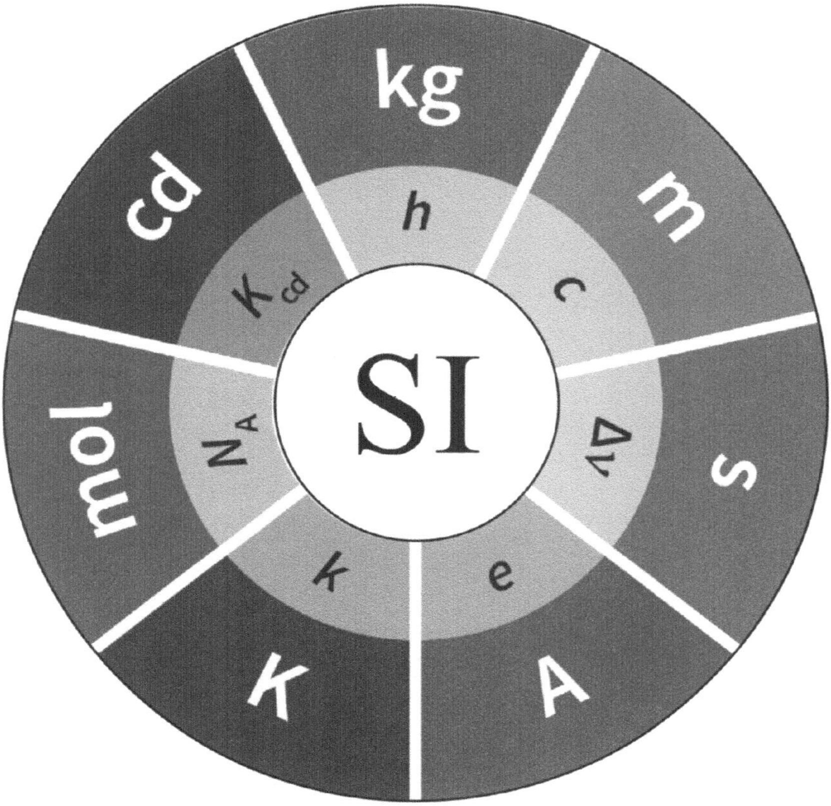

FIGURE 1.5 Logo of the SI, illustrating the relationship between the units and the fundamental constants.

in trade and legislation (for example, pricing, speeding offenses) is mandatory. Most countries legally demand the SI to be used in science and trade; however, in the USA, this is not fully implemented and even varies from state to state; also in other countries, the situation can be a little complicated, so care must be taken. Because of the importance of the proper use of recognized units in dimensional metrology, much attention is given to the SI system here. In brochures from BIPM (SI Brochure, BIPM 2019) and NMIs, this is explained in more details and with more examples.

1.4.3 Accepted Non-SI Units

Some units are so common that their use within the SI system is allowed. These are listed in Table 1.8. Units that are more or less accepted but for which use is

TABLE 1.4
Examples of Derived SI Units

Derived Quantity	Typical Symbol	Derived Unit
Area	A	m^2
Volume	V	m^3
Speed, velocity	v	$m\ s^{-1}$
Acceleration	a	$m\ s^{-2}$
Wavenumber	σ	m^{-1}
Density, mass density	ρ	$kg\ m^{-3}$
Surface density	ρ_A	$kg\ m^{-2}$
Specific volume	v	$m^3\ kg^{-1}$
Current density	j	$A\ m^{-2}$
Magnetic field strength	H	$A\ m^{-1}$
Mass concentration	ρ, γ	$kg\ m^{-3}$
Luminance	L_v	$cd\ m^{-2}$

TABLE 1.5
Derived SI Units with Their Own Name and Symbol

Quantity		SI Unit		
Name	Symbol	Name	Symbol	Expressed in Other SI (Base) Units
Plane angle	α, β	radian	rad	$M\ m^{-1}$
Solid angle	Ω,ω	steradian	sr	$m^2\ m^{-2}$
Frequency	ν, f	hertz	Hz	s^{-1}
Force	F	newton	N	$Kg\ m\ s^{-2}$
Pressure, stress	p	pascal	Pa	$N\ m^{-2} = kg\ m^{-1}\ s^{-2}$
Energy, work	E, W	joule	J	$N\ m = kg\ m^2\ s^{-2}$
Power	P	watt	W	$J\ s^{-1} = kg\ m^2\ s^{-3}$
Electric charge	Q	coulomb	C	$A\ s$
Electric potential	U, V	volt	V	$W\ A^{-1} = kg\ m^2\ s^{-3}\ A^{-1}$
Capacitance	C	farad	F	$C\ V^{-1} = kg^{-1}\ s^4\ A^2$
Electric resistance	R	ohm	Ω	$V\ A^{-1} = kg\ m^2\ s^{-3}\ A^{-2}$
Celsius temperature	t	degrees Celsius	°C	K

TABLE 1.6
Examples of Derived SI Units Whose Names and Symbols Include SI Units with Special Names and Symbols

Quantity		Derived SI Unit		
Name	**Symbol**	**Name of Unit**	**Unit**	**In Base SI Units**
Moment of force	M	newton meter	N m	m^2 kg s^{-2}
Surface tension	σ	newton per meter	N m^{-1}	Kg s^{-2}
Angular velocity	ω	radian per second	rad s^{-1}	s^{-1}

TABLE 1.7
Prefixes of SI Units

Prefix	Symbol	Factor	Name	Prefix	Symbol	Factor
quetta	Q	10^{30}	nonillion	deci	d	10^{-1}
ronna	R	10^{27}	octillion	centi	c	10^{-2}
yotta	Y	10^{24}	septillion	milli	m	10^{-3}
zetta	Z	10^{21}	sextillion	micro	μ	10^{-6}
Exa	E	10^{18}	quintillion	nano	n	10^{-9}
Peta	P	10^{15}	quadrillion	pico	p	10^{-12}
Tera	T	10^{12}	trillion	femto	f	10^{-15}
Giga	G	10^{9}	billion	atto	a	10^{-18}
mega	M	10^{6}	million	zepto	z	10^{-21}
Kilo	K	10^{3}	thousand	yocto	y	10^{-24}
hecto	H	10^{2}	hundred	ronto	r	10^{-27}
Deca	Da	10	ten	quecto	q	10^{-30}

TABLE 1.8
Accepted Non-SI Units

Name	Symbol	Value in SI Units
minute	min	1 min = 60 s
hour	h	1 h = 60 min = 3600 s
day	d	1 d = 24 h = 86 400 s
degree	°	$1° = (\pi/180)$ rad
minute	'	$1' = (1/60)° = (\pi/10\,800)$ rad
second	"	$'' = (1/60)' = (1/3600)° = (\pi/648\,000)$ rad ≈ 5 μrad
hectare	ha	1 ha = 1 hm^2 = 10^4 m^2
liter	l, L	1 L = 1 dm^3 = 10^{-3} m^3

discouraged include the are (Dutch unit for 10 m × 10 m area), bar (1 bar = 10^5 Pa), ångstrom (1 Å = 10^{-10} m = 0.1 nm) and micron (1 micron = 1 μm). In dimensional metrology, the 'micron' is still rather commonly used in daily conversations, while 'micrometer' is commonly used for a measuring instrument that should be called a screw gage.

Imperial Units
There are several systems of imperial units. These are outside the scope of the SI but are still frequently used, which can be confusing at least, life-threatening at worst. It is important to remember the inch in length: 1 in = 25.4 mm (exactly, with zero uncertainty). In the aviation industry, almost exclusively inch sizes are used, and elsewhere (screen displays, bicycle tires, screw threads), inch sizes are still implicitly or explicitly used. For pressure, the pound force per square inch (psi) is common: 1 lbf/in^2 ≈ 6.895 kPa. In addition, peculiar units exist, such as 'foot seawater' for pressure, and the resistance of wires expressed in ohm per thousand feet. Caution needs to be applied whenever non-SI units are being used.

1.4.4 Standardized Units

In science and technology, units are also in use that cannot, or cannot easily, be traced back to SI units because measurements depend on the method used. If this method is fixed in, for example, an ISO specification standard, measurements can still be comparable. An example is the measurement of hardness. Various units have been used for this purpose: Rockwell, Brinell, Vickers, Shore, IHRD and others. The quantity 300 HB (H) (50) according to ISO 6506:1981 means: in a workpiece with a hard metal ball with a force of 50 × 9.814 N an indentation is made; the ratio force/(surface area of the plastic indentation) was 300. The SI unit should, therefore, actually be pascal, but because the value depends on the force and the specific indentation body, a unit is used that implies a specific standardized measurement method.

1.4.5 Correct Use of SI Units

Below are some rules and examples of correct and incorrect use of SI units.

Decimal Marker
The decimal marker shall be either the point on the line or the comma on the line. The decimal marker chosen should be that which is customary in the language and context concerned. The point is more common in calculators, computer output, etc. Thousands may be separated by a (typically small) space, not by a period or comma. A general recommendation is to use either the point or comma consistently in a document, certificate or table.

Display of Uncertainty

The standard uncertainty that is associated with a quantity x is displayed as $u(x)$. If the length of a gage block l = 300.000 24 mm with a standard uncertainty of $u(l)$ = 0.15 µm, this can be summarized as l = (300.000 24 ± 0.000 15) mm or l = 300.00024 (15) mm. With an expanded uncertainty $U(x)$, the coverage probability p and the coverage factor k must also be documented (see Chapter 3 on uncertainties for more details on the terms used here). Uncertainties are given in to two decimal places, maximum, and the quantity is given in the corresponding decimals.

Technical Drawings

Essential for dimensional metrology is the indication of sizes on technical drawings.

Metric dimensions on drawings use a comma for the decimal marker and generally follow these rules:

1. Where the dimension is less than one millimeter, a zero precedes the comma.
2. Where the dimension is a whole number, neither the comma nor a zero are shown.
3. Where the dimension exceeds a whole number by a fraction of one millimeter, the last digit to the right of the comma is not followed by a zero (for example, 12,7 or 12,07 – not 12,070).
4. A small space separates each group of three digits either side of the decimal marker position (using a comma or point for this purpose is deprecated).

There is just one exception to the millimeter rule: surface texture. The values indicated on drawings for surface texture are not in millimeters, but in micrometers, which is one thousandth of a millimeter. Surface texture values are often below the micrometer (µm) level.

For the imperial (inch) system, these are the rules generally followed in dimensioning:

1. Normally a point is used for the decimal marker.
2. A zero is not used before the decimal point for values less than one inch.
3. A dimension is expressed to the same number of decimal places as its size limits by adding zeros to the right of the decimal point as necessary.

In the imperial system, the numbers applying to surface texture are in millionths of an inch, as in 18 which means 0.000 018 inch (or 18 µ inch).

Names of Units

Names of units are written as singular, abbreviated as symbols if numbers are used in a sentence; the abbreviation is without a point except at the end of a sentence.

There are also rules for breaking down and writing products and quotients. The examples below are for illustrative purposes. This mainly concerns spaces, uppercase letters, multiples, points and the like.

Correct:
The distance is three kilometer. The distance is 3 km. The diameter is d = 16 mm. The unit of moment is N·m. The unit of moment is N m. The acceleration is 3 m/s^2 or 3 m·s^{-2}. The volume is 51 mm × 51 mm × 45 mm. The distance is 63.2 m ± 0.1 m, either (63.2 ± 0.1) m or 63.2(1) m. The maximum allowable voltage U_{max} is U_{max} = 1000 V. The measuring range of the microscope is 225 nm to 2400 nm. The angle of the corner is 89°59'48.5" or 89,996 806° The temperature is 15 °C or 288.15 K. The mass concentration of lead is 5 ng/l. The pendulum contains an object with a mass of 5 kg. The electrical voltage is three volt.

Wrong:
The distance is three km. The distance is 3km. The distance is three kilometers. The unit of moment is Nm. The acceleration is 3 m/s/s. The content is 51 × 51 × 45 mm. The distance is 63.2 m ± 0.1. The maximum allowable voltage U is U = 1000 V_{max}. The measuring range of the microscope is 225 to 2400 nm. The angle of the corner is 89° 59' 48.5". The temperature is 15 °C or 288.15 K The mass concentration is 5 ng lead/*l*. The electrical voltage is 3 Volts.

Roman and Italic
Table 1.9 shows which symbols and units should be displayed in Roman or in italics.

TABLE 1.9
Formatting of Numbers, Quantities and Symbols

Subject	Typeface	Example
Numbers in digits	Roman	3.1415
Symbol of number	italic	*N*
Symbol of defined number	Roman	π = 3.14 e = 2.7 i, j (imaginary)
Vectors and tensors	italic, bold	***a; B***
Symbols of physical quantities	italic	*m* (mass), *C* (capacity)
Symbol of free to choose function or operator	italic	*f*(*x*); *y''*
Symbol for function or operator with fixed meaning	Roman	∂; Δ; lim; sin; Re; div
Sign for unit and prefix	Roman	m; μF
Symbols for chemistry and atomic physics	Roman	$\mathrm{H_2SO_4}$, $\mathrm{e^-}$ (electron)

TABLE 1.10
Good and Bad Practice in Table Column Headings

Bad	Bad	Good	Good
h	h	h	h/m
[10^{-7} m]	(mm)		
6.123	$6.123 \cdot 10^{-4}$	0.612 3 µm	$6.123 \cdot 10^{-7}$

Format of Table Headings
Although it is common practice to write units between parenthesis or square brackets, this is not good practice. Good practice is summarized in Table 1.10.

Graph Axes
Correct display is, for example, 'h in mm' of 'h/mm' so that the tick marks are labeled only with numbers. Sometimes one of the tick marks is replaced by the unit, which is also acceptable.

Note that the use of brackets, for example, (mm) or [mm] is widespread in publications but this is not mathematically correct and is better avoided.

1.5 CLASSIFICATION OF MEASUREMENT PROCESSES

Various classifications of measurement processes are possible, which partially overlap each other. Though it is not intended to give a comprehensive overview, some of the classifications are useful, because they provide a better insight into the operation of measuring instruments and their properties.

1.5.1 Direct–Indirect

A measurement process is considered as 'direct' if the quantity to be determined is directly compared with the corresponding reference standard. An example of this is measuring a length with a tape measure. Almost always, a measuring instrument contains some sort of conversion. This can be an amplification system or the use of an auxiliary variable. In this case, this is called an 'indirect measurement process'.

An example of indirect measurement process is measuring temperature with a mercury thermometer. As a result of the temperature increase, the mercury expands in the thermometer. Due to the difference in expansion between mercury and glass, the height of the mercury column increases, which can be read on the temperature scale. The temperature is thus determined by measuring the volume change of the mercury by reading the length change of the mercury column. Another example is the determination of mass with an equal-arm balance as depicted in Figure 1.6. To illustrate that this principle is both basic and very

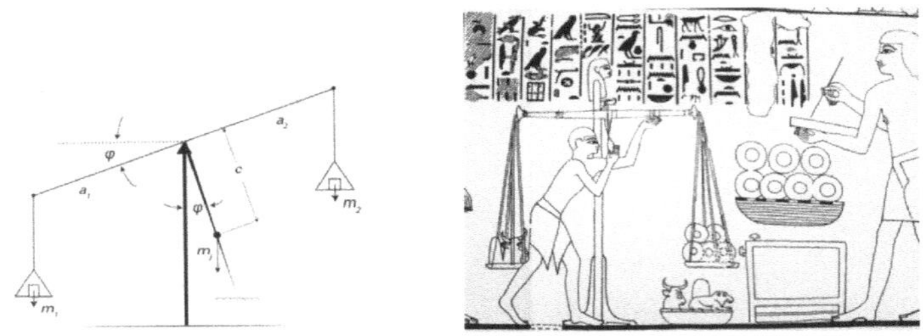

FIGURE 1.6 Mass determination using an equal-arm balance as a schematic (left) and as depicted in the tomb of Rekh-mi-re, about 1440 BC. Note the recording of the measurement result. $a_{1,2}$: arm lengths, $m_{1,2}$ mass pieces, m_j equivalent mass of dial at arm length c, φ rotation angle.

old, a corresponding figure from an Egyptian tomb is depicted on the right-hand side of the figure. Because of gravity, forces act on the masses. These forces cause a resulting torque on the yoke of the balance, causing it to turn until an equilibrium state is reached. The resulting torque is then zero. The mass difference $\Delta m = m_1 - m_2$ can be determined from the measured angle from the equilibrium equations:

$$m_1 g a_1 \cos\varphi = m_2 g a_2 \cos\varphi + m_j g c \cos\varphi, \tag{1.3}$$

where, for small angles φ, it can be written that $\sin(\varphi) \approx \varphi$ and $\cos(\varphi) \approx 1$, and with $a_1 = a_2 = a$, it follows that

$$\Delta m = m_1 - m_2 = \frac{m_j \cdot c \cdot \varphi}{a}. \tag{1.4}$$

This means that the indication angle φ is proportional to the mass difference between m_1 and m_2 and in turn that the mass difference can be determined in an indirect way by reading the angle φ.

1.5.2 Static–Dynamic

A measuring process is called static if the value of the measurand can be considered constant in the course of the measurement. A measurement, on the other hand, is dynamic if the value of the variable changes over time – this also means that the measurement is not repeatable. Examples of this are discussed in the following classifications.

Note that the terms static and dynamic refer to the measured quantities, not to the measurements themselves.

1.5.3 Analogue–Digital Measurement Instrument

The terms analogue and digital relate to the presentation of the value determined by an instrument. In the case of an analogue instrument, the value presented is a continuous function of the measured values. In the case of a digital instrument, only discrete values are displayed. Examples of analogue instruments are the aforementioned equal-arm balance or a moving coil meter. Many instruments have a digital character, especially those that display the measured values in digits on a display. Here, the measured value always 'jumps' by one digit, that is, by one step of the lowest (latest) number. Commonly, a digital caliper has a digital step of 0.01 mm and a digital micrometer (screw gage) has steps of 0.001 mm.

1.5.4 Indication and Null Measurement Method

An indication measurement determines the magnitude of a measurand because the object to be measured gives an indication, which can be read on a scale. An example of this is a moving coil meter for measuring potential difference (voltage) or a quantity that is transformed to voltage. Figure 1.7 shows such a meter. The meter consists of a rotating coil placed in a magnetic field, which is connected to a torsion spring. In the static case, if the coil with the pointer rigidly attached is in equilibrium and thus indicates a constant value, the Lorentz force caused by the current applies, thus

$$F = nBib \tag{1.5}$$

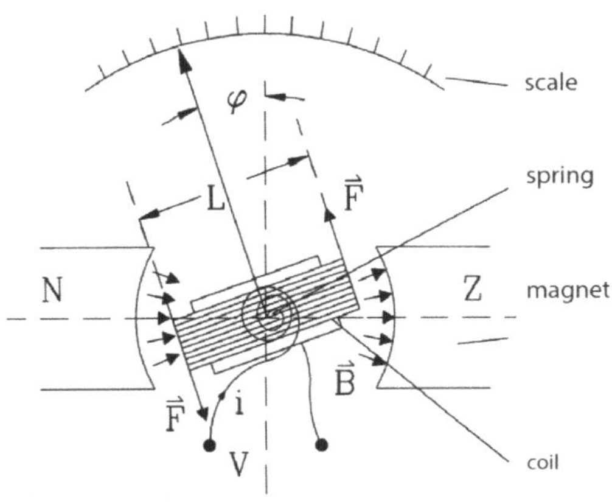

FIGURE 1.7 Schematic of a moving coil meter as an example of an indication measurement. N-Z: magnetic poles, φ: angular deviation, L: dial length, $\overline{B}$ magnetic field, $\overline{F}$ Lorentz force, I electric current.

and

$$i=\frac{V}{R_s} \tag{1.6}$$

where F is the Lorentz force, n is the number of windings of the coil, b is the diameter of the coil, B is the magnetic induction, I is the current, V is the voltage across the coil and R_s is the resistance of the coil. The Lorentz force causes a torque given by

$$M = FL, \tag{1.7}$$

where M is the torque and L is the length of coil. By rotating the coil, an opposing torque is generated:

$$M_v = -c_v \varphi, \tag{1.8}$$

where M_v is the torque, c_v is the spring constant and φ is the angular deviation.

In the static case, there is equilibrium, $\Sigma_i M_i = 0$,

$$M + M_v = 0 \tag{1.9}$$

Substitution of equations (1.7) and (1.8) into (1.9) gives:

$$FL = c_v \varphi. \tag{1.10}$$

Rewriting and substituting equations (1.5) and (1.6) then gives

$$\varphi=\frac{nBVbL}{c_v R_s}, \tag{1.11}$$

which can be written as

$$\varphi = c_t V, \tag{1.12}$$

where c_t is a constant. This means that the measurement value φ is linearly proportional to the measured voltage and the voltage can be obtained indirectly by reading the angle φ.

In this measurement, there is a start-up phenomenon because the current jumps from 0 to i at $t = 0$. This generates a torque that counters the torque generated by the Lorentz force. The resulting time-dependent moment $M(t)$ gives an angular acceleration $\frac{d^2\varphi}{dt^2}$, so the differential equation that describes the start-up phenomenon can be written as

$$M(t) = I_0 \frac{d^2\varphi}{dt^2}, \tag{1.13}$$

where I_0 is the angular moment of inertia of the coil. From this equation, the dynamic behavior of the measurement can be determined; in a similar manner to that of a damped oscillator.

In the 'null-measurement method', the value of the object to be measured is determined by means of an instrument with which the object is brought into balance, by adjusting the quantities of known values. An example of this method is the equal-arm balance of Figure 1.6. In the case that it is balanced by, for example, putting additional known masses on one side of the balance until the indication angle φ is zero.

1.5.5 Substitution and Difference Measuring Methods

In the 'substitution measurement method', the object to be measured is replaced by a reference object or objects that makes the instrument measure the same quantities; however, these reference objects have known values, such that their effect on the instrument is equal to that of the object. An example of this is Mendeleev's method of weighing (Figure 1.8). In this example, the mass m to be determined is replaced by mass pieces m_3, m_4 and m_5. These mass pieces are chosen in such a way that the balance is accurately balanced again. So,

$$m = m_3 + m_4 + m_5. \tag{1.14}$$

An important advantage of this measuring method is that significant systematic deviations in the balance, for example, if one arm of the balance is longer than the other, or frictional influences in the pivot point, have almost no influence on the result (there will still be some random error).

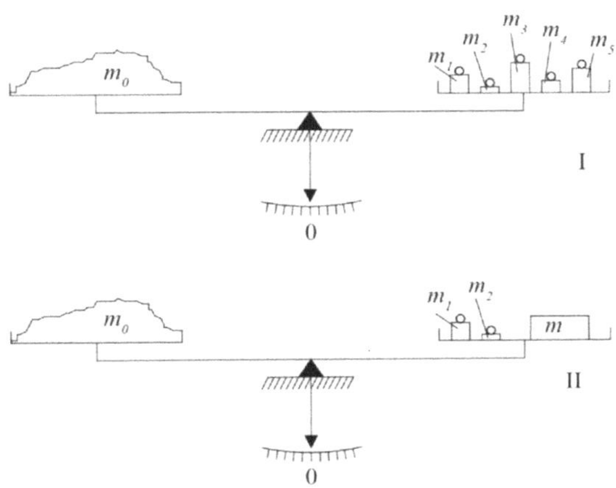

FIGURE 1.8 Weighing method of Mendeleev m_0: counter mass piece, m: unknown mass piece, $m_{1..5}$ known mass pieces.

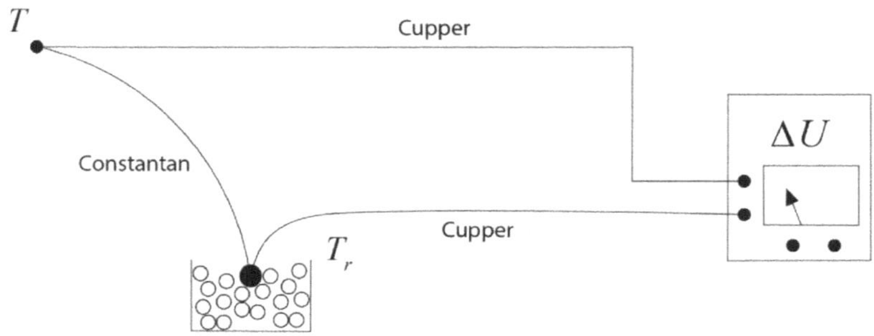

FIGURE 1.9 Thermocouple as an example of a difference measuring method. T temperature T_r reference temperature, ΔU: voltage.

In the 'difference measurement method', the object is compared with the same kind of quantity, of known magnitude which, however, differs a little from the value to be measured. The difference between the two values is measured. An example of this is the measurement of temperature with respect to a reference value (melting ice, 0 °C) with the aid of a thermocouple (see Figure 1.9). The operation of a thermocouple is based on the physical principle that if wires of different material (in this case, copper and constantan) are connected together in a circuit, and if there is a temperature difference between their ends, a potential difference between these points can be measured, thus

$$\Delta U = S\left(T - T_r\right). \tag{1.15}$$

and the temperature T can be written as

$$T = T_r + \frac{\Delta U}{S} \tag{1.16}$$

Where ΔU is the potential difference, T is the temperature to be measured, T_r is the reference temperature and S is the sensitivity constant: $S \approx 40\ \mu V\, K^{-1}$. Another example is a temperature measurement system using resistance thermometers (see Figure 1.10). Temperature measurements using resistors are based on the fact that the resistance changes when the temperature changes. In the example, four resistors are included in a so-called Wheatstone bridge arrangement. One of the resistors is temperature dependent. If the temperature changes, this resistance value also changes, which in turn results in a changing potential difference ΔU. The following equation applies to the relative difference in voltage

$$\frac{\Delta U}{U_0} \approx \frac{1}{4}\frac{dR\left(T\right)}{R_0}, \tag{1.17}$$

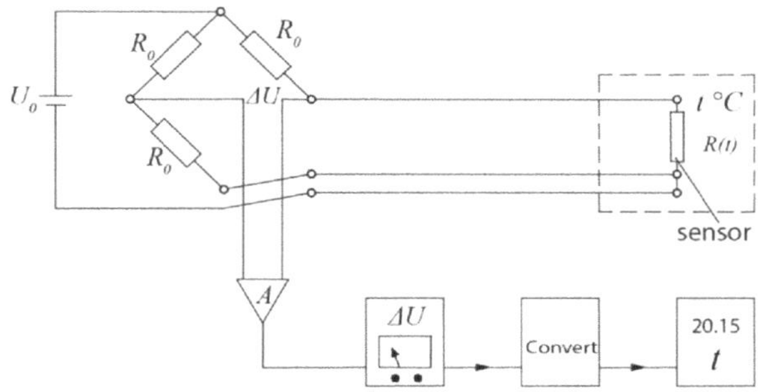

FIGURE 1.10 Temperature measurement system with resistance thermometers U_0: Voltage source, R_0: fixed, nominally equal resistors. $R(t)$: variable resistor, t temperature, A: amplifier, ΔU voltage difference.

where U_0 is the supply voltage, ΔU is the differential voltage, $dR(T)$ is the change in resistance and R_0 is the reference resistance. The relationship between resistance and temperature is usually quadratic, so that

$$R(T) = R_0\left(1 + aT + bT^2\right), \tag{1.18}$$

where a and b are constants depending on the material characteristics of the resistor. From this equation, T can be determined.

1.5.6 Reversal Methods

Methods that use an orientation change in a measurement that reverses the sign of a systematic deviation are called 'reversal methods', 'error separation' or 'absolute testing'. In the case of the nominal equal-arm balance in Figure 1.6, a difference in arm length can be corrected for by exchanging mass pieces m_1 and m_2 and repeating the measurement. In geometrical measurements of shape and angle, a reversal method, or an extension thereof, is often necessary to achieve traceability. An example is a straightness measurement compared to a reference. This is shown schematically in Figure 1.11. The process consists of measuring the distance variations between the reference and the object under test. However, the reference itself has a straightness deviation as well. Half of the difference between measurement 1 and measurement 2 gives the straightness deviation of the object in which the straightness deviation of the reference is effectively eliminated. Accordingly, it can be written that

$$Object(z) = \frac{M_1 - M_2}{2} \text{ and } Reference(z) = \frac{M_1 + M_2}{2}. \tag{1.19}$$

A comprehensive and systematic treatment of reversal and related methods in dimensional metrology is given by Evans (Evans et al. 1996).

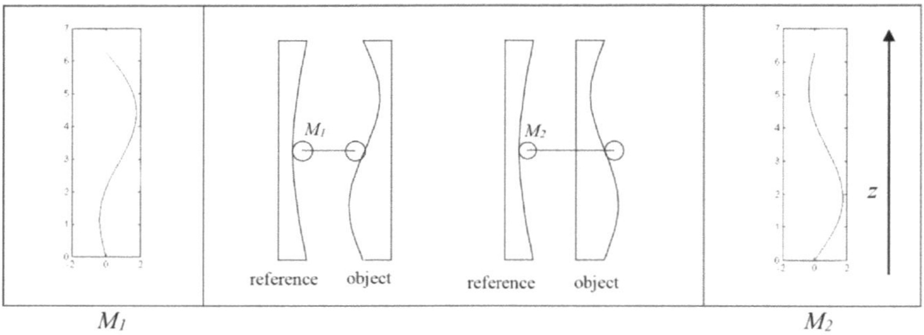

FIGURE 1.11 Schematic view of an absolute straightness measurement by a reversal method.

1.6 PROPERTIES: RESOLUTION, SENSITIVITY, AND LINEARITY

Various indications are used when specifying the behavior of a measuring instrument. The most generally used characteristics are given here with an explanation of their most usual meaning.

1.6.1 Resolution

Resolution, also referred to as resolving power or discrimination, is the smallest input signal change that is still usefully noticeable in the output signal of the measuring instrument. If with a digital readout the last decimal is stable, the last decimal unit can be used as a resolution. Care must be taken not to confuse metrological resolution and the resolution used in optics (see Sections 3.4.3 and 7.9.3.3).

1.6.2 Sensitivity

The static sensitivity of a measuring instrument can be defined as being the slope of the calibration curve that describes the relationship between the input signal and the output signal of the instrument. Referring to Figure 1.12, the static sensitivity S is defined as

$$S = \frac{\partial U}{\partial I}. \tag{1.20}$$

The static sensitivity is, therefore, not necessarily constant, but can be a function of the input signal. However, unlike dynamic sensitivity, static sensitivity is not frequency or time-dependent. The maximum measuring range $U_{\max}$ is often referred to as the 'range' or 'span'.

1.6.3 Linearity

A measurement system is linear if the relation between input and output signal can be described by

$$U = \alpha I, \tag{1.21}$$

where α is a constant value, called the 'calibration coefficient' or 'amplification coefficient'. This implies that the sensitivity is constant. Figure 1.13 shows both a linear and a nonlinear relationship between the actual input value and the indicated

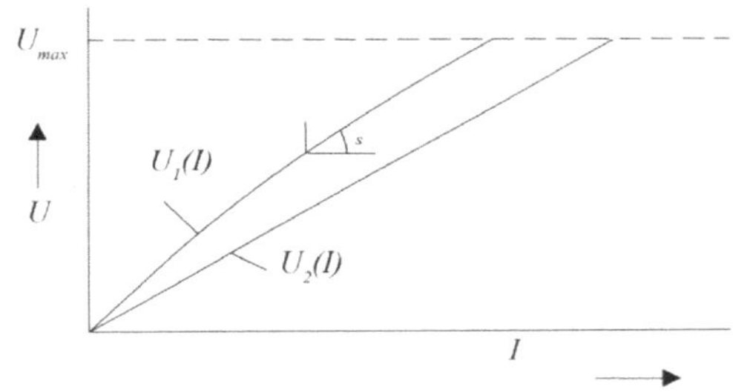

FIGURE 1.12 Relationship between input signal I and output signal U, illustrating a different sensitivity.

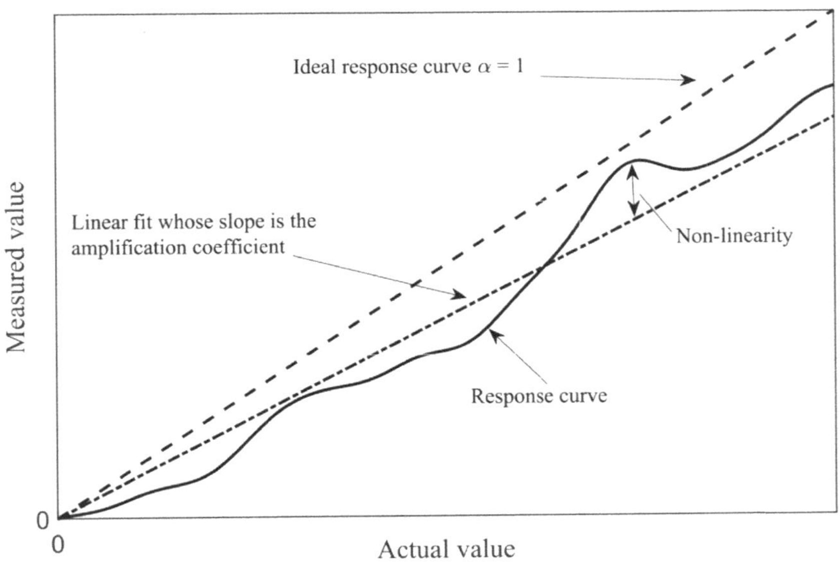

FIGURE 1.13 Illustration of nonlinearity.

measured value as the output signal. The nonlinearity of the measuring instrument is indicated as an absolute value, a percentage of the measuring range or as a percentage of the reading. A nonlinear instrument is not necessarily less accurate than a linear instrument, provided that the calibration curve is adequately applied as a correction (adjustment) value on the measured value or the output signal.

1.6.4 Threshold

The response value T is that value of the input signal, measured from zero, below which no output signal of the measuring instrument is detectable. This is shown schematically in Figure 1.14.

1.6.5 Zero Drift

Zero-point drift occurs in many measuring instruments. This means that the output signal drifts with an amount of ΔU while the input signal is $I = 0$, or stable. This situation is shown in Figure 1.15. In dimensional metrology, zero-drift is commonly caused by temperature effects or mechanical instabilities during the measurement.

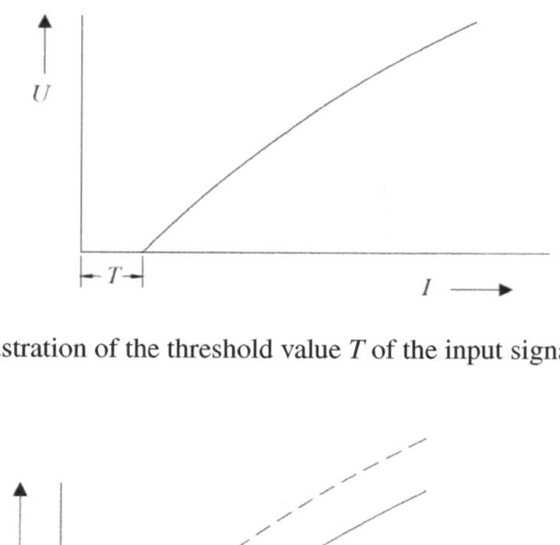

FIGURE 1.14 Illustration of the threshold value T of the input signal.

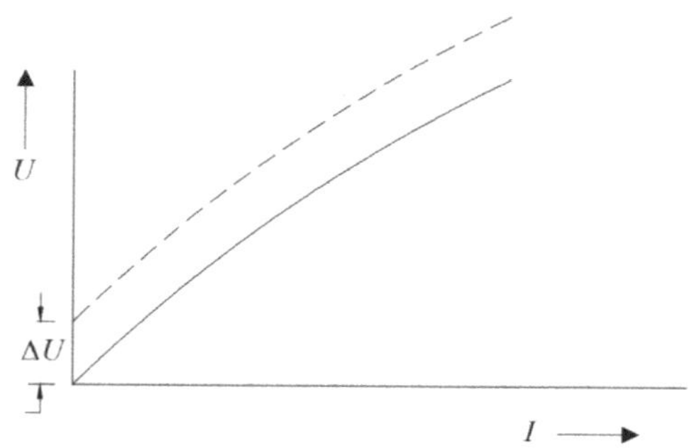

FIGURE 1.15 Illustration of the zero-point drift ΔU of the output signal.

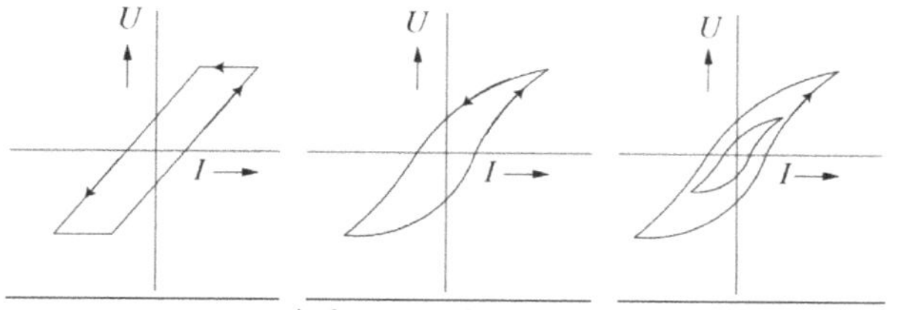

FIGURE 1.16 Illustration of different types of hysteresis.

1.6.6 Hysteresis

There are various definitions of the term hysteresis. Here, the hysteresis is defined in the output signal. Hysteresis in the output signal means that a difference ΔU in the output signal of a measuring instrument occurs at certain values I of the input signal, when the measuring range around I is moved along in a closed loop. Figure 1.16 illustrates the different ways a hysteresis can occur. The hysteresis can be constant (Figure 1.16a), a function of the input signal (Figure 1.16b) or dependent on the previous history (Figure 1.16c). The latter has to do with the size of the measurement area that is swept through, so where the zero point and the reversal point are selected. Hysteresis occurs in play, friction and different material behaviors, such as with piezoelectric and ferromagnetic materials.

1.7 UNCERTAINTY AND TRACEABILITY

Traceability ensures that a measurement can be traced back to a primary standard, via an unbroken chain of calibrations (comparisons) that always contribute some uncertainty. Simply stated: uncertainty is the interval of values within which the true value of a quantity will be located. Table 1.11 gives an example of how the traceability of a measurement using a caliper can be arranged.

The caliper is calibrated against a reference standard, in this case a 20 mm gage block of class 2 according to ISO 3650:1998. This gage block is calibrated against a class K gage block. These comparisons continue until the meter definition is reached. Temperature also plays a role: lengths are defined at 20 °C according to ISO 1:2022. The calibration of the grade K gage block against a stabilized laser and the calibration of the stabilized laser against a primary standard usually takes place in a NMI. The calibration of class 2 gage blocks against a reference, class K gage block usually takes place in a laboratory that is accredited according to ISO 17025:2017. Table 1.10 illustrates how metrology organizations, traceability, primary standards, accreditation, uncertainty calculations and ISO standards all play a role in achieving traceability. An example of the traceability of a caliper is given in Table 1.11. Uncertainty and traceability are interdependent concepts: if a

TABLE 1.11
Illustration of the Traceability Chain of a Caliper Measurement

Measured Item	Standard	Instrument	Location	Uncertainty
Manufactured part	Caliper	Caliper	Shop floor	0.1 mm
Caliper	Gage block class 2	Caliper	Measurement room	0.05 mm
Gage block class 2	Gage block class K	Gage block comparator	Accredited laboratory	0.05 μm
Gage block class K	Stabilized He-Ne laser	Gage block interferometer	NMI	0.02 μm
Stabilized He-Ne-laser	Iodine stabilized He-Ne laser	Laser frequency comparison	NMI	0.02 nm
Iodine stabilized He-Ne laser	Meter definition	BIPM documents	(universal)	2 fm

measurement is not traceable, no uncertainty can be calculated, and if one is unable to calculate an uncertainty, a measurement cannot be called traceable.

1.8 SHORT HISTORY OF THE METER

To illustrate the scientific and practical issues that can be encountered in the definition of standards and units, we give a brief history of the meter. Egypt already had a system of measures and weights, which were maintained by servants of the Pharaoh. The improper treatment of standards, for example, the failure to re-calibrate mass standards on time, was punishable by decapitation (nowadays an accreditation organization such as UKAS will issue a 'noncompliance form' if it finds nonconformance at a laboratory and threatens with withdrawal of the accreditation, a lot milder, but still a 'punishment'). The basic unit of length was the Royal Cubit, based on the length of the forearm of the Pharaoh, a standard of about 523 mm made of stone or gilded wood, divided into smaller units. Famous examples of the uncertainties reached in constructions from that time are the pyramids. For example, the pyramid of Cheops, built 2500 years BC, has the following longitudinal and angular deviations: <40 mm length deviation over 230 m; <1' (0.017°) deviation from 90°, <2.8' (0.06°) deviation from the north–south direction; <20 mm deviation from the horizontal plane. Figure 1.17 shows one of the oldest known representations of geometric measurements: an Egyptian wall

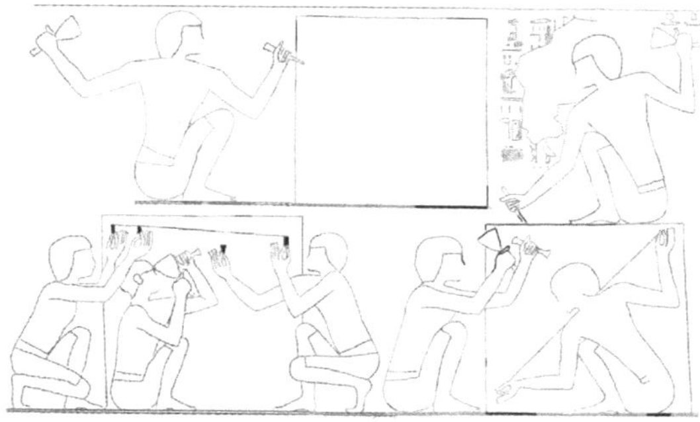

FIGURE 1.17 One of the oldest known representations of geometric measurements: Egyptian wall painting in the tomb of Rekh-mi-re in Thebes, built 1440 BC.

painting in the tomb of Rekh-mi-re in Thebes, built 1440 BC. It depicts stone cutters preparing a block for building purposes (top). A wire is tightened over the surface using two gages; with a third gauge – equal in length with the other two – the constant distance of the surface to the wire (straightness) is checked and the stone cutter adjusts the surface accordingly (left). By doing this, in several directions an overall flatness is achieved (right) (De Garies Davis 1943). Until the end of the 18th century, the longitudinal standards were based on subdivided scales, mostly derived from body measurements (thumb, inch, pouce, Zoll), elbow (el, aune, Elle), foot (voet, pied, Fuß), of leading persons of the political systems (pharaoh, emperor, king, ruler). There were also other approaches, such as the definition of an average foot. This large variety of lengths standards impeded trade. In France, it was decided to relate the length to a stable reference, not dependent on human dimensions: the length of the earth meridian. The necessary triangulation measurements between the cities of Dunkirk, Paris and Barcelona were carried out by Delambre and Méchain up to 1795. The 1/10 000 000 part of the quarter meridian was defined as the new standard of length: the meter (literally: 'he who measures'). From this meridian definition, a standard meter was derived, the so-called 'archive meter'; an 'end standard': a bar with parallel planes on both sides of which the distance between these measuring planes defined the meter. After some time, it was found that a mistake was made in the correction of the earth's ellipticity and the meter should actually be 0.2 mm shorter. For practical reasons, however, the archive meter was kept as a standard. In addition to this error, there was also some dissatisfaction about the fact that the reference was an end standard, while in practice line standards were more used. There was also a fear that the measuring surfaces would wear out due to the required reference measurements. After a few attempts, a new stable line scale was found (Pt-Ir alloy: 90% platinum, 10% iridium) produced by Johnson Matthey in London, with an X-shaped cross section to minimize torque effects, to be supported on the Airy points (see Section 2.6). During the 1st General Conference of Weights and Measures (CGPM) in 1889, the length of this international meter prototype, or to be more precise, the distance of its measurement lines at 0 °C, was defined as the length unit, traceable to the former archive meter by copying this length as well as possible. All participating countries received a copy of this meter whose deviation from the international meter prototype, colloquially called the 'meter in Paris', had to be determined periodically, a situation that still applied until 2018 for the kilogram. Around 1915, the ångström was already defined on the basis of interferometry and light wavelengths, but it was still very difficult to measure lengths up to 1 m with interferometry. It took until 1960 until the material-based definition of the length unit was replaced by a wavelength-based definition at the 11th CGPM. It was decided to use the orange line of the krypton-86 isotope ($\lambda \approx 606$ nm) as the new vacuum wavelength standard, based on stable gas discharge light sources with long coherence length, that allowed for improvements in interferometric precision. With this spectral line a gage block of 1 m length could just be measured. Based on this definition, high-precision length measurements with relative uncertainties of about 10^{-8} could be performed. In 1960, the laser was also invented, which then offered new possibilities for interferometry. It also became clear that the speed of light is highly constant within each measurable uncertainty, that is, independent of the speed of the source and of the frequency, and that the definition of the meter was the main impediment in the measurement of the speed of light. Figure 1.18 illustrates the development in the accuracy of

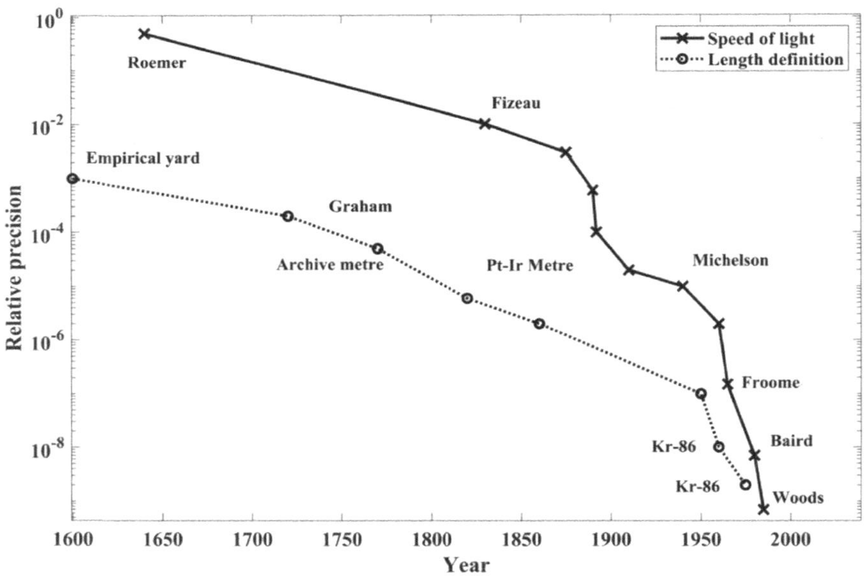

FIGURE 1.18 Development in the precision of the length measurement and in the measurement of the velocity of light.

the length standards and the velocity of light determination in the course of time. In 1983, it was decided to change the definition of the length unit again. It was decided to reverse the situation and use the best-known value of the speed of light in vacuum (c = 299 792 458 m/s) for the definition of the length unit. This definition of the unit of length can be used directly for long-distance measurements based on 'time of flight', such as, long-range metrology and astronomy ('light-years'). For applications in a normal laboratory and production environment, because time-of-flight measurements do not offer the required accuracy over smaller lengths, interferometric length measurements based on recommended stabilized light sources offer the smallest uncertainties. For high-precision length interferometry, the recommended stabilized light sources are defined in the so-called 'Mise en Pratique'. An example is the iodine-stabilized red helium-neon-, commonly abbreviated as HeNe, laser at $\lambda \approx 633$ nm, stabilized on an iodine line, with a relative uncertainty of the vacuum wavelength of $2.1 \cdot 10^{-11}$. The spectral lines of gas discharge lamps that are used in gage block interferometers as a standard are also defined and can still be regarded as primary standards. The unstabilized red HeNe laser at $\lambda \approx 633$ nm can be used as a primary standard for interferometric length measurements if its relative uncertainty of $1.5 \cdot 10^{-6}$ is sufficient for the given application, such as, the shape measurement of precision surfaces by interferometry. The method of defining a constant of nature to derive standards became a model for the redefinition of the entire SI system in 2019. In the history of the meter, some issues can be observed that are still an issue in contemporary metrology: the end standard versus the line standard; the use of material measures versus measuring with light wavelengths; and the question of temperature (now 20 °C for length measurements, but 23 °C for, for example, electrical measurements). The accuracy of length measurements is still being improved both in technology (for example, in wafer steppers for making chips) and in science. Figure 1.19 is given as an extreme example: in the detection of gravitational waves, relative length changes of 10^{-21} over a distance of 4 km are measured, which corresponds to absolute length changes of $4 \cdot 10^{-18}$ m.

1.9 METROLOGICAL DICTIONARY

In this section, some important metrological concepts are defined, based on the Vocabulaire International de Métrologie (popularly called the 'VIM') (BIPM 2012). These concepts are not intended to be memorized, but used as a reference. Also, it is good to know that terms such as 'accuracy', 'precision' and 'measurement' are defined, and lengthy discussions can be shortcut by consulting this reference. More importantly, if we are all using agreed terms and definitions, then there is much less likelihood of mistakes and differences in manufacturing, engineering, trade and other collaborative endeavors. This does not mean that all discussions are over; the VIM changes every few years, but not by much.

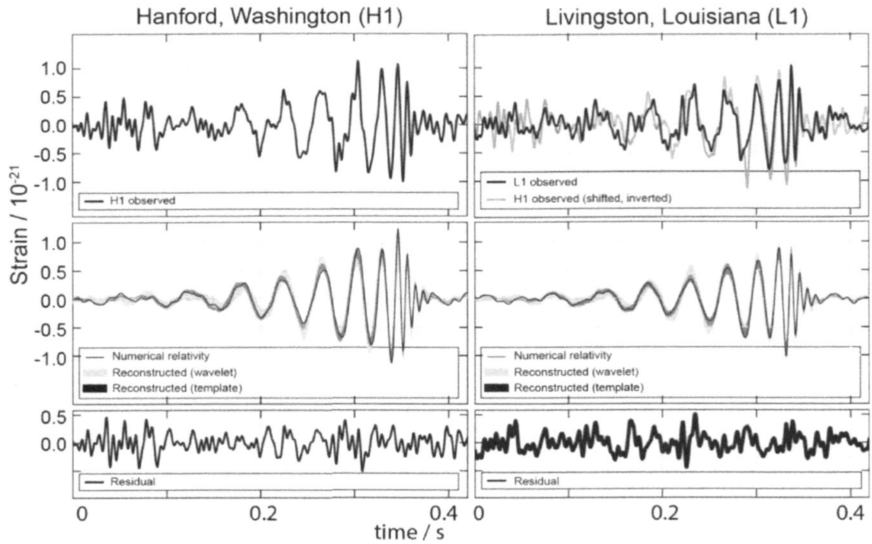

FIGURE 1.19 Interferometric measurement of a gravitational wave. The length of the horizontal scale is 0.2 s (after Abbott et al. 2017).

Measurement Accuracy
Closeness of agreement between a measured value and a true value of a measurand.

Note 1: The concept measurement accuracy is qualitative and is not given a numerical value. A measurement is said to be more accurate when it offers a smaller measurement error.

Note 2: Measurement accuracy is sometimes understood as closeness of agreement between measured quantity values that are being attributed to the measurand.

Measurement Error
Measured value minus a reference value.

Note 1: The concept of measurement error can be used when there is a reference value to refer to, which occurs if a calibration is made by means of a measurement standard with a much smaller measurement uncertainty.

Note 2: Measurement error should not be confused with production error, a failure or mistake.

Measurement Unit
Quantity, defined and adopted by convention, with which any other quantity of the same kind can be compared to express the ratio of the two quantities as a number.

Sensitivity
Quotient of the change in an indication of a measuring system and the corresponding change in a value of a quantity being measured.

Quantity
Property of a phenomenon, body or substance, where the property has a magnitude that can be expressed as a number and a reference.

Examples:

1. Quantities in a general sense: length, time, mass, temperature, resistance, molar concentration.
2. Special quantities: length of a bar, resistance of a wire.

Precision
Closeness of agreement between indications or measured values obtained by replicate measurements on the same or similar objects.

Note 1: Measurement precision is usually expressed numerically by measures of imprecision, such as standard deviation, variance or coefficient of variation.

Note 2: Measurement precision is used to define repeatability and reproducibility.

Repeatability
Precision under a number of repeatability conditions.
These repeatability conditions include:

- same measurement method;
- same observer;
- same instrument, used under same conditions;
- same location; and
- repeated measurements in a short period of time.

Reproducibility
Measurement precision under reproducibility conditions.
These reproducibility condition may include:

- different locations;
- different operators;
- different measuring systems; and
- replicate measurements on the same or similar objects.

A specification should give the changed conditions.

Metrological Traceability

Property of a measurement result whereby the result can be related to a reference through a documented chain of calibrations, each contributing to the measurement uncertainty.

Note: For this definition, a reference can be a definition of a measurement unit through its practical realization (i.e., a primary standard), a measurement procedure that includes the measurement unit (for example, an ISO standard that defines a hardness unit) or a measurement standard.

Calibration

Operation that, under specified conditions, in a first step, establishes a relation between the quantity values with measurement uncertainties provided by measurement standards and corresponding indications with associated measurement uncertainties. In a second step, this information can be used to establish a relation for obtaining a measurement result from an indication.

Note 1: A calibration may be expressed by a statement, calibration function, calibration diagram, calibration curve or calibration table. In some cases, it may consist of an additive or multiplicative correction of the indication with an associated measurement uncertainty.

Note 2: Calibration should not be confused with adjustment of a measuring system, often mistakenly called 'self-calibration', nor with verification of calibration.

Note 3: Often, the first step alone in the above definition is perceived as being calibration.

Measurement Result

Value or values being attributed to a measurand.

Note 1: A measurement result is generally expressed as a single value and a measurement uncertainty. If the measurement uncertainty is considered to be negligible for some purpose, the measurement result may be expressed as a single measured quantity value. In many fields, this is the common way of expressing a measurement result.

Note 2: A measurement result can mean an indication, an uncorrected result or a corrected result, according to the context.

Measurement Standard

Realization of the definition of a given quantity, with stated value and associated measurement uncertainty, used as a reference.

Examples:

- 1 kg mass standard;
- 100 Ω standard resistor;

- standard-ampere meter; and
- cesium-frequency standard.

Measurement
Process of experimentally obtaining one or more values that can reasonably be attributed to a quantity.

Metrology
Science of measurement and its application.

Note: Metrology includes all theoretical and practical aspects of measurement, whatever the measurement uncertainty and field of application.

Nominal Value
Rounded or approximate value of a characterizing quantity of a measuring instrument or measuring system that provides guidance for its appropriate use.

Example 1: 100 Ω as the nominal value marked on a standard resistor.
Example 2: 3.998 mm as the nominal diameter indicated on a setting ring.

Measurement Uncertainty
Nonnegative parameter characterizing the dispersion of the values being attributed to a measurand, based on the information used.

Note 1: Measurement uncertainty includes components arising from systematic effects, such as components associated with corrections and the assigned values of measurement standards, as well as the definitional uncertainty. Sometimes estimated systematic effects are not corrected for but, instead, associated measurement uncertainty components are incorporated.

Note 2: The parameter may be, for example, a standard deviation called standard measurement uncertainty (or a specified multiple of it), or the half-width of an interval, having a stated coverage probability.

Note 3: Measurement uncertainty comprises, in general, many components. Some of these may be evaluated by Type A evaluation of measurement uncertainty from the statistical distribution of the quantity values from series of measurements and can be characterized by standard deviations. The other components, which may be evaluated by Type B evaluation of measurement uncertainty, can also be characterized by standard deviations, evaluated from probability density functions based on experience or other information.

Primary Standard
A standard that has been designated or is widely recognized as a standard with the highest metrological quality and whose value is accepted without reference to other standards of the same quantity.

Examples:

1. The vacuum wavelength of the light emitted by an iodine-stabilized He-Ne laser is considered a primary length standard.
2. A pressure balance with calibrated mass pieces, a calibrated effective surface, at a location with a known gravitational acceleration, is considered a primary pressure standard.

The term *primary standard* is equally valid for base quantities and derived quantities.

Resolution
Smallest change in a quantity being measured that causes a perceptible change in the corresponding indication

Systematic Error
Component of measurement error that in replicate measurements remains constant or varies in a predictable manner

Note: Systematic measurement error, and its causes, can be known or unknown. A correction can be applied to compensate for a known systematic measurement error.

Random Error
Component of measurement error that in replicate measurements varies in an unpredictable manner.

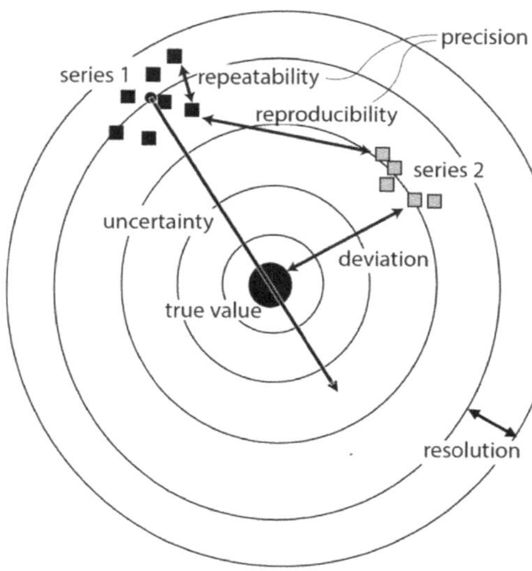

FIGURE 1.20 Schematic overview of a few metrological terms.

Note: Random measurement errors of a set of replicate measurements form a distribution that can be summarized by its expectation value, which is generally assumed to be zero, and its standard deviation. Figure 1.20 gives a schematic overview of some of these concepts.

EXERCISES

1. Which statements are correct?
   a. The kilogram is a base unit of the SI system
   b. The light intensity, expressed in cd/m^2, is a derived SI unit
   c. The atmosphere is a derived SI unit with an own name
   d. The siemens is a derived SI unit with an own name and symbol
   e. The poise is a non-SI unit for magnetic field strength
   f. 1 μmm = 10^{-9} m
   g. Comparing the length of an object with a gage block with the same nominal value can be classified as a substitution method
   h. The repeatability is equal to the reproducibility with as different condition the time at subsequent connecting time intervals
   i. Units are put after quantities in square brackets
   j. When detecting gravity waves, displacements of the order of attometers are detected.
2. Wheatstone bridge
   One of the resistors in a Wheatstone bridge changes its value. Give the sensitivity S of the measured voltage ΔU for this change.
3. Measurement method
   The method of resistance measurement with a Wheatstone bridge, in which the bridge is balanced by means of a variable resistance, is in fact:
   a. A substitution method
   b. A null method
   c. A difference method
   d. A calibration
4. Resistance thermometer
   The relationship between the resistance and the temperature of a resistance thermometer is:

   $$R(T) = R_0 + dR_0(T) = R_0(1 + a \cdot T + b \cdot T^2).$$

   Give the sensitivity of the resistance for the temperature.
5. Magnetic field
   When measuring a magnetic field of an electro-magnet, the current is slowly increased from zero to half of the maximum, the magnetic field strength is recorded, then the current is increased to the maximum and then

reduced to half of the maximum. The magnetic field strength is recorded again and it is noted that it is somewhat larger than in the first measurement. This is an example of:

a. sensitivity drift
b. zero-point drift
c. hysteresis
d. negative threshold

6. Equal-arm balance

When weighing a mass piece against a mass standard of exactly 1 kg on a (nominally) equal-arm balance, the balance indicates a difference of +5 mg. When exchanging the mass piece and the mass standard, the indication of the mass difference does not change.

From this it can be concluded that:

a. The mass piece has a mass that is 1 kg within the measurement uncertainty
b. This balance is not equal-arm and so not suitable for mass measurements
c. The mass of the mass piece is 1 kg + 5 mg
d. The mass of the mass piece is 1 kg + 1 kg + 10 mg

7. Equal-arm balance

When weighing a mass piece m against a mass standard m_r of exactly 1 kg on a (nominally) equal-arm balance, the balance indicates a difference of $M_1 = m - m_r = +5$ mg.

When exchanging the mass piece and the mass standard the indication of the mass difference changes to $M_2 = m - m_r = -11$ mg.

a. Draw the balance and give the equilibrium equation
b. Give the equations as a function of masses and arm length for both situations
c. Give the mass m
d. Give the (relative) difference in arm length

8. Straightness

Take in Figure 1.11 as vertical scale the length of the measured object in centimeters and as horizontal axis the measured values in micrometers. Calculate the form of the reference and the measured object from the first and second measurement in steps of 5 mm along the reference.

9. Linearity

A height gage is calibrated using a laser interferometer as a reference. The calibration results are obtained, as given in Table 1.12.

Determine the amplification coefficient and the nonlinearity of the height gage.

TABLE 1.12
Calibration Result of a Height Gage

Laser Interferometer Reading/mm	Height Gage Indication/mm
0	0
99.99	100
199.98	200
299.99	300
399.94	400
499.95	500
599.94	600

REFERENCES

Abbott, B.P. et al., 2017, LIGO Scientific Collaboration and Virgo Collaboration *Phys. Rev. Lett.* **118(22)**:221101.

BIPM, 2019, SI Brochure, www.bipm.org/en/publications/si-brochure accessed 2025-01-02.

BIPM, IEC, IFCC, ILAC, ISO, IUPAC, IUPAP and OIML, 2012, International Vocabulary of Metrology – Basic and General Concepts and Associated Terms (VIM 3rd edition), Bureau International des Poids et Mésures, JCGM 200 www.bipm.org/utils/common/documents/jcgm/JCGM_200_2012.pdf accessed 2025-01-02.

Evans, C.J., Hocken, R.J. and Tyler Estler, W., 1996, Reversal, Redundancy, Error separation, and 'Absolute Testing'. *Ann. CIRP* **45(2):**617–634.

Garies Davies, N. de, 1943, *The Tomb of Rekh-Mi-Re at Thebes* Vol II. New York, The Metropolitan Museum of Art – Egyptian Expedition.

ISO 1, 2022, Geometrical Product Specifications (GPS) – Standard Reference Temperature for the Specification of Geometrical and Dimensional Properties. Geneva, International Organization for Standardization.

ISO 3650, 1998, Geometrical Product Specifications (GPS) – Length Standards – Gauge Blocks. Geneva, International Organization for Standardization.

ISO 6506-1, 2014, Metallic Materials – Brinell Hardness Test – Part1: Test Method. Geneva, International Organization for Standardization.

ISO 10360-2, 2009, Geometrical Product Specifications (GPS) – Acceptance and Reverification Tests for Coordinate Measuring Machines (CMM) – Part 2: CMMs Used for Measuring Linear Dimensions. Geneva, International Organization for Standardization.

ISO 16610-21, 2011, Geometrical Product Specifications (GPS) – Filtration – Part 61: Linear Areal Filters – Gaussian Filters. Geneva, International Organization for Standardization.

ISO 17025, 2017, General Requirements for the Competence of Testing and Calibration Laboratories. Geneva, International Organization for Standardization.

ISO 21920-2, 2021, Geometrical Product Specifications (GPS) – Surface Texture: Profile Part 2: Terms, Definitions and Surface Texture Parameters. Geneva, International Organization for Standardization.

Leach, R.K., Ferrucci, M. and Haitjema, H., 2019, Dimensional Metrology. In: Chatti, S., Laperrière, L., Reinhart, G. and Tolio, T. *CIRP Encyclopaedia of Production Engineering* 2nd edition. Berlin, Springer-Verlag.

Leach, R.K. and Smith, S.T., 2018, *Basics of Precision Engineering*. Boca Raton, CRC Press.

2 General Principles of Dimensional Metrology

2.1 THE METER DEFINITION AND 'MISE EN PRATIQUE'

In dimensional metrology, the traceability to the unit of length, the meter, is of essential importance. As stated in Chapter 1, the unit length has the speed of light as its defining constant.

The speed of light in vacuum *c* is 299 792 458 meters per second. With this, the full definition of the meter is, since 2019 (BIPM 2019-1):

The meter, symbol m, is the SI unit of length. It is defined by taking the fixed numerical value of the speed of light in vacuum *c* to be 299 792 458 when expressed in the unit m s^{-1}, where the second is defined in terms of the cesium frequency $\Delta\nu_{Cs}$.

This definition implies that 'The meter is the length of the path travelled by light in vacuum during a time interval of 1/299 792 458 of a second', as it was stated in the definition of the meter that has been in place since 1983.

When using a caliper, one might wonder how a relationship with this definition can be established in daily length measurements: in a workshop, it is seldom seen that somebody tries to measure a photon passing by using a stopwatch. Therefore, in parallel to establishing the formal definitions, a document has been prepared that describe the various routes of traceability for dimensional measurements at various scales. The title of this document is the '*Mise en pratique* for the definition of the meter in the SI' (BIPM 2019-2); where the use of French refers to the original position of the BIPM, near Paris, where once the meter was maintained as a physical length bar. The *mise en pratique* distinguishes primary and secondary methods for establishing traceability, where the primary methods use, directly or indirectly, the meter definition, and the indirect methods use reference values for physical constants with an established uncertainty. A further elaboration on the practical aspects of these definitions is given by Schödel (Schödel et al. 2021) and is summarized in the next sections.

DOI: 10.1201/9780367821746-2

2.1.1 Primary Methods for the Practical Realization of the Definition of the Meter

The fundamental equation underlying the most recent definition of the meter is a direct relationship between a length, a time interval and the speed of light:

$$l = c \cdot \Delta t, \tag{2.1}$$

in which c is the fixed value for the speed of light in vacuum, $c = 299\,792\,458$ m s^{-1}, and Δt is the traveling time of the light along a geometrical path of length l. Realization of the length unit, at a primary level, is thus linked to measurement of the light traveling time; this may be achieved directly with high relative accuracy for long ranges (for example lunar ranging) but at typical macroscale ranges, the indirect traveling time measurement offers higher accuracy, due to the challenge of measuring a very short time interval. Accordingly, the definition of the meter can be realized in practice by either direct or indirect measurement of the light traveling time.

2.1.1.1 Direct Measurement of the Light Traveling Time (Time of Flight Measurement)

The direct measurement of the traveling time of light requires some form of modulation of the light to generate fiducial features used for the timing process. Such a modulation generates a superposition of light waves forming a wave packet. The path length of propagation of such a wave packet, for example, of a laser pulse, can be determined. This principle is used for long distances, for example, the distance to the moon. The attainable accuracies for length measurements in the range of typical dimensional measurements (<10 m) are too limited to be practical, so technologies based on this principle are not treated here.

2.1.1.2 Indirect Measurement of Light Traveling Time: Optical Interferometry

For the realization of lengths below a few meters, and also for the most accurate realization of length in general, interferometric techniques are preferable. Optical interferometry is a measurement method based on the interference of light. Using this method, distances or displacements l can be expressed as N multiples (interference fringes) of the used wavelength λ divided by two, and fractions thereof expressed as a phase difference $\Delta\varphi$:

$$l = \frac{\lambda}{2}\left(N + \frac{\Delta\varphi}{2\pi}\right) = \frac{c}{2fn}\left(N + \frac{\Delta\varphi}{2\pi}\right) = \frac{1}{2}\frac{c}{n}\Delta t, \tag{2.2}$$

where f is the frequency of the light and n is the refractive index of the medium in which the measurement takes place, commonly air. Knowing the frequency of the light f, which indirectly defines the traveling time in equations (2.1) and (2.2), is an essential requirement for the realization of the unit of length. For highest demands on the accuracy of the light frequency measurement, a light source can be synchronized to a primary frequency standard by, for example, a frequency comb technique (Hall 2006). A frequency comb can make the transfer from radio frequencies (RF) of approximately 9 GHz ($\lambda \approx 0.3$ m) of the primary time realization to optical frequencies of about 600 THz ($\lambda \approx 0.5$ µm, green light) that are useful for dimensional metrology. This is achieved by pulsing a laser having a wide spectrum with a frequency that is derived from a time realization. The resulting spectrum of this laser light consists of a series of peaks, the separation of which is given by the pulse frequency; this is the 'comb' that gives this technology its name. This principle is illustrated in Figure 2.1; a pulsed laser gives pulses of 'white' light that cover the visible spectrum. The spectrum, that is the Fourier transform of such a pulse train, is a set of frequencies equally spaced by the repetition frequency f_{rep} with an overall frequency shift called the carrier envelope offset frequency f_{ceo}. Both f_{rep} and f_{ceo} are typically in the RF spectral region and, in principle, can be measured with traditional electronics. Therefore, frequency combs connect

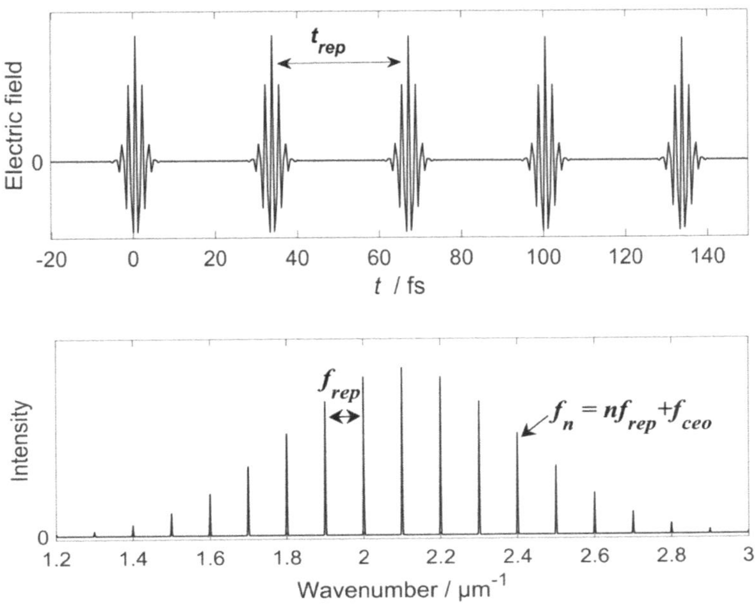

FIGURE 2.1 Relationship between the electric field of a mode-locked laser and its Fourier transform: the frequency spectrum. The time between pulses is the repetition period t_{rep} and is related to the spacing between the comb teeth f_{rep} through $f_{\text{rep}} = 1/t_{\text{rep}}$.

the optical and RF frequency ranges, since the optical frequency of the individual comb modes can be written as:

$$f_n = nf_{\text{rep}} + f_{\text{ceo}}. \tag{2.3}$$

Here, the index n denotes the comb tooth number and f_n the frequency of an individual comb mode. Equation (2.3) corresponds to a ruler in frequency space with potentially millions of tick marks. Therefore, the frequency comb can serve as a 'meter stick' in frequency space, whereby an unknown optical source – such as a stabilized laser – can be compared to the comb and its frequency determined. As an alternative to using a comb, one of the recommended values of standard frequencies can be used that are summarized in the CPM list of recommended frequency standard values (BIPM 2019-2). A few examples in decreasing uncertainty are given in the following.

EXAMPLE 1: IODINE STABILIZED LASER

When a red He-Ne iodine-stabilized laser $(\lambda \approx 633)$with an internal iodine cell is stabilized using the third harmonic detection technique to the so-called f-dip, the frequency is given by f = 473 612 353 604 kHz and the vacuum wavelength by λ = 632.991 212 58 nm, both with a relative standard uncertainty of 2.1 × 10^{-11} subject to the conditions: cell-wall temperature (25 ± 5) °C, cold-finger temperature (15.0 ± 0.2) °C and peak-to-peak modulation width (6.0 ± 0.3) MHz. A schematic of an iodine-stabilized laser is given in Figure 2.2. An iodine-stabilized laser is a common primary standard. It can be used to calibrate the frequency of laser sources as they are commonly used in displacement laser interferometer systems; see Section 4.4.

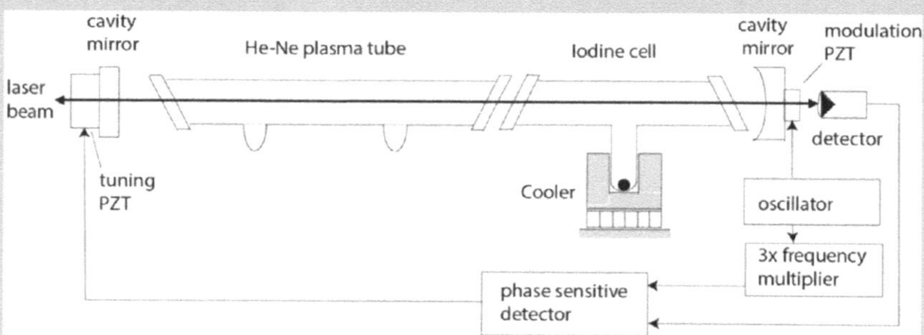

FIGURE 2.2 Typical scheme of an iodine-stabilized laser. The modulation piezo-electric transducer (PZT) gives a sinusoidal modulation to a cavity mirror that corresponds with a modulation of 6 MHz in the laser frequency. When a dip in the iodine absorption spectrum is detected the laser frequency can be stabilized to this dip using a feedback loop to tune the PZT.

EXAMPLE 2: SPECTRAL LAMP

The red spectral line of a spectral lamp based on ^{114}Cd can be assumed to have a vacuum wavelength of 644.024 8 nm with a relative standard uncertainty of 2×10^{-8}. Spectral lamps are in use for gage-block interferometers (see Section 4.6) and in spectroscopy.

EXAMPLE 3: UNSTABILIZED RED HE-NE LASER

An unstabilized 'red' He-Ne laser can be assumed to have a vacuum wavelength of $\lambda = 632.990\,8$ nm with a relative standard uncertainty of 1.5×10^{-6}.

Many instruments are operated with He-Ne lasers, where the accuracy of the wavelength is less critical, for example, Fizeau flatness interferometers (see Section 4.5.2). For users, it may be convenient to know that calibration is not necessary or useful if the laser is not stabilized and a relative uncertainty of higher than 1.5×10^{-6} is not needed for the application.

The accuracy of interferometry is limited by fringe-counting and interpolation techniques for small distances and, in air, by the effect of the refractive index of the air in which the laser operates. This is treated in further detail in Chapter 4.

2.1.2 Secondary Methods of Realizing the Meter for Dimensional Nanometrology

At scales relevant to current dimensional nanometrology, the methods based on interferometry in the visible spectrum are limited by fringe-counting and interpolation of interference fringes. Therefore, for nanometrology applications, it may be advantageous to refer to the lattice spacing of the Si atom. The Si {220} lattice spacing $d_{220} = 192.015\,571\,4 \times 10^{-12}$ m may be used as a secondary realization of the definition of the meter using the following techniques and uncertainty limits (BIPM 2019-2).

a. Measurement of a displacement by reference to the d_{220} lattice plane, using an X-ray interferometer (REF). An uncertainty of 10 pm can be assumed for a 10 μm displacement. A correction must be applied to the lattice spacing to take into account impurities within the crystal.
b. Calibration of transmission electron microscopy (TEM) magnification by reference to a single crystal silicon artifact, where the crystal lattice is visible in the field of view of the TEM and the size or width of the single crystalline nanostructure can thus be determined by counting the number of

lattice planes in the nanostructure. By this method, expanded uncertainties below 1 nm for the widths of line structures smaller than 200 nm can be achieved.

c. Measurement of step height standard artifacts manufactured from single crystal silicon, where the height range of multiple monoatomic steps is currently limited to 10 nm and the expanded uncertainties of the monoatomic step heights are of the order 5 pm under UHV conditions and 15 pm under ambient conditions.

2.2 REFERENCE TEMPERATURE

The form and size of physical objects is temperature-dependent due to the thermal expansion of materials with temperature. Therefore, it is logical that a temperature is defined where measurements of the dimensions of objects are taken. The definition of this temperature has a long and interesting history; just for the purpose of this book, it is relevant to know that length measurements are assumed to be taken at 20 °C, and where this is not possible, the deviation from 20 °C must be taken into account in the stated uncertainty and a correction should be made if relevant. The importance of this standard temperature is reflected in the reference number ISO standards: it is number 1: ISO 1. Its formulation has slightly evolved through the years, from the original statement of the CIPM in 1931: 'As the normal temperature for adjustment of industrial standards, the Committee adopts the temperature of 20 °C'. In the most recent version (ISO 1:2022), taking into account the revision of the SI system in 2018, it is:

> The standard reference temperature value for the specification of geometrical and dimensional properties shall be fixed at $t_{90} = 20\ ^{\circ}\mathrm{C}$, where t_{90} is the International Celsius Temperature according to the ITS-90, the International Temperature Scale of 1990.

The ITS-90 is a practical approximation and does not provide the exact thermodynamic temperature according to the definition (see Chapter 1), giving deviations at the milli-kelvin level. Still, this practical approach is taken because all thermometers used in dimensional metrology and in engineering worldwide are calibrated to ITS-90. Note that the unit degree Celsius (°C) is linked to the temperature T by a fixed shift of scale according to $t = T - 273.15\,\mathrm{K}$, where T is the absolute temperature and t the temperature in degrees Celsius.

In corrections and uncertainty evaluations, it is common to assume a linear dependence between the length and the temperature of the object, where the object material is characterized by the coefficient of linear expansion (CTE) α defined as:

$$\alpha = \frac{1}{L}\frac{\partial L}{\partial T} \tag{2.4}$$

where α is the CTE, L is the measured length or displacement and T is the temperature. The length of an object at temperature T, L_T in terms of the length L_{20} at the reference temperature T_{ref} = 20 °C, can be written as:

$$L_T = L_{20} + \int_{T_{\mathrm{ref}}}^{T} \alpha L(T)\,dT = L_{20} + \alpha L_T T - \alpha L_{20} T_{\mathrm{ref}} \approx L_{20}\left(1 + \alpha\left(T - 20\,^{\circ}\mathrm{C}\right)\right). \tag{2.5}$$

Here, the approximation $\frac{L_T}{L_{20}} \approx 1$ is made in the second term which is normally justified as the expansion coefficient and the magnitude of the expansion are relatively small. Similarly, equation (2.5) can be written as:

$$L_{20} = L_T + \int_{T}^{T_{\mathrm{ref}}} \alpha L(T)\,dT = L_T + \alpha L_{20} T_{\mathrm{ref}} - \alpha L_T T \approx L_T\left(1 - \alpha\left(T - 20\,^{\circ}\mathrm{C}\right)\right). \tag{2.6}$$

Equations (2.5) and (2.6) are the most-used equations in dimensional metrology for the evaluation of temperature corrections and uncertainties in measured dimensions due to temperature.

Table 2.1 gives the CTE of some common materials. See also Doiron and Beers (2014). As a rule of thumb, the value of steel ($\alpha \approx 10^{-5}$/K) can be phrased as: a steel object expands about 10 μm per ºC per meter of its length. In uncertainty evaluations, it is common to attribute a standard uncertainty of $0.5 \cdot 10^{-6}$/K to the values in Table 2.1, if the CTE is not explicitly specified. When length measurements must be carried out with high precision at temperatures significantly different from 20 °C, it may be necessary to incorporate nonlinear terms in the expansion calculation. The second-order expansion coefficient is defined as:

$$\beta = \frac{1}{2L}\frac{\partial^2 L}{\partial T^2}. \tag{2.7}$$

For metals and ceramics, this value is typically $\beta = 6 \cdot 10^{-9}$ K^{-2} (Schödel 2008). Length indicating measuring instruments and standards are calibrated at 20 °C or the results given in the calibration certificate are corrected to 20 °C. A typical situation in dimensional metrology is given in Figure 2.3, where an object having a temperature T_x and linear expansion coefficient α_x is measured using a line standard (ruler) having a temperature T_s and linear expansion coefficient α_s. The actual measured length is L, where the scale indication is $L_{s,20}$. The measured length at 20 °C is $L_{x,20} = L\left(1 - \alpha_x\left(T_x - 20\,^{\circ}\mathrm{C}\right)\right)$, with $L = L_{s,20}\left(1 + \alpha_s\left(T_s - 20\,^{\circ}\mathrm{C}\right)\right)$,

TABLE 2.1
Coefficient of Linear Expansion (CTE) of Some Common Materials (Doiron and Beers 2014)

Material	CTE $\alpha/(10^{-6}/\text{K})$ at 20 °C
Carbon steel	10.8
Stainless steel	10.1 to 17.3
Steel	11.0 to 13.0
Glass	8.5
Fused quartz	0.59
Optical glass – BK7	7.1
Aluminum	23.1
Brass	19
Copper	17
Diamond	1
Invar	1.2
Iron	11.8
Nickel	13
Silicon	2.56
Granite	6.3
Tungsten carbide	4.23
Chrome carbide	8.4
Zirconia oxide (ceramic gage block)	9.2
'Zerodur' (glass ceramic)	0.007 to 0.1

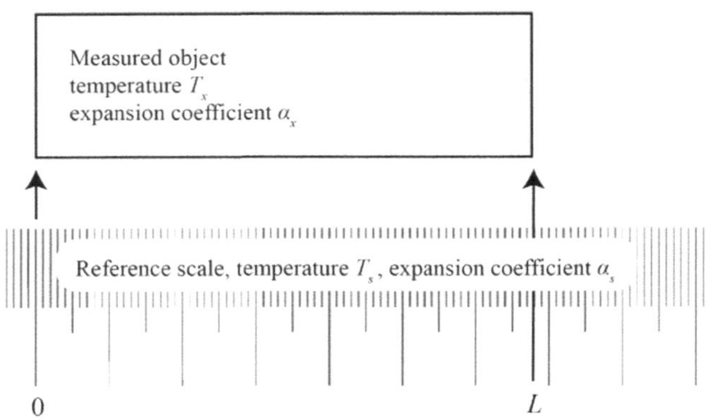

FIGURE 2.3 Typical measurement situation in dimensional metrology: a measurement instrument containing a reference scale at temperature T_s having linear expansion coefficient α_s takes a measurement of an object with linear expansion coefficient α having a temperature T_x. The measured length indicated on the scale is L.

giving

$$L_{x,20} = L_{s,20}\left(1+\alpha_s\left(T_s - 20\,^\circ\mathrm{C}\right)\right)\left(1-\alpha_x\left(T_x - 20\,^\circ\mathrm{C}\right)\right)$$
$$\approx L_{s,20}\left(1+\left(\alpha_x - \alpha_s\right)\cdot 20\,^\circ\mathrm{C} + \alpha_s T_s - \alpha_x T_x\right), \tag{2.8}$$

where higher-order terms in αT are omitted. Equation (2.8) can be re-written as:

$$L_{x,20} = L_{s,20}\left(1+\frac{\alpha_x+\alpha_s}{2}\left(T_x - T_s\right)+\left(\alpha_x - \alpha_s\right)\left(\frac{T_x+T_s}{2} - 20\,^\circ\mathrm{C}\right)\right)$$
$$= L_{s,20}(1+\bar{\alpha}\Delta T + \Delta\alpha\left(\bar{T} - 20\ ^\circ\mathrm{C}\right) \tag{2.9}$$

where $\bar{\alpha}$ and Δ indicate the mean and difference of the temperatures and expansion coefficients, respectively. From equation (2.9), some elementary principles of the temperature effects on dimensional measurements can be derived:

- The difference in expansion coefficient has no effect when the temperature is 20 °C.
- The mean expansion coefficient has no effect when the temperatures are equal.
- The deviation of the mean temperature from 20 ℃ has no effect when the expansion coefficients are equal.
- The temperature difference is less relevant when the mean expansion coefficient is small.

Ideally, the object and the scale have the same temperature which is 20 °C. This is the reason why dimensional calibration laboratories are preferable operated at 20 °C. If a temperature close to 20 °C cannot be achieved, it is still important to make the object and scale have the same temperature. If the object and the reference scale also have the same expansion coefficient, the measurement gives directly the dimension of the object at 20 °C. This is the reason why gage blocks are commonly made of steel: when comparing a steel object to an assembly of gage blocks having the same size and temperature, the measured size of the object is the size at 20 °C, even if the actual temperature deviates somewhat (for example, by a few degrees Celsius) from 20 °C .

2.3 REFERENCE PRESSURE

Although it is rarely relevant in practice, it is noteworthy that practical length measurement are assumed to take place in ambient air at a pressure of 1013.25 hPa. This is stated in ISO 3650:1998 that defines the lengths of gauge blocks and, where gauge blocks are commonly used as references for other length measurements, this applies to length measurement in practice in general. As the relative length changes are in the 10^{-8} range, they can be ignored under normal atmospheric conditions. For example, a steel object having a length l will be approximately $2 \cdot 10^{-7} \cdot l$ longer in vacuum relative to atmospheric conditions (Schödel 2008).

2.4 THE ABBE PRINCIPLE AND ABBE AND COSINE ERROR

The Abbe principle (Leach 2015) is named after the German physicist Ernst Karl Abbe (1840–1905) who formulated the principle in 1890 as the following (Abbe 1890):

> If errors of parallax are to be avoided, the measuring system must be placed co-axially (in line with) the line in which displacement (giving length) is to be measured on the work-piece.

This principle is illustrated in Figure 2.4, where the length L of an object is measured using a length scale. In position A, the measured length is co-axial with the displacement of the measurement indication (the arrow). In position B, this is not the case. A misalignment in position B, where the arrow is rotated over an angle θ, gives an Abbe error δ_A given by:

$$\delta_A = O \tan \theta \approx O\theta. \tag{2.10}$$

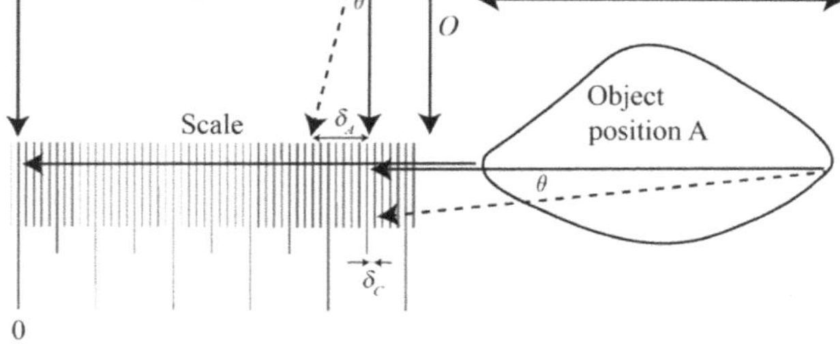

FIGURE 2.4 Illustration of the Abbe and cosine error. In position A, if there is an angular error θ when the indicator moves from one side of the object to the other, there will an error δ_C that is called the cosine error. In position B, if there is an angular error θ when the indicator moves from one side of the object to the other, there will an error δ_A that is called the Abbe error.

The linear dependence on the offset O and the angle θ makes this a first-order effect. The Abbe error can be reduced by making either the offset O or the rotation angle θ small. If these reductions cannot be achieved in a sufficient way, the alternative is to measure O and θ and apply a correction according to equation (2.9). This is reflected in the re-formulation of the Abbe principle by Brian (Brian 1979):

> The displacement measuring system should be in line with the functional point whose displacement is to be measured. If this is not possible, either the slideways that transfer the displacement must be free of angular motion or angular motion data must be used to calculate the consequences of the offset.

In position A, the relevant dimension of the object is aligned to the measurement axis. A misalignment of the measurement system with an angle θ, as indicated by the arrow in Figure 2.4, will still give a measurement error, given by

$$\delta_C = L - L\cos\theta \approx L - L\left(1 - \frac{1}{2}\theta^2\right) \approx \frac{1}{2}L\theta^2. \tag{2.11}$$

The error in equation (2.11) is generally referred to as the cosine error. Because of the quadratic term (for small angles), this is a second-order effect. This quadratic function can have advantages; for example, reducing the angle by a factor of two reduces the cosine error by a factor of four. On the other hand, the cosine error can be difficult to estimate for small displacements, as the length range available to estimate θ may be small.

2.5 THE EPPENSTEIN PRINCIPLE

The Eppenstein principle is named after the German physicist Martin Otto Eppenstein (1876–1942) who designed an optical length measuring instrument in 1922 (Eppenstein 1924). The basic principle is that Abbe errors are automatically compensated in the design of the readout of a measured length. This is illustrated in Figure 2.5 that schematically gives the readout of a measuring machine based on the Eppenstein principle. The Abbe offset must be equal to the focal length f of the illumination unit. When observed from above, for example, using a traveling microscope, the illumination will always be at the same reading point independent of the angular error θ. The term Eppenstein principle can be more generally used for systems where Abbe errors are inherently compensated.

2.6 BESSEL POINTS AND AIRY POINTS

A kinematic support (Leach and Smith 2018) of a one-dimensional beam requires two support points. If the length of a beam is to be measured it must be supported in such a way that the length measurement is as close as possible to the beam in a 'free state'. This is accomplished by supporting the beam at the Bessel points,

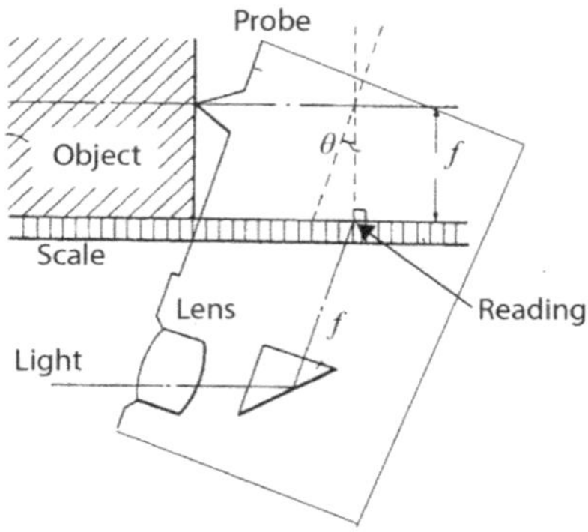

FIGURE 2.5 Illustration of the Eppenstein principle. When the Abbe offset is equal to the focal length f of the lens, the scale will be illuminated at the reading point, independent of the rotation error θ of the probe assembly.

which give a minimum length deviation, or the Airy points, which keep the end faces parallel (this is useful for gage blocks and for any object with parallel end faces) (Lewis 1994). The Airy points are symmetrically arranged around the center of the object of length L and are separated by a distance equal to

$$d_{\text{Airy}} = \frac{L}{\sqrt{3}} \approx 0.5774 \cdot L. \tag{2.12}$$

The Bessel points are symmetrically arranged around the center of the object of length L and are separated by a distance equal to

$$d_{\text{Bessel}} = 0.5594 \cdot L. \tag{2.13}$$

The Airy and Bessel points are illustrated in Figure 2.6. In addition, there are the support points for minimum sag, separated by $0.5536 \cdot L$. In practice, the difference between the Bessel and Airy points is negligible.

Support Points on 3D Objects

The principles of the support points can be extended to 3D objects, especially surface plates and optical flats. For surface plates, it is common to keep the Bessel points at 0.56 multiplied by the length of the respective faces; for round surfaces, such as optical flats, the sag is minimized by positioning the supporting points on a circle with a diameter of 0.67 times the diameter of the workpiece to be supported; see Figure 2.7.

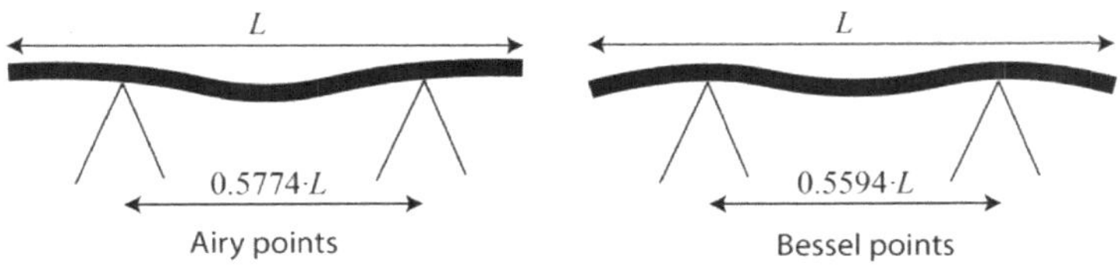

FIGURE 2.6 Illustration of the supporting points for a beam keeping the end points parallel (Airy points, left) and giving a minimum length change (Bessel points, right).

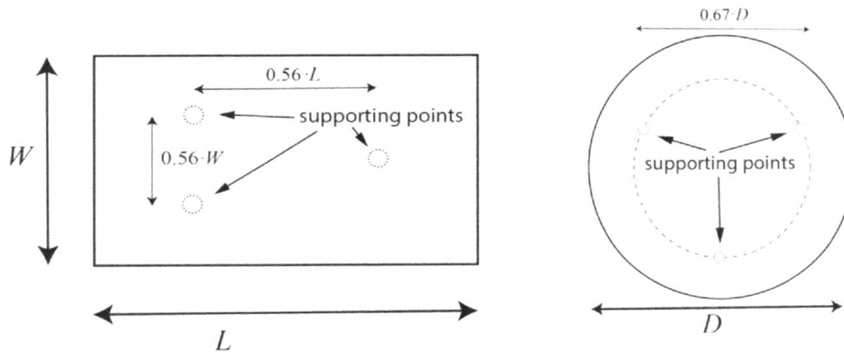

FIGURE 2.7 Top view on rectangular (left) and circular (right) workpieces with the position of the support points indicated. The supporting points are positioned symmetrically at 0.56 multiplied by the length and width of a rectangular workpiece and equidistant at 0.67 multiplied by the diameter of a round surface.

2.7 GO AND NO-GO GAGES, TAYLOR'S PRINCIPLE

Where this book deals mainly with dimensional measurement rather than testing, there is one basic principle in testing dimensional features that cannot be overlooked. This is the Taylor's principle (Kverneland 2007) for limit gages. Limit gages are typically used for a fast check of the dimensional features of workpieces. Limit gages have a 'Go' and a 'No-Go' feature. The Go and No-Go principle of gauging is that the Go-end of the gages must fit to the feature of the component being tested. The No-Go end must not fit into the same feature.

The result of this Go–No-Go test can be:

- Approve: the component is within tolerance. This means the 'Go' side of the gage did fit **and** the No-Go side did not fit.
- Disapprove: the component is not within tolerance. This means the 'Go' side of the gage did not fit **or** the No-Go side did fit.

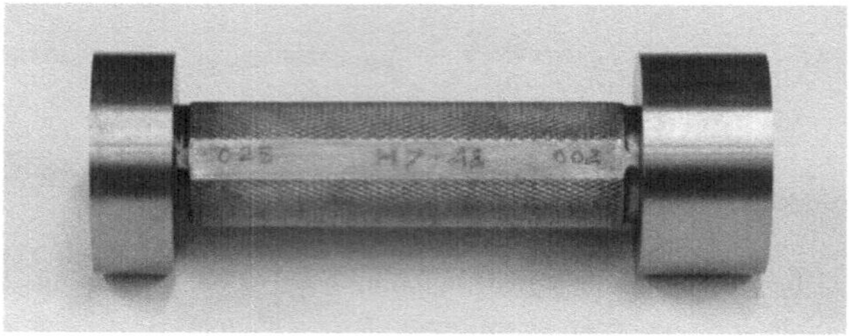

FIGURE 2.8 Picture of a typical cylindrical limit gage with a 'Go' side on the right and a 'No-Go' side on the left. This example is for testing a 48 mm diameter hole according to tolerance class H7 according to ISO 286-1:2010.

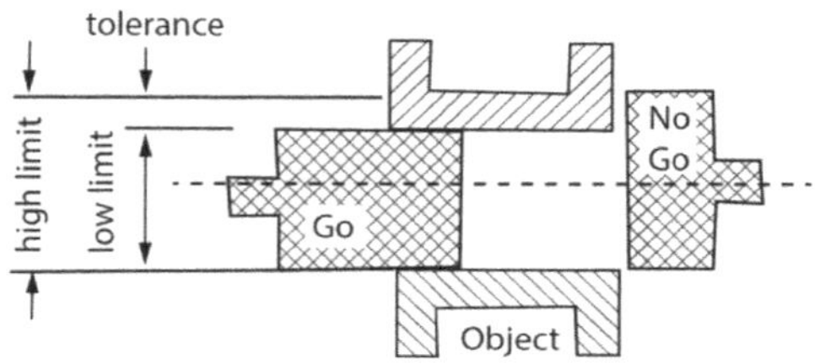

FIGURE 2.9 Illustration of the Taylor's principle: the hole in the object has tolerances limited by a high limit and a low limit. The Go-gage should be able to enter the hole over its full length, the No-Go gage should not be able to fit in the hole at all.

The dimensions of the Go and No-Go ends of gages are determined from the limits stated on the dimension of the component to be gaged. The dimension of the Go end is equal to the minimum permissible dimension and that of the No-Go end is equal to the maximum permissible dimension. Figure 2.8 shows a typical gage with a Go and a No-Go end. In practice, the Go-side may be indicated by a green color and the No-Go end by a red color. The Taylor's principle states that the Go gage should be designed in such a way that it will check the maximum material condition of as many dimensions as possible. A No-Go gage should inspect the minimum material condition of only one dimension at a time. According to this principle, a Go plug gage should have a full circular section and be of the full length of the hole that is to be tested. In addition, to control the diameter at any one point, the gage also tests the straightness or parallelism of the hole over its full length. The Taylor's principle is illustrated in Figure 2.9 for a hole checked by a plug gage.

2.8 TRACEABILITY OF ANGULAR MEASUREMENTS

In the overview of primary standards in Chapter 1, a standard of angle was not defined. This is because an angle can often be expressed as a ratio of two lengths

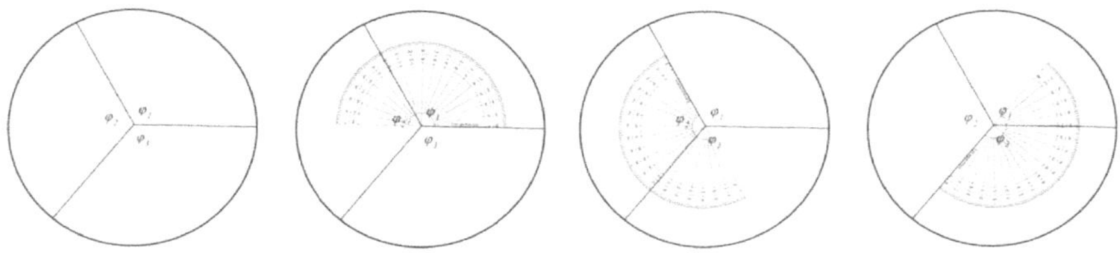

a. b. c. d.

FIGURE 2.10 Illustration of the inherent traceability of angular measurements. Within a circle, three angles are measured for which $\varphi_1 + \varphi_2 + \varphi_3 = 360°$ must hold. This requirement makes the measurement traceable without the need for any external reference.

or, more importantly, the fact can be used that, for a full circle, the angles add up to 360° or 2π radians. Figure 2.10 gives an example where a circle is divided into three pie slices characterized by three angles φ_1, φ_2 and φ_3. These angles are measured by a protractor as shown in Figures 2.10b, c and d.

The results of these measurements are respectively $\varphi_1 = 120.4°$, $\varphi_2 = 112.1°$ and $\varphi_3 = 129.3°$.

These results include the errors due to the scale itself, alignment and smaller effects. The traceability to the full circle is now imposed by the fact that the three angles **must** add up to 360°: $\varphi_1 + \varphi_2 + \varphi_3 = 360°$. The angles as measured add up to 361.8°, so obviously there is an error of 1.8°. Dividing this error over the three measured angles gives as final, traceable, result:

$\varphi_1 = 119.8°, \varphi_2 = 111.5°$ and $\varphi_3 = 128.7°$. This can also be formulated as a general least-squares problem, see Section A.2. To achieve this final result, no external reference was needed; even the protractor does not need any calibration. In fact, this method calibrates the protractor so long as the scale divisions are monotonic and consistent.

2.9 TRACEABILITY OF FORM ERRORS – ERROR SEPARATION AND REVERSAL METHODS

As with standards of angle, there is no defined standard of form to which all forms of that kind should be traceable. This is because, for any type of form error, there are error separation methods available that separate the measurement errors from the form errors and make absolute measurements traceable to the 'perfect form' possible. An overview of these techniques was given by Evans (Evans et al. 1996). This principle does not make form standards useless or obsolete. On the contrary, many are useful reference standards in calibration laboratories: reference flats, cylinders, straightedges, etc. The principal point made here is that ideal forms such as flats, spheres, straightedges, and so on do not need to be compared to some form of absolute reference in order to be traceable. As an example, a method to calibrate straightness using three nominally straight objects is illustrated in Figure 2.11.

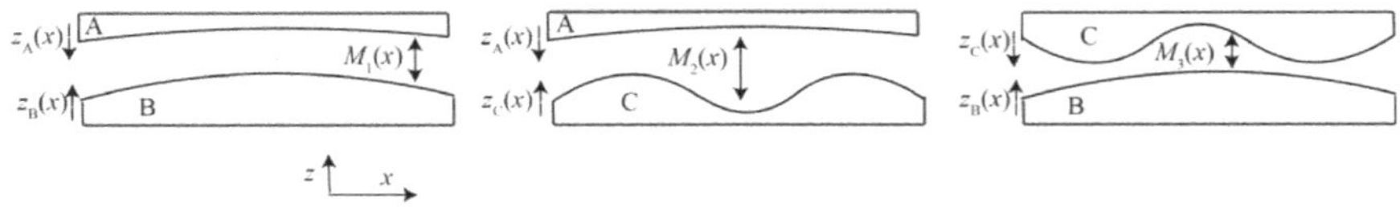

FIGURE 2.11 Absolute straightness measurements where the straightness profiles of three objects are compared to each other.

The straightness profile of objects A, B and C are denoted by $z_A(x)$, $z_B(x)$ and $z_C(x)$, respectively. The objects are positioned opposite to each other and the mutual difference, where one object is rotated 180° around the x-axis, is measured using a length measuring device. The measurements are denoted by $M_1(x)$, $M_2(x)$ and $M_3(x)$. The measurements can, omitting a constant term, be written in matrix form as:

$$\begin{pmatrix} M_1(x) \\ M_2(x) \\ M_3(x) \end{pmatrix} = \begin{pmatrix} -1 & -1 & 0 \\ 0 & -1 & -1 \\ -1 & 0 & -1 \end{pmatrix} \begin{pmatrix} z_A(x) \\ z_B(x) \\ z_C(x) \end{pmatrix}, \tag{2.14}$$

with a solution given by

$$\begin{pmatrix} z_A(x) \\ z_B(x) \\ z_C(x) \end{pmatrix} = \begin{pmatrix} -0.5 & 0.5 & -0.5 \\ -0.5 & -0.5 & 0.5 \\ 0.5 & -0.5 & -0.5 \end{pmatrix} \begin{pmatrix} M_1(x) \\ M_2(x) \\ M_3(x) \end{pmatrix}. \tag{2.15}$$

This gives the absolute straightness profiles $z_A(x)$, $z_B(x)$ and $z_C(x)$ expressed as combinations of the measurements $M_A(x)$, $M_B(x)$ and $M_C(x)$, without the need of an external absolute straightness reference.

2.10 INDENTATION IN MECHANICAL MEASUREMENTS

When a mechanical measurement is carried out where probing is involved, some indentation will take place that changes the measured value (Puttock and Thwaite 1969). Two elementary cases of bodies in contact that give an indentation are given in Figure 2.12. The indentation/deformation d as indicated on the left-hand side of Figure 2.12 is given by the Hertz equations:

$$d = \left(\frac{9F^2}{2E^{*2}R} \right)^{\frac{1}{3}} \tag{2.16}$$

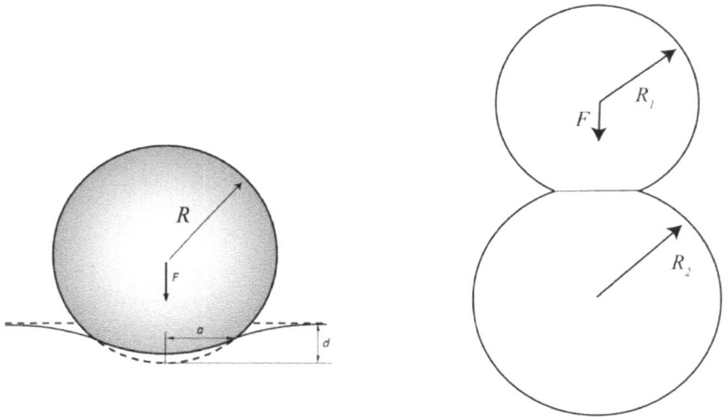

FIGURE 2.12 Indentation of a ball on a semi-infinite flat surface (left) and indentation of two balls (right) in contact.

TABLE 2.2
Poisson's Ratio and Elastic Modulus of Some Materials

Material	Poisson's Ratio	Elastic Modulus/GPa
Tungsten carbide	0.28	724
Steel 52100	0.29	205
Aluminum	0.35	72
Glass (light borate crown)	0.27	46
Ruby (synthetic sapphire)	0.25	38
Diamond (synthetic)	0.20	860
Ceramic (ZrO_2)	0.23	200

where R is the radius of the ball, F is the applied force and E^* is the effective elastic modulus given by

$$E^* = \left(\frac{1-\nu_1^2}{E_1} + \frac{1-\nu_2^2}{E_2} \right)^{-1} \quad (2.17)$$

and E_1, E_2 are the elastic moduli and ν_1, ν_2 are the Poissons's ratios associated with the respective bodies. Some typical values of E and ν are given in Table 2.2.

2.10.1 Contact between Two Spheres

For contact between two spheres of radii R_1 and R_2, as depicted in the right-hand side of Figure 2.12, equation (2.16) still holds, except that the radius R is defined as an effective radius given by the harmonic sum of R_1 and R_2

$$R = \left(\frac{1}{R_1} + \frac{1}{R_2} \right)^{-1}, \tag{2.18}$$

where it can be noted that the case for the ball and a semi-infinite flat is retained with $R_1 = R$ and $R_2 = \infty$.

2.10.2 Other Cases of Mechanical Contact

There are many other cases, such as two crossed cylinders, a ball against cylinder, a cylinder against a flat surface for a given length *L*, a sphere in a vee-groove and more. These calculations can be found in the literature and an internet page: https://emtoolbox.nist.gov/Elastic/Photo.asp

If in a measurement, the measurement force can be varied, an extrapolation to zero force is possible by plotting the displacement as a function of $F^{2/3}$ and calculating the intercept for $F = 0$. In Section A.4.1, it is illustrated how this can be calculated and how the uncertainty may propagate. The indentation can be doubled by increasing the force by a factor of 2.8, or it can be halved by decreasing the force by a factor of 2.8.

2.10.3 Some Typical Examples of Indentations

This section gives some examples that are common in practice. In many cases, the indentation is taken into account as the instrument varies the measurement force, or adjusts for it, for example, a coordinate measuring machine (CMM) measuring a reference sphere. In other cases, the measuring force approximately cancels as a reference is measured with the same force and indentation effect, for example, when a length measurement of a steel object is compared to a steel gage block. Also, the measurement force may have a negligible effect on a measurement result as it gives a constant measurement bias that is irrelevant when the measurement is evaluated, for example, in surface texture, roundness and straightness measurements.

Example 1: Surface texture measurement with a 2 μm radius diamond stylus on a flat steel surface, measurement force 0.75 mN: indentation 0.1 μm.

Example 2: Surface texture measurement with 5 μm radius diamond stylus on a flat steel surface, measurement force 4 mN: indentation 0.05 μm.

Example 3: CMM measurement; spherical aluminum oxide (ruby) probe, radius 1 mm, on 12.7 mm radius ceramic reference sphere; measurement force 0.5 N: indentation 0.2 μm.

Example 4: CMM measurement; spherical aluminum oxide (ruby) probe, radius 0.5 mm on flat aluminum surface, measurement force 0.5 N: indentation 0.4 μm

Example 5: Gage block comparator; spherical tungsten carbide probe, 3 mm radius; measuring force 0.8 N. Measurement of (flat) steel gage block: indentation 0.16 μm. Measurement of tungsten carbide gage block: indentation 0.09 μm.

Example 6: Measurement wire of 2 mm diameter, measuring an M5 screw thread that is approximated by a 60° vee-groove, measurement force 1.5 N: indentation 0.7 μm.

Note that these examples refer to a static measurement force in the elastic compression regions of the Hertzian curve.

EXERCISES

1. Laser frequency measurement
   The frequency of a laser interferometer is measured in comparison to an iodine-stabilized laser that is considered as a primary standard, operating under the conditions mentioned on the BIPM website. The calibrated laser interferometer system has as a nominal vacuum wavelength λ_{nom} = 632.991 370 nm. The iodine-stabilized laser is stabilized on the so-called d-dip that has a defined value of f_d = 473 612 379.8 MHz. The frequency difference between both lasers is recorded as a function of time and is displayed in Figure 2.13.
   a. Find and check the defined reference frequency on the BIPM-website www.bipm.org/en/: select the web-pages: technical –practical realization – recommended values of standard frequencies-wavelength – wavelength ≈ 633 nm.

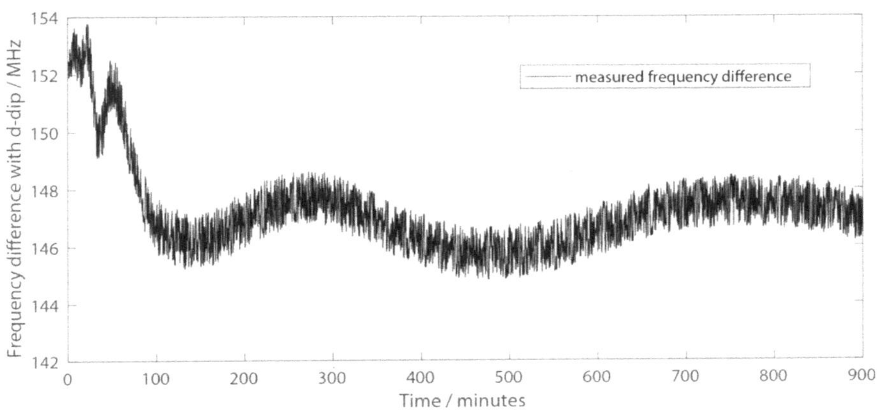

FIGURE 2.13 Measured frequency difference between a laser interferometer and an iodine-stabilized laser.

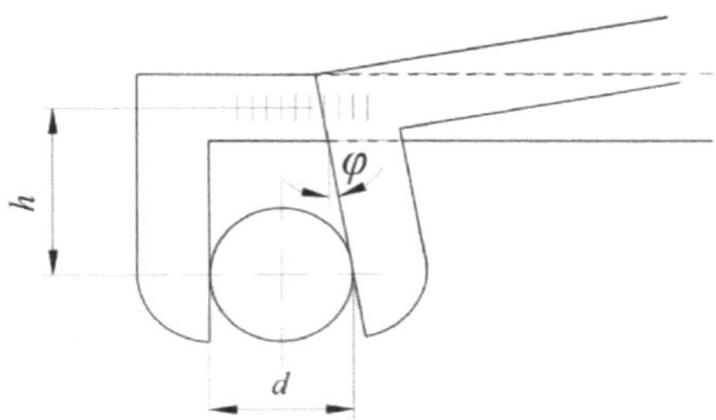

FIGURE 2.14 Sketch of a cylinder measurement using a Vernier caliper with an angular deviation.

Consider the standard uncertainty in this reference frequency on the vertical scale in Figure 2.13 (for example, in terms of a plot line thickness).

b. Calculate the nominal frequency of the laser interferometer system.
c. Draw the expected frequency difference as a horizontal line in Figure 2.13.
d. Calculate the relative short-term and long-term stability of the laser.
e. Consider the maximum deviation from the nominal frequency and calculate the corresponding vacuum wavelength.
f. Suppose a length of 1 m is measured using this laser interferometer. Note that a laser interferometer calculates a measured length from a number of wavelength counts multiplied by half its nominal wavelength. Calculate the maximum deviation based on the maximum deviation of the vacuum wavelength from its nominal value and argue whether the system will measure the 1 m length too long or too short.

2. Temperature correction
Consider Figure 2.3. The scale is calibrated at 20 °C and has no deviations. The scale indicates a measured length L = 100.125 mm. The glass scale has a temperature T_s = 21 °C and an expansion coefficient $\alpha_s = 5.0 \cdot 10^{-6}$ / K. The steel object has a temperature T_x = 22 °C and an expansion coefficient $\alpha_x = 11.5 \cdot 10^{-6}$ / K. Calculate the length of the object at 20 °C.
3. Abbe error and cosine error
Consider Figure 2.4. An object with a length L = 100 mm is measured using a scale without deviations. The object is measured in position A with an angular deviation of 40". The object is measured in position B with an angular deviation of 40" and an offset O = 51.5 mm.
Calculate the deviations δ_A and δ_C in the measurement.
4. First- and second-order Abbe error
Figure 2.14 shows how the diameter d of a cylinder is measured using a caliper having an angular deviation φ.
a. Determine the first-order deviation δ of the measurement, that is, the deviation that is linear in φ.

b. Determine δ until up to second order.
c. Determine the deviation δ for $h = 30$ mm, $d = 20$ mm and $\varphi = 1°$ using the results from a. and b.
Note that this is a classical calculation in dimensional metrology and that the second-order term often is not required in practice (as it is negligible compared to other influence factors).

5. Straightness measurement using three objects
Consider Figure 2.11. The measurement values taken from three combinations of the three straightedges are given in Table 2.3. Calculate the straightness profiles $z_A(x)$, $z_B(x)$ and $z_C(x)$.
6. Indentation
a. A 6 mm diameter ruby ball is measured using a screw gage (commonly called a micrometer), having tungsten carbide measurement faces, see Figure 2.15. The measurement force is 5 N. Calculate the indentation.
b. A 5 mm ball of unknown material is measured using a measurement bench having flat steel measurement faces. Applying a force of 2 N gives a measured length $L_{2N} = 5.000\,28$ mm. When reducing the force

TABLE 2.3
Measurement Values *M(x)* as a Function of the Location (*x*-Position) of the Straightness Measurement

x/mm	0	25	50	75	100
$M_1(x)$/µm	0	1.7	2.5	1.7	0
$M_2(x)$/µm	0	1	–0.5	1	0
$M_3(x)$/µm	0	0	2	0	0

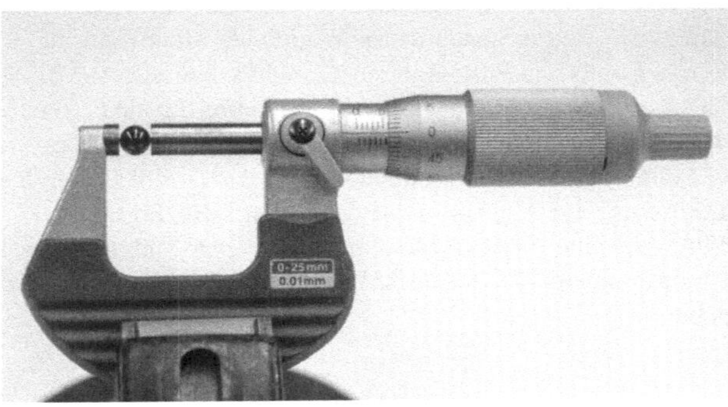

FIGURE 2.15 Ball measurement using a screw gage.

to 1 N, the measured length is L_{2N} = 5.000 56 mm. Calculate the ball diameter, extrapolating the length to a measurement force of 0 N.

c. Derive a general equation $L = f\left(L_F, L_{2F}\right)$ for the length of an object derived from two measurements: a first measurement using a certain force giving a measured length L_F and a second measurement using twice this force L_{2F}.

REFERENCES

Abbe, E., 1890, Messapparate für Physiker [Measuring Instruments for Physicists]. *Zeitschrift fur Instrumentenkunde [J. Scien. Instrum.]* **10**:446–448 (in German).

BIPM, 2019-1, The International System of Units (SI Brochure) 9th edition. www.bipm.org/en/publications/si-brochure accessed November 2024

BIPM, 2019-2, The International System of Units (SI Brochure), Appendix 2: Practical Realizations of the Definitions of Some Important Units – Mises en Pratique, Mise en Pratique for the Definition of the Metre in the SI 9th edition. https://bipm.org/utils/en/pdf/si-mep/SI-App2-metre.pdf

Bryan, J.B., 1979, The Abbé Principle Revisited: An Updated Interpretation. *Precis. Eng.* **1**:129–132.

Doiron, T. and Beers, J., 2014, *The Gauge Block Handbook, NIST Monography 180.* CreateSpace Independent Publishing Platform. www.nist.gov/system/files/documents/calibrations/mono180.pdf

Eppenstein, O.S.E, 1924, Measuring Machine, US patent US1589796A.

Evans, C.J., Hocken, R.J. and Estler, W.T., 1996, Self-Calibration: Reversal, Redundancy, Error Separation and 'Absolute Testing'. *Ann. CIRP* **45(2)**:617–634.

Hall, J.L., 2006, Nobel Lecture: Defining and Measuring Optical Frequencies (2006). *Rev. Mod. Phys.* **78(4)**1279:1295.

ISO 1, 2022, Geometrical Product Specifications (GPS) – Standard Reference Temperature for the Specification of Geometrical and Dimensional Properties. Geneva, International Organization for Standardization.

ISO 286-1, 2010, Geometrical Product Specifications (GPS) – ISO Code System for Tolerances on Linear Sizes – Part 1: Basis of Tolerances, Deviations and Fits. Geneva, International Organization for Standardization.

ISO 3560, 1998, Geometrical Product Specifications (GPS) – Length Standards – Gauge Blocks. Geneva, International Organization for Standardization.

Kverneland, K.O., 2007, Metric Standards for Worldwide Manufacturing, Chapter 6: The ISO System of Limits and Fits – Tolerances and Deviations. ASME Press.

Leach, R.K., 2015, Abbe Error/Offset. In: Laperrière, L. and Reinhart, G. *CIRP Encyclopaedia of Production Engineering.* Berlin, Springer-Verlag.

Leach, R.K. and Smith, S.T., 2018, Introduction to Precision. In: Leach, R.K. and Smith, S.T. *Basics of Precision Engineering.* Boca Raton, CRC Press.

Lewis, A.J., 1994, Absolute Length Measurement Using Multiple-Wavelength Phase-Stepping Interferometry, PhD Thesis, University of London, Appendix C: Flexing of Length bars.

Puttock, M.J. and Thwaite, E.G., 1969, *Elastic Compression of Spheres and Cylinders at Point and Line Contact,* National Standards Laboratory Technical Paper 15, CSIRO, Melbourne, Australia. See also https://emtoolbox.nist.gov/Elastic/Photo.asp

Schödel, R., 2008, Ultra-High Accuracy Thermal Expansion Measurements with PTB's Precision Interferometer. *Measure. Sci. Tech.* **19(8):**084003.

Schödel, R., Yacoot, A. and Lewis, A., 2021, The New Mise en Pratique for the Metre – A Review of Approaches for the Practical Realization of Traceable Length Metrology from 10^{-11}m to 10^{13} m. *Metrol.* **58 (5)**:052002.

3 Measurement Uncertainty

3.1 THE GUM UNCERTAINTY FRAMEWORK

The *Guide to the Expression of Uncertainty in Measurement* (GUM) is considered the generally accepted basis of the evaluation of uncertainty (JCGM 100 2008). In 2008, Supplement 1 was published entitled *Propagation of Distributions Using a Monte Carlo Method* (JCGM 101 2008). The applications of this framework are documented in the guideline of the European Accreditation: EA/4-02 *Evaluation of the Uncertainty of Measurement in Calibration* (EA 2011), the specification standard ISO 14253-2:2011 *Guidance for Estimation of Uncertainty in Geometric Product Specification (GPS) Measurement, in Calibration of Measurement Equipment*, and in *Guidelines for Evaluating and Expressing the Uncertainty of NIST Measurement Results* (Taylor and Kuyatt 1994). This chapter is based on these documents and on an earlier elaboration of this subject (Haitjema 2021); however, this chapter uses an example that is characteristic for dimensional metrology and elaborates more on comparison methods and evaluations.

3.2 SYSTEMATIC AND RANDOM ERRORS; TYPE A AND B UNCERTAINTIES

In the context of measurement uncertainty, the term 'error' means a deviation from a 'true' or average value; this should not be confused with a failure or a mistake. Historically, errors in measurement processes are classified as either systematic or random. In this context, a systematic error is a deviation from the true value in a measurement that remains constant in repeated measurements, and a random error is a component of a measurement error that varies in an unpredictable manner when the measurement is repeated. In uncertainty evaluations, it is considered more useful to classify uncertainty evaluations as 'type A' and 'type B'. Type A refers to the evaluation of a component of measurement uncertainty by a statistical analysis of measured values obtained under defined measurement conditions. This is related to, but not the same as, random errors. Examples of type A evaluations are:

DOI: 10.1201/9780367821746-3

- The standard deviation in a series of measurements.
- The standard deviation in the mean of a series of measurements.
- The standard deviation from a determined calibration curve.

It is common to express the uncertainty resulting from a type A evaluation as a standard deviation or the standard deviation in the mean measured value. Type B evaluation encompasses any other method of uncertainty evaluation. Examples are:

- The calibration uncertainty of a used physical reference specimen.
- The published uncertainty of a physical constant.
- The known uncertainty of equations that approximate quantities, for example, for the local gravitational acceleration as a function of altitude and latitude.
- A known but uncorrected systematic error.
- The resolution of a stable measurement reading.
- The knowledge or assumption that a quantity that is not measured but affects the measurand is within certain limits or conforms to specifications, for example, the temperature in a stable room where geometrical measurements are performed.

Type B evaluations are commonly related to systematic errors. Such errors can be expressed as a deviation, which may be represented by a standard deviation. Therefore, both type A and type B errors are quantified as standard uncertainties. This commonly implies a standard deviation or a standard deviation in a mean value in a type A evaluation and an expected deviation in a type B evaluation.

3.3 DIMENSIONAL MEASUREMENT EXAMPLE: THREAD MEASUREMENT USING MEASURING WIRES

As an example to illustrate several aspects of uncertainty evaluation, a screw thread measurement is considered. Figure 3.1 shows the schema and a photograph of the

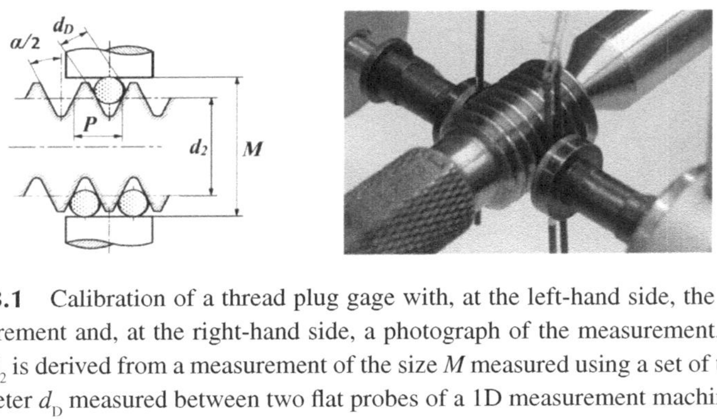

FIGURE 3.1 Calibration of a thread plug gage with, at the left-hand side, the schema of the measurement and, at the right-hand side, a photograph of the measurement. The pitch diameter d_2 is derived from a measurement of the size M measured using a set of three wires with diameter d_D measured between two flat probes of a 1D measurement machine.

measurement of an M18 × 2.5 thread plug gage on which three measuring wires are positioned. The pitch diameter has to be determined from the measured diameter M, defined as the distance between the center of the probing elements touching the thread on both sides; the pitch P; the thread angle α and the diameter d_D of the measuring wires. From a geometrical analysis, one obtains for a symmetrical thread (Euramet 2012):

$$d_2 = M + \frac{P}{2\tan(\alpha/2)} - d_D\left(1 + \frac{1}{\sin(\alpha/2)}\right) - A_1 + A_2, \tag{3.1}$$

where A_1 is the rake correction and A_2 is the correction for the measurement force. A_1 and A_2 will be neglected in this analysis. Also, temperature effects are neglected.

3.4 SINGLE VALUE UNCERTAINTY EVALUATION

The basic quantities for the evaluation of a single value are the mean value, the standard uncertainty in the measured value and the standard uncertainty in the mean measured value.

In this example, in equation (3.1), evaluations of the variables M, P, α and d_D are needed for an evaluation of the value of d_2, the standard uncertainty in d_2, $u(d_2)$ and the expanded uncertainty of d_2, $U(d_2)$. Of these quantities, commonly only M is obtained by a repeated measurement, enabling a type A evaluation. Table 3.1 gives an example of five measured values of M, obtained from a 1D measuring machine.

TABLE 3.1
Pitch Diameter Measurement as an Illustration of the Calculation of the Mean and Standard Deviation of a Measurement Series

	Measured Value	Difference from Mean	Squared Difference from Mean
Measurement Number i	M_i/mm	$\frac{\delta_i}{\text{mm}} = M_i - \bar{M}$	$\delta_i^2/\mu\text{m}^2$
1	18.3314	–0.0046	21.2
2	18.3332	–0.0028	7.8
3	18.3347	–0.0013	1.7
4	18.3412	+0.0052	27.0
$n = 5$	18.3395	+0.0035	12.3
Sum	91.6800	0	70.0

3.4.1 Type A Evaluation: Statistical Evaluation

From a number of repeated measurements M_i, the following quantities can be derived:

The mean value $\bar{M}$, given by:

$$\bar{M} = \frac{\sum_{i=1}^{n} M_i}{n} = \frac{\sum_{i=1}^{5} M_i}{5} = \frac{91.6798}{5} = 18.3360 \text{ mm}, \quad (3.2)$$

the experimental standard deviation s, given by:

$$s = \sqrt{\frac{\sum_{i=1}^{n} \left(M_i - \bar{M}\right)^2}{n-1}} = \sqrt{\frac{\sum_{i=1}^{5} \delta_i^2}{4}} = \sqrt{\frac{70.0}{4}} = 4.2\ \mu\text{m}, \quad (3.3)$$

and the experimental standard deviation in the mean value s_m, given by:

$$s_m = s / \sqrt{n} = \sqrt{\frac{\sum_{i=1}^{n} \left(M_i - \bar{M}\right)^2}{n(n-1)}} = \sqrt{\frac{70.0}{20}} = 1.9\ \mu\text{m}. \quad (3.4)$$

Note that both s and s_m are given to two significant figures (digits) and that the significant figures in which the mean value is given is adapted to this. The mean value $\bar{M}$ is the best approximation for the expected value $\langle M \rangle$ that would be obtained after a (hypothetical) infinite number of measurements. The experimental standard deviation s is the best approximation of the standard deviation σ that would be obtained after a (hypothetical) infinite number of measurements.

The standard deviation σ characterizes the expected deviation of each individual measurement M_i from the expected value $\langle M \rangle$. The experimental standard deviation in the mean s_m is the best approximation of the standard deviation in the mean $\sigma_m = \frac{\sigma}{\sqrt{n}}$ that characterizes the expected deviation of the obtained mean value $\bar{M}$ from five measurements ($n = 5$) from the expected value M. A less known and used quantity is the 'standard deviation in the standard deviation' s_s that characterizes the expected deviation of the experimentally obtained standard deviation s from the expected standard deviation σ. It is given by (Squires 2008):

$$s_s = \frac{s}{\sqrt{2n-2}} = 1.5\ \mu\text{m}. \quad (3.5)$$

This equation is given for completeness and as an illustration for the fact that an accurate determination of the standard deviation takes quite some measurements.

TABLE 3.2
Relationship between the Statistical Parameters and Their Approximation by a Finite Number of Measurements

Expected 'True' Value	Approximation	Expected Deviation
M	M_i	s
M	$\bar{M}$	s_m
σ	s	s_s

An overview of what can be concluded when taking a series of measurements is given in Table 3.2.

Taking more measurements has the effect that the mean value $\bar{M}$ will be closer to the expected value M, as expressed by the decrease of the standard deviation in the mean s_m, and the experimental standard deviation s will be a better approximation of the theoretical standard deviation σ. It must be noted that the decrease of s_{m} with the number of measurements is rather slow: to halve the standard deviation in the mean, four times as many measurements need to be taken. However, if measurements are taken within a time interval that is too short to enable the instrument to reach a new independent value, the measurements will be correlated, making the standard deviation meaningless. Taking many measurements over a prolonged period of time has the disadvantage that drift may occur, apart from the trivial disadvantage that it just takes time that is not always available and does not always outweigh the improved accuracy, especially not when the measurement uncertainty has type B components that have the same magnitude or are even larger. If a certain distribution of measurement values can be assumed, the probability P that a measured value will be located in the interval $\left[\langle M\rangle - k\sigma;\ \langle M\rangle + k\sigma\right]$, where k is a coverage factor, can be calculated. For example, it is known that for a Gaussian distribution $P(\left[\langle M\rangle - \sigma;\ \langle M\rangle + \sigma\right]=68\%$ and $P(\left[\langle M\rangle - 2\sigma;\ \langle M\rangle + 2\sigma\right] = 95.5\%$. The probability that $\langle M\rangle$ is located within the interval $\left[\bar{M} - ks;\ \bar{M} + ks\right]$ depends on both the factor k and the number of measurements n taken. In order to take the limited amount of measurements into account for a Gaussian distribution, the experimentally determined standard deviation s must be multiplied with a factor of t, the so-called Student's-t factor, in order to achieve the same probability as when the standard deviation σ is multiplied with a factor z. This can be written as:

$$P\left(M \pm z(P)\cdot\sigma\right) = P\left(\bar{M} \pm t(n,P)\cdot s\right). \tag{3.6}$$

This Student's-t factor depends both on the number of measurements and the probability, while the factor z depends on the probability only. The factor should be applied instead of the factor z when there is a relatively small number of measurements, typically for $n < 10$. The t-factor for $n < 10$ and a probability

TABLE 3.3
Student's *t*-Factor as a Function of the Number of Measurements, Providing the Probability $P\left(\bar{M} \pm t_{95\%,n} \cdot s\right) = 95\%$ That the Value of $\langle M \rangle$ is inside this Interval

$n = \upsilon + 1$	2	3	4	5	6	7	8	9	12	18	60	∞
$t_{95\%,\upsilon}$	13	4.3	3.2	2.8	2.6	2.5	2.4	2.3	2.2	2.1	2.0	1.96

$P = 95\%$ is given in Table 3.3. The symbol υ stands for the number of degrees of freedom, that is, $\upsilon = n - 1$ when an average is calculated from a series of n measurements. Application of the student's t-factor may be relevant if the limited repeatability or reproducibility of the measurement is a major contribution to the final uncertainty.

An important point of consideration is the question as to whether the standard deviation or the standard deviation in the mean must be taken as the standard uncertainty in the measured value of M. In the example above, it can be assumed that the variation is caused by the measurement, its alignment, the parallelism of the probes, etc., and that the value of d_2 itself is not statistically varying or not well defined. This justifies the consideration of the mean of the measurements, that is, the standard deviation can be divided by $\sqrt{n}$. This is generally the case when gages or other well-defined material objects are measured. For the calibration of indicating instruments, the situation can be different; the variation in the measured value can be caused by the measuring instrument itself. When in later use a single measurement is taken, the standard deviation is the number that characterizes its variation, so this value should be used in further uncertainty evaluation. For the example above, the standard deviation s_m in the mean value can be taken as the standard uncertainty u in the value of M in equation (3.1), giving $M = 18.3360$ mm and $u(M) = s/\sqrt{5} = 1.9$ µm. For the distribution, a Gaussian distribution can be assumed.

3.4.2 Type B Evaluations

As in equation (3.1), the values of the pitch P, the thread angle α and the diameter d of the measuring wires are not measured, its values and uncertainties must be obtained by other means. This is a type B evaluation that is illustrated below for the values and uncertainty of the pitch, tread angle and wire diameter in different ways.

Pitch P

The nominal pitch of the caliper is 2.5 mm. In the specification standard ISO 1502:1996, the specifications of metric screw threads are given. Table 7 in this standard states that the pitch tolerance T_p applies to the maximum measured error, whether it occurs as a cumulative progressive error over the total length of the thread, as a periodic error or as a local error of pitch. This implies that the deviation in the pitch P, as indicated in Figure 3.1 and in equation (3.1), will never be larger than 5 μm; assuming that the caliper conforms to this standard. However, no statement can be made about its probability of being closer to the edges or closer to the nominal value. Therefore, for this uncertainty, a rectangular (or uniform) distribution is assumed ranging from 2.5 mm – 5 μm to 2.5 mm + 5 μm. In general, the standard uncertainty of a quantity having a tolerance $\pm a$, assuming a rectangular distribution is given by:

$$u(P) = \frac{a}{\sqrt{3}} = \frac{5\ \mu\text{m}}{\sqrt{3}} = 2.9\ \mu\text{m}. \tag{3.7}$$

Thread Angle α

The nominal angle α of the thread is 60°. In ISO 1502:1996, the specifications for metric screw threads are given. Table 6 in this standard states that the tolerances for each half flank angle $\alpha/2$ = 30° are $T_{\alpha/2}$ = ±10 arc minutes. This means that the tolerance for the total flank angle will be twice this value. Similarly, as in the case of the pitch, a rectangular distribution can be assumed for every half thread angle. However, the thread angle is the sum of two half thread angles. Where the uncertainties in the half threads are assumed to have a rectangular distribution, the distribution of the sum of two half angles will be triangular with limits 60° – 20' and 60° + 20'. The standard uncertainty of a quantity having a tolerance $\pm a$, assuming a triangular distribution, is given by:

$$u(\alpha) = \frac{a}{\sqrt{6}} = \frac{20'}{\sqrt{6}} = 8.2' = 0.13°. \tag{3.8}$$

Measurement Wire Diameter d_D

The measurement wires have been calibrated by an accredited laboratory. The calibration certificate states: 'the three wires have an equal diameter of d_D = (1.3500 ± 0.0005) mm'. As an explanation of the uncertainty, it is stated: 'The reported uncertainty is based on a standard uncertainty multiplied by a coverage factor $k = 2$, which provides a confidence level of approximately 95%. The standard uncertainty has been determined in accordance with EA 04/02'. This means that the standard uncertainty can be taken as half of the stated uncertainty, and the use of k = 2 with a 95% confidence level suggests that a Gaussian distribution was assumed. This means that as value and standard uncertainty for the measurement wires can be taken: d_D = 1.35000 mm and $u(d_D)$ = 0.00025 mm = 0.25 μm.

3.4.3 Standard Uncertainty, Resolution and Repeatability

In the VIM, a distinction is made between 'resolution' that is defined as 'smallest change in a quantity being measured that causes a perceptible change

FIGURE 3.2 Digital caliper with a stable reading of 29.97 mm.

in the corresponding indication' and the 'resolution in a displaying device' that is defined as 'smallest difference between displayed indications that can be meaningfully distinguished'. If the uncertainty in a quantity is limited by the resolution of an instrument, this should be reflected in a value for the standard uncertainty due to this resolution. A few guidelines for typical cases are given here.

Stable Digital Readout

When taking repeated measurements using a digital caliper, this may give the same results displayed digitally. This gives a standard deviation of zero, but a standard uncertainty should be attributed to the measurand due to the limited resolution. In the example given in Figure 3.2, it can be assumed that the measured value will be between 29.965 mm and 29.975 mm with a uniform likelihood. Therefore, it is common to assume a rectangular uncertainty distribution ranging from 29.965 mm to 29.975 mm that gives a standard uncertainty of $0.01\ \text{mm}/(2\sqrt{3})=2.9\ \mu\text{m}$. In general, a stable digital readout with a resolution r can be given a standard uncertainty of $u = r/(2\sqrt{3})$.

Varying Digital Readout

When a digital readout is not stable, it makes no sense to consider the last decimal that is displayed. An example is four consecutive readings of a laser interferometer, where the measured distance is subject to turbulence and vibration (see Figure 3.3). The last displayed digit corresponds to 1 nm, but when it comes to resolution, the 'meaningfully' aspect of the definition becomes relevant: the last digit displaying the nanometers is not meaningful. In this case, the mean value of 18.24587 mm and the standard deviation of 0.00032 mm are useful to report, giving a standard uncertainty of 320 nm, not $1/(2\sqrt{3})$ nm.

FIGURE 3.3 Photographs of a laser interferometer display continuously measuring the same displacement.

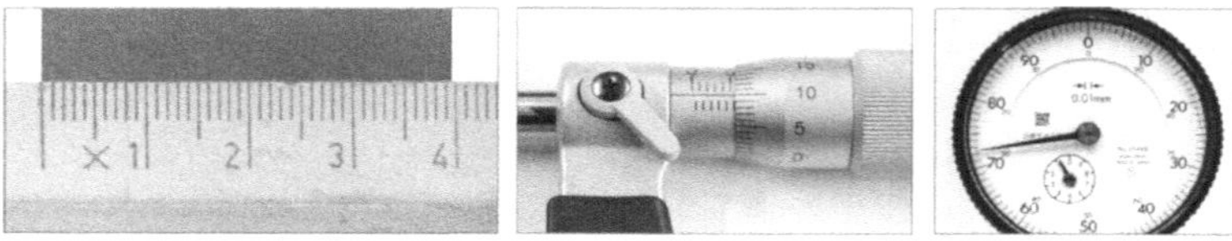

FIGURE 3.4 Examples of reading and interpolation of a scale. Left: reading of a ruler, in this example 39.7 mm, standard uncertainty 0.2 mm. Center: reading of a screw micrometer, in this example 5.60 mm, standard uncertainty 2 μm. Right: reading of a dial gage, in this example 0.724 mm, standard uncertainty 2 μm.

Reading and Interpolation of a Scale

When an analogue scale is read, the indication will generally be between two scale lines. It is more or less agreed that the human eye can interpolate to about 1/5th of the reading resolution. This can be taken as well as the standard uncertainty. In Figure 3.4, a few cases are illustrated.

Reading of a Vernier

Some instruments are equipped with a Vernier that allows a better interpolation between scale lines than an estimation between two lines. This uses the fact that the human eye is very sensitive to symmetry and the straightness of lines. The most commonly encountered Vernier is the Vernier of a Vernier caliper. Mostly the discrimination between two Vernier lines is just possible; therefore, the Vernier resolution is commonly taken as the standard uncertainty. This is illustrated in Figure 3.5 together with the Vernier of a bevel protractor.

Combined Vernier/Indication Reading

There are systems where lines can be brought to coincidence in an eyepiece, where the reading of the displaced eyepiece scale can be used as an interpolation-aid. It may appear attractive to take the resolution of the latter scale as a standard uncertainty, but in this case, the repeatability of the reading must also be taken in to account. As an example, the reading of an autocollimator is taken, where an interpolation between two lines is possible by shifting these until a scale line coincides with a line of the reflected cross through the autocollimator. Then the dial that makes this displacement can be read, with a scale in arc seconds. This

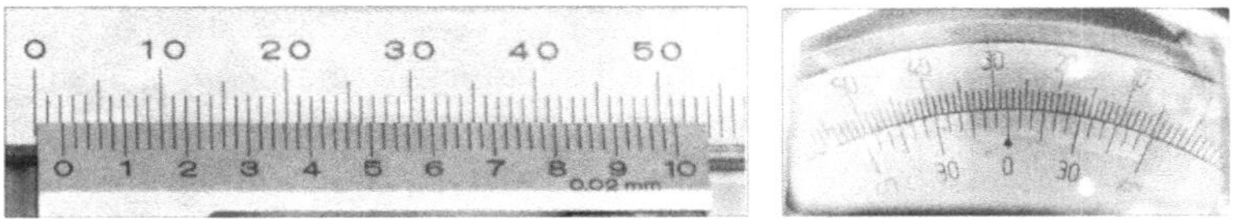

FIGURE 3.5 Examples of readings of a Vernier. Left: Vernier caliper, reading 2.22 mm, resolution and standard uncertainty 0.02 mm. Right: Bevel protractor, reading 27°45', resolution and standard uncertainty 5' (5 arc minutes). Note the improvement over just interpolating the zero-indication of the Vernier.

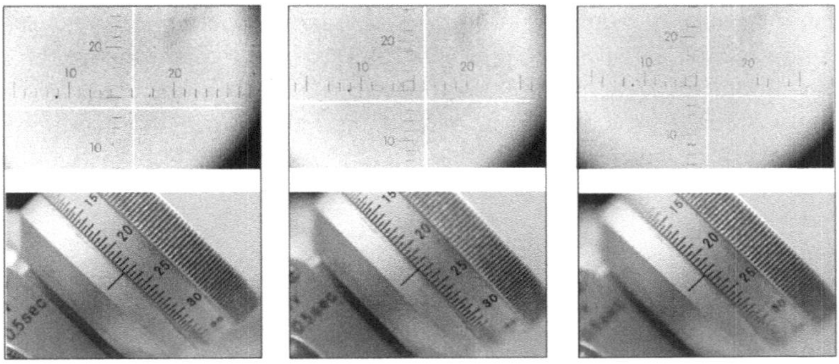

FIGURE 3.6 Three consecutive adjustments of the bright horizontal line to coincide with the 14' line of the vertical scale. The view in the eyepiece is displayed on top and below the corresponding dial position is displayed.

is illustrated in Figure 3.6 by displaying the view where the 14' vertical scale line is displaced to make it coincide with the horizontal bright line of the cross that is back-reflected in the autocollimator, with the corresponding reading of the dial. The dial can be read with a resolution of 0.5 arc seconds that would lead to a standard uncertainty of 0.1 arc seconds when just the dial is read and interpolated. However, the adjustment of the scale depends on the viewer and is not perfectly reproducible, leading to different readings when reproduced. In this case, the respective readings of the dial are 22.7, 22.5 and 22.2 arc seconds, giving a standard deviation of 0.25 arc seconds, which can be taken as the standard uncertainty of this reading. The mean value is 14'22.47".

3.4.4 Uncertainty due to the Definition of a Quantity

The way a quantity is defined may imply an uncertainty or a reduction of uncertainty. Already mentioned are a single reading of a display or the mean of several readings of a display. In dimensional measurements, the definition is often implied in ISO specification standards. For example, ISO 3650:1998 defines the central length of a gage block as 'length of a gage block taken at the center point of the free measuring face'. This is an essential different from a definition like 'the length of a gage block taken at any point of its measuring face'. In the latter case, the variation in length contributes to the uncertainty in the quantity value of the length.

The same applies to the diameter of a hole. Defining this as just 'the diameter' implies that this diameter can be measured in any direction and in any depth of the hole, implying that the variation in diameter has to be determined as a function of depth and orientation. Defining the diameter of a setting ring as 'the diameter in the center of the ring, measured in the direction of the text indicating the nominal diameter' overcomes this ambiguity. Another case is the pitch diameter of a screw caliper as treated as an example in this chapter. For simplifying the calibration and enabling better measurement comparison, a more specific parameter than the 'pitch diameter' as given in equation (3.1) is the 'simple pitch diameter'. In the simple pitch diameter, the pitch P is supposed to have its nominal value, without uncertainty. Actually, measuring the pitch P and attributing an uncertainty to it means more effort, a different value and a larger uncertainty in the pitch diameter d_2, compared to the simple pitch diameter d_2.

3.5 PROPAGATION OF UNCERTAINTIES

A measurement result is often used to calculate a different quantity than the one that is measured. That other quantity may be obtained using known relationships, typically in the form of mathematical models or numeric simulations, and its uncertainty is related to the uncertainty in the original measurement, that is, the uncertainty of the measurement propagates through the equation of the model and appears as an uncertainty in the other quantity. This is illustrated by the example given in Section 3.3. The measurements and estimations of the size M, the pitch P, the angle α and the wire diameter d_D are used for the calculation in the pitch diameter d_2. The uncertainties in the measured diameter will propagate into an uncertainty of the calculated pitch diameter. This propagation can be carried out in a number of ways that are illustrated in the following sections.

3.5.1 Direct Calculation

The method of direct calculation is straightforward and may appear rather unsophisticated. However, in Monte-Carlo methods (see Section 3.5.3), direct calculation is also used. Direct calculation may be the only possibility if the measured quantity is subject to a calculation for which there is no direct expression, for example, as performed by a computer program using an iterative or unknown algorithm. In general, the functional relationship, also called the model function, is given by

$$y = f(x_1, x_2 .. x_n), \tag{3.9}$$

where y is the output estimate of the measurand Y and x_i is the input estimate of the input quantity X_i. For the direct calculation of the standard uncertainty in y, $u(y)$ as a result of the standard uncertainty in x_i, the variation of the output quantity is considered as a function of the variation of the input quantity, and $u(y)$ is given by

$$u(y)_i = \left| f(..x_i + u(x_i)..) - f(x_1, x_2 .. x_n) \right|. \tag{3.10}$$

Omitting the correction factors A_1 and A_2, the model function for the thread measurement (see equation (3.1)) is:

$$d_2(M, P, \alpha, d_D) = M + \frac{P}{2\tan(\alpha/2)} - d_D\left(1 + \frac{1}{\sin(\alpha/2)}\right). \tag{3.11}$$

Substitution of the measured and found values, M = 18.336 mm, P = 2.5 mm, $\alpha = 60°$ and d_D = 1.35 mm gives d_2/mm = 18.3360 + 2.1651 – 4.0500 = 16.4511. In Section 3.4.2, the standard uncertainties have been determined as $u(M)$ = 1.9 μm; $u(P)$ = 2.9 μm; $u(\alpha)$ = 0.13° and $u(d_D)$ = 0.25 μm, respectively.

The standard uncertainty in d_2 as a result of the standard uncertainty in M is given by

$$\begin{aligned} u(d_2)_M &= \left| d_2(M + u(M), P, \alpha, d_D) - d_2(M, P, \alpha, d_D) \right| \\ &= \left| 16.4530 - 16.4511 \right| \text{mm} = 1.9\ \mu\text{m}. \end{aligned} \tag{3.12}$$

Similarly, the standard uncertainties in d_2 as a result of the standard uncertainties in P, α and d_D are given by:

$$u(d_2)_P = \left| d_2(M, P + u(P), \alpha, d_D) - d_2(M, P, \alpha, d_D) \right| = 2.5\ \mu\text{m}, \tag{3.13}$$

$$u(d_2)_\alpha = \left| d_2(M, P, \alpha + u(\alpha), d_D) - d_2(M, P, \alpha, d_D) \right| = 0.37\ \mu\text{m}, \tag{3.14}$$

and

$$u(d_2)_{d_D} = \left| d_2(M, P, \alpha, d_D + u(d_D)) - d_2(M, P, \alpha, d_D) \right| = 0.75\ \mu\text{m}. \tag{3.15}$$

This method is straightforward and appropriate but gives little insight. For example, it is not immediately clear what happens with the uncertainty in the pitch diameter when the uncertainty in the wire diameter doubles.

3.5.2 Calculation Using Partial Derivatives

More insight into the propagation of uncertainty is obtained when using the partial derivative of the function $y = f(x)$. This can be derived from the first two terms of the Taylor series expansion of $f(x+u(x))$ in equation (3.10), assuming $u(x) \ll x$:

$$\begin{aligned} u(y)_i &= \left|f(x_1..x_i + u(x_i)..x_n) - f(x_1, x_2..x_n)\right| \\ &= \left|f(x_1, x_2..x_n) + \frac{\partial f(x_1, x_2..x_n)}{\partial x_i}u(x_i) - f(x_1, x_2..x_n)\right| \\ &= \left|\frac{\partial f(x_1, x_2..x_n)}{\partial x_i}\right|u(x_i). \end{aligned} \tag{3.16}$$

The partial derivative $\frac{\partial f(x_i)}{\partial x_i}$ is called the sensitivity coefficient. It can be useful to substitute the original function $y = f(x)$ in the derivative where possible. Using the derivative, the uncertainties can be written as:

$$u(d_2)_M = \left|\frac{\partial d_2(M, P, \alpha, d_D)}{\partial M}\right|u(M) = 1 \cdot u(M) = 1.9\ \mu\text{m}, \tag{3.17}$$

$$u(d_2)_P = \left|\frac{\partial d_2(M, P, \alpha, d_D)}{\partial P}\right|u(P) = \frac{u(P)}{|2\tan(\alpha/2)|} = 0.87 \cdot u(P) = 2.5\ \mu\text{m}, \tag{3.18}$$

$$\begin{aligned} u(d_2)_\alpha &= \left|\frac{\partial d_2(M, P, \alpha, d_D)}{\partial \alpha}\right|u(\alpha) \\ &= \left|\frac{d_D\cos(\alpha/2) - P/2}{2\sin^2(\alpha/2)}\right|u(\alpha) = 161\frac{\mu\text{m}}{\text{rad}} \cdot u(\alpha) = 0.37\ \mu\text{m}, \end{aligned} \tag{3.19}$$

$$\begin{aligned} u(d_2)_{d_D} &= \left|\frac{\partial d_2(M, P, \alpha, d_D)}{\partial d_D}\right|u(d_D) = \left(1 + \frac{1}{\sin(\alpha/2)}\right)u(d_D) \\ &= 3 \cdot u(d_D) = 0.75\ \mu\text{m}. \end{aligned} \tag{3.20}$$

For classes of functional relationships $y = f(x)$, general rules for the uncertainty propagation can be derived. An overview is given in Table 3.4.

3.5.3 Calculation Using a Monte-Carlo Method

With a Monte-Carlo method, both the propagation of standard uncertainties and the propagation of uncertainty distributions can be calculated. The Monte-Carlo method

TABLE 3.4
Propagation of Uncertainties for Some Functional Relationships

Function $y = f(x)$ c Is a Constant	Uncertainty Propagation	In Words
$y = x + c$	$u(y) = u(x)$	Uncertainty is unchanged
$y = c \cdot x$	$\frac{u(y)}{y} = \frac{u(x)}{x}$	Relative uncertainty is unchanged
$y = x^c$	$\frac{u(y)}{y} = c \cdot \frac{u(x)}{x}$	Relative uncertainty is multiplied with a factor of c
$y = \log(x)$	$u(y) = \frac{u(x)}{x}$	Absolute uncertainty in y becomes relative uncertainty in x
$y = e^x$	$\frac{u(y)}{y} = x$	Relative uncertainty in y becomes absolute uncertainty in x

consists of the generation of a large number of simulated measurements according to a defined distribution, applying the calculation using the model function for each individual simulated measurement and considering the distribution of the output values. Simulations are made using the input variables ξ_i that are given by

$$\zeta_j = x + u(x) \cdot z_j \tag{3.21}$$

where x is the best estimate of the input quantity, usually the mean, $u(x)$ is the standard uncertainty of x and z_j is a random number, taken from a distribution with a mean value of zero and a standard deviation of unity. Most mathematical software contains a function to generate Gaussian distributed random numbers. If a random number generator gives uniformly distributed random numbers r_j between 0 and 1, then a random number z_i with zero mean value and unit standard deviation with a rectangular distribution is obtained from

$$z_j = (r_j - 0.5) \cdot 2\sqrt{3}. \tag{3.22}$$

The simulations give a distribution of the output deviations $\delta_j(x_i)$ that is given by:

$$\delta_j(x_i) = f(x_1..x_i + u(x_i) \cdot z_j..x_n) - f(x_1..x_n). \tag{3.23}$$

The root mean square of the output deviations can be used to estimate the standard uncertainty in y:

$$u(y)_{xi} = \sqrt{\frac{\sum_{j=1}^{K} \delta_j^2(x_i)}{K}}, \tag{3.24}$$

with K the number of simulations. The coverage interval that contains the true value Y with a specified probability can be obtained from the distribution of $\delta_j(x_i)$. This is illustrated for the influence of the thread angle on the pitch diameter. The thread angle is the sum of the two half-angles with a rectangular distribution. Simulated thread angles can be written as:

$$\begin{aligned} \alpha &= (29^0 50' + r_{1,i} \cdot 20') + (29^0 50' + r_{2,i} \cdot 20'), \\ &= (29.83^0 + r_{1,i} \cdot 0.34°) + \left(29.83° + r_{2,i} \cdot 0.34°\right), \end{aligned} \tag{3.25}$$

where the i independent random numbers r are taken from a rectangular distribution between 0 and 1. A Monte-Carlo evaluation of the distribution of d_2 due to the distribution of α using 10 000 simulations can be written in MATLAB code as follows.

```
>> M = 18.336; P = 2.5; dD = 1.35; % constants in mm
>> ad = 29.83 + 0.34*rand(1,10000) + 29.83 + 0.34*rand(1,10000); % simulated
       angles in degrees
>> ar = ad*pi/180; % convert angles to radians
>> d2m = 16.4511; % measured value
>> ud2dist = M + P./(2*tan(ar/2)) – dD*(1 + 1./sin(ar/2)) – d2m; % distribution
       of d2
```

Figure 3.7 gives the distribution of the simulated angles and how this transfers to a distribution of the calculated pitch diameter d_2 around its measured value. It shows how the rectangular distributions of both half-angles (not shown) combine

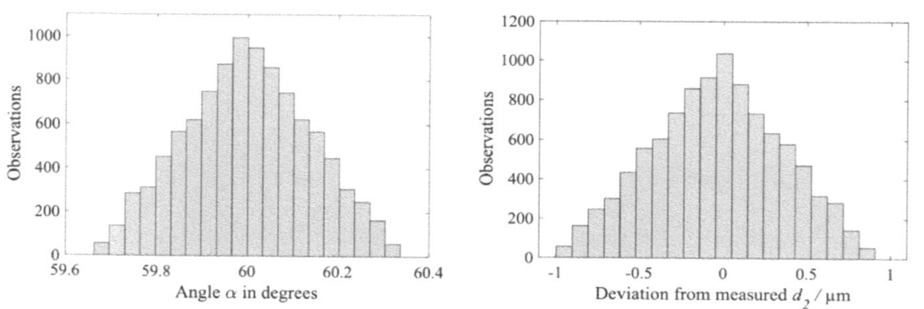

FIGURE 3.7 Distribution of simulated angles (left) and its propagation to a distribution of d_2 values (right).

to the triangular distribution of the angle α that transfers to a triangular distribution of the deviations of the d_2 values. Once this distribution is obtained, the standard uncertainty can be calculated using equation (3.24), giving $u_\alpha(d_2) = 0.37\ \mu\text{m}$, as found earlier with both the direct calculation and derivative methods. Moreover, confidence intervals can be directly derived from this distribution. For example, the 95% confidence interval can be derived by considering for which value 97.5% of the simulated values are larger on one side, and for which value 97.5% of the simulated values are smaller on the other side. In MATLAB, the function 'prctile' is available; in this case the command >>prctile(ud2dist, [2.5 97.5]) will give the 95% confidence interval, here [–0.80, 0.70] µm, but this may vary somewhat between simulations. This section illustrates the Monte-Carlo concept as it is described in the *JCGM Guide* (JCGM 2008), where it is considered that in uncertainty evaluations, distributions are propagated rather than single measurement values and standard uncertainties.

3.6 MEASUREMENTS WITH MULTIPLE VARIABLES

In Section 3.5, the uncertainty propagation for single variables was treated. However, in the example, and also in practice, there are often multiple variables that contribute to an uncertainty in a measured quantity. The uncertainty analysis for such cases is treated in this section.

3.6.1 Propagation of Standard Uncertainties

The propagation of standard uncertainties can be carried out numerically, as in Section 3.5.1; however, for some more general considerations, the propagation similar to that in Section 3.5.2, using partial derivatives, is considered. The expression for the combined variance $u^2(y)$ associated with the result of measurements of multiple variables x_i is obtained using an extension of equation (3.15) to multiple variables, and is given by

$$
\begin{aligned}
u^2(y) &= \sum_{i=1}^{N}\sum_{j=1}^{N} \frac{\partial f}{\partial x_i}\frac{\partial f}{\partial x_j} u(x_i, x_j) \\
&= \sum_{i=1}^{N}\left(\frac{\partial f}{\partial x_i}\right)^2 u^2(x_i) + 2\sum_{i=1}^{N-1}\sum_{j=i+1}^{N} \frac{\partial f}{\partial x_j}\frac{\partial f}{\partial x_i} u(x_i)\cdot u(x_j)\cdot r(x_i, x_j)
\end{aligned}
\tag{3.26}
$$

where $r(x_i, x_j)$ is the correlation coefficient between variables x_i and x_j that is given by

$$
r(x_i, x_j) = \frac{u(x_i, x_j)}{u(x_i)\cdot u(x_j)}. \tag{3.27}
$$

The value of $r(x_i, x_j)$ may vary between −1 (full negative correlation) through 0 (no correlation) to +1 (full positive correlation). Uncorrelated input quantities, for which $r(x_i, x_j) = 0$, mean that a change in one quantity does not imply an expected change in another. In many cases, this assumption is justified. Taking the screw thread measurement as an example, there is no reason to assume that if the pitch P is near the upper boundary of the specification, the pitch angle would be at the lower (or upper) side, so these quantities can be assumed to be uncorrelated. In the case of uncorrelated quantities, equation (3.25) reduces to:

$$u^2(y) = \sum_{i=1}^{N} \left(\frac{\partial f}{\partial x_i} \right)^2 u^2(x_i). \tag{3.28}$$

Equation (3.28) is the essential equation for the propagation of standard uncertainties that is the basis of the large majority of uncertainty evaluations. The partial derivatives can be omitted by applying the direct calculation similar to equation (3.10):

$$u^2(y) = \sum_{i=1}^{N} \left(f\left(x_i + u(x_i)\right) - f(x_i) \right)^2. \tag{3.29}$$

For the example of the thread measurement, equation (3.28) can be written as:

$$\begin{aligned} u^2(d_2) &= \left(\frac{\partial d_2}{\partial M} \right)^2 u^2(M) + \left(\frac{\partial d_2}{\partial P} \right)^2 u^2(P) + \left(\frac{\partial d_2}{\partial \alpha} \right)^2 u^2(\alpha) + \left(\frac{\partial d_2}{\partial d_D} \right)^2 u^2(d_D) \\ &= u^2(M) + \left(\frac{1}{2\tan(\alpha/2)} \right)^2 u^2(P) \\ &\quad + \left(\frac{d_D \cos(\alpha/2) - P/2}{2\sin^2(\alpha/2)} \right)^2 u^2(\alpha) + \left(1 + \frac{1}{\sin(\alpha/2)} \right)^2 u^2(d_D). \end{aligned} \tag{3.30}$$

Here, the partial derivatives given in equations (3.17) to (3.20) have been used. The standard uncertainty $u(d_2)$ can now be evaluated numerically using the numerical results as given in equations (3.17) to (3.20):

$$u(d_2) = \sqrt{1.9^2 + 2.5^2 + 0.37^2 + 0.25^2}\ \mu\text{m} = 3.2\ \mu\text{m}. \tag{3.31}$$

3.6.2 Uncertainty Budgeting

Although equation (3.28) (or equation (3.29)) is essentially all that is needed for the majority of uncertainty evaluations, insight can be obtained more easily if the evaluation is arranged in a so-called 'uncertainty budget'. An uncertainty budget is an ordered arrangement of the quantities, estimates, standard uncertainties, sensitivity coefficients and uncertainty contributions used in the uncertainty analysis of a measurement. The uncertainty budget is the representation of equation (3.28) in the form of a table, as shown in Table 3.5. Sometimes the distribution used is included as an additional column, but this is omitted in Table 3.5. For the example of the thread measurement, using the numerical values given in equations (3.16) to (3.19), the uncertainty budget is as given in Table 3.6.

TABLE 3.5
Schema of an Ordered Arrangement of All Aspects of the Uncertainty Analysis of a Measurement, Also Known as an Uncertainty Budget

Quantity X_i	Estimate x_i	Standard Uncertainty $u(x_i)$	Sensitivity Coefficient $c_i = \frac{\partial f(y)}{\partial x_i}$	Contribution to Standard Uncertainty $u_i(y) = c_i \cdot x_i$
X_1	x_1	$u(x_1)$	c_1	$u_1(y)$
X_2	x_2	$u(x_2)$	c_2	$u_2(y)$
:	:	:	:	
X_N	x_N	$u(x_N)$	c_N	$u_N(y)$
Y	y			$u(y) = \sqrt{\sum_{i=1}^{N} u_i(y)}$

TABLE 3.6
Example of an Uncertainty Budget for a Screw Thread Measurement

Quantity X_i	Estimate x_i	Standard Uncertainty $u(x_i)$	Sensitivity Coefficient $c_i = \frac{\partial f(y)}{\partial x_i}$	Contribution to Standard Uncertainty $u_i(y) = c_i \cdot x_i$
M	18.336 mm	1.9 μm	1	1.9 μm
P	2.5 mm	2.9 μm	0.87	2.5 μm
α	60°	0.13°	2.8 μm/°	0.37 μm
d_D	1.35 mm	0.25 μm	3	0.75 μm
d_2	16.4511 mm			$u(d_2)$=3.2 μm

From Table 3.6, it can directly be observed that the uncertainty in the pitch P is the main contributor to the standard uncertainty in the pitch diameter, so if the uncertainty is to be reduced, this is the first parameter where the determination needs to be improved, for example, by a direct measurement, or by defining d_2 as the simple pitch diameter which omits the uncertainty in P, giving a standard uncertainty in d_2 of $u(d_2) = 2.1$ µm.

3.6.3 Confidence Intervals and Degrees of Freedom

In Section 3.4.1, it was illustrated how a standard deviation can be used to define an uncertainty based on a confidence interval. Similar to the Student's t-factor used in Section 3.4.1, it can be stated more generally that a standard uncertainty must be multiplied by a factor k to obtain an uncertainty that can be associated with a defined confidence interval, thus

$$U(y) = ku(y) \tag{3.32}$$

where $U(y)$ is the expanded measurement uncertainty. In general, it is assumed that taking $k = 2$ leads to a confidence interval of at least 95% (actually $k = 1.96$ gives a confidence interval of 95%, and $k = 2$ gives a confidence interval of 95.5% for a Gaussian distribution). This is based on assuming that most quantities will have a Gaussian distribution, and even if this is not the case, the combination of different kinds of distributions will often result in a Gaussian or Gaussian-like distribution. This observation is known as the central limit theorem. In the example of the pitch diameter d_2, the measurement result could be reported as follows: 'The measured pitch diameter is $d_2 = (16.4511 \pm 0.0032)$ mm, where the reported expanded measurement uncertainty is the standard uncertainty multiplied by the coverage factor $k = 2$, which for a Gaussian distribution corresponds to a coverage probability of approximately 95%'. For most cases, such an evaluation is sufficient. When a type A uncertainty with a low number of degrees of freedom is the major contributor to the uncertainty, taking $k > 2$ may be needed. For such a case, the degrees of freedom for the different uncertainty contributors can be combined to an 'effective degrees of freedom' for which the t-distribution and its confidence interval can be used, see Table 3.3. The effective degrees of freedom depends on the relative magnitude of the uncertainties and their degrees of freedom, and may be approximated by the so-called Welch-Satterthwaite equation:

$$\upsilon_{eff} = \frac{u^4(y)}{\sum_{i=1}^{N} \frac{u_i^4(y)}{v_i}}, \tag{3.33}$$

where $u(y)$ is the combined standard uncertainty and u_i and v_i are the contributions and their degrees of freedom, respectively. The degrees of freedom, also called the number of free parameters, is the number of measurements minus the number of parameters that is derived from them before a statistical analysis is made. For example, when taking an average out of N measurements, the degrees of freedom is $\upsilon = N-1$. When a least-squares plane is fitted through a number of height coordinates $z(x,y)$, an intercept is calculated plus gradients in the x- and y-direction, making the degrees of freedom $\upsilon = N-3$. For type-B uncertainties, it is common to take the degrees of freedom as infinite: $\upsilon = \infty$. Then the expanded uncertainty is given by $U = ku(y) = t(v_{eff})u(y)$, using the t-factor corresponding to the effective degrees of freedom for a 95% confidence interval. When the type-B uncertainties are the dominant factors in an uncertainty budget, the difference with taking $k = 2$ is commonly negligible. When the uncertainty distribution and the related confidence intervals need a closer consideration, the Monte-Carlo method can be used. This is explained in the following section.

3.6.4 Propagation of Probability Distributions

The Monte-Carlo method introduced in Section 3.5.3 can be extended to multiple input quantities x_i. This approach has the advantage that each variable can be given its own uncertainty distribution and that the confidence intervals of the output estimate y can be calculated directly from its distribution. The name 'Monte-Carlo' means in general that a statistical distribution is generated using random numbers. In uncertainty evaluations, this means that using the functional relationship as given in equation (3.11), many simulations evaluating y are carried out using

$$y_k = f\left(x_1 + u(x_1)\cdot r_{1,k} \,\ldots\ldots\ldots\, x_N + u(x_N)\cdot r_{N,k}\right) \tag{3.34}$$

where y_k is the output value for the kth simulation, k is the simulation number, $k = 1\ldots K$, where K is the total number of simulations, x_i are the input values and $r_{i,k}$ is a random number with zero mean and unit standard deviation that has a distribution according to the input variable x_i. The simulations give a distribution of the output deviations δ_k that is given by:

$$\delta_k = y_k - f(x_1..x_n). \tag{3.35}$$

These simulated values can be used for evaluating the distribution of y; for example, the standard uncertainty in y, $u(y)$, can be evaluated from

$$u(y) = \sqrt{\frac{\sum_{k=1}^{K} \delta_k^2}{K}}. \tag{3.36}$$

This evaluation becomes more reliable with more simulations (larger K), similar to how the standard deviation becomes more reliable with more measurements;

see equation (3.5). The distribution of the values of d_2 can be used to determine confidence intervals. The Monte-Carlo method is illustrated here for the thread measurement that was used as an example in the previous sections. The input quantities and their distribution are discussed in Section 3.5.3 and summarized in Table 3.7. The difference with Section 3.5.3 is that the other parameters are also varied. Figure 3.8 gives the distribution of the quantity δ as defined in equation (3.35) for the screw thread measurement. It can be observed that the distribution does not look purely Gaussian; this is because the rectangular distributed pitch P dominates the distribution of δ. The standard uncertainty calculated from this

TABLE 3.7
Calculation of the Uncertainty of the Pitch Diameter d_2 of a Screw Thread Caliper Using a Monte-Carlo Method

Quantity X_i	Estimate x_i	Standard ncertainty $u(x_i)$	Distribution	MATLAB Code for 10 000 Simulations
M	18.336 mm	1.9 μm	Gaussian	M=18.336+ 0.0019*randn(1,10000);
P	2.5 mm	2.9 μm	Rectangular	P=2.495+0.010*rand(1,10000);
α	60°	0.13°	Triangular	ad=59.66+0.34*rand(1,10000)+ 0.34*rand(1,10000); ar=ad*pi/ 180; % convert angles to radians;
d_D	1.35 mm	0.25 μm	Gaussian	dD=1.35+0.00025*randn(1,10000);
d_2	16.4511 mm			delta=M+P./(2.*tan(ar/2))-dD.*(1+ 1./sin(ar/2))-16.4511;

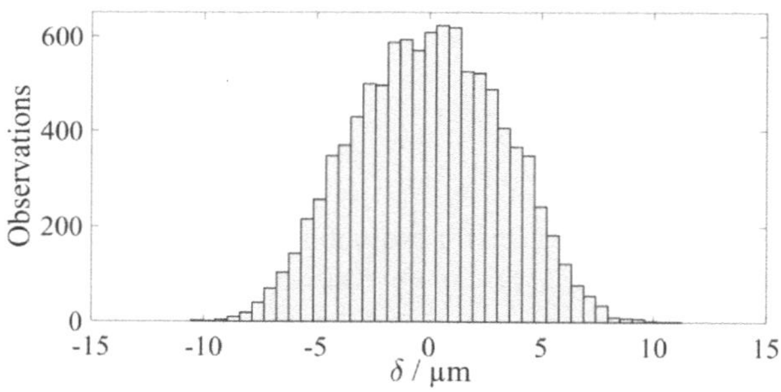

FIGURE 3.8 Distribution of the measurement deviation δ as the result of a Monte-Carlo simulation of a screw thread measurement.

distribution using equation (3.36) is $u(d_2) = 3.3$ μm, which is consistent with the result in Table 3.6 using the uncertainty budget. The 95% confidence interval is given by $d_2 = (16.4511 \pm 0.0061)$ mm, where the uncertainty $U = 6.1$ μm is somewhat narrower than $U = 6.6$ μm that would be expected using $k = 2$, because of the somewhat dominant influence of the rectangular distribution.

Although the results from Monte-Carlo simulations may be similar to a more traditional uncertainty budget calculation in many cases, there are apparent advantages to propagating uncertainty distributions by Monte-Carlo simulations. These advantages are as follows.

- The Monte-Carlo method in fact uses the direct calculation method as in equation (3.10), so the calculation of partial derivatives is omitted.
- Correlations of input quantities can be incorporated into the simulations by correlating (combining) the random numbers used to characterize the distributions involved.
- The standard uncertainty and confidence intervals are readily obtained; considerations about the appropriate k-factor can be omitted.
- When testing against specifications, the probability of meeting the specifications can readily be calculated.
- The Monte-Carlo method is considered as a more rigorous method of uncertainty calculations and may provide more insight of how the individual measurements will influence the values and subsequent distributions coming from a particular process (JCGM 2008).

3.7 SIGNIFICANT FIGURES AND DECIMAL PLACES

In the reporting of measurement results and uncertainties, the significant figures and decimal places used must be considered. A digit is significant if it is required to express the numerical value of a quantity. In the expression $l = 1200$ mm, it is not possible to tell whether the last two zeroes are significant or only indicate the magnitude of the numerical value of the length l. However, in the expression $l = 1.200$ m, the two zeroes are assumed to be significant because if they were not, the value of l would have been written $l = 1.2$ m. For this reason, it must be considered what prefixes are used when expressing a dimensional quantity.

In reporting uncertainties, standard deviations and standard uncertainties, it is common to express these in two significant figures at maximum. Tolerances may be given with more or less significant figures where relevant. The mean quantity value is given in the same significant figures so that the uncertainty or tolerance can be added or subtracted without losing significant figures. A few examples are given in Table 3.8, see also the NIST guidelines (Taylor and Kuyatt 1994) and Section 1.4.5.

TABLE 3.8
Significant Figures in the Reporting of a Room Temperature Specification and in the Reporting of a Pitch Diameter

Wrong Way of Reporting	Right Way of Reporting
$T = (20 \pm 0.2)$ °C	$T = (20.0 \pm 0.2)$ °C
$d_2 = (16.4511 \pm 0.006)$ mm	$d_2 = (16.4511 \pm 0.0061)$ mm
$d_2 = (16.45112 \pm 0.00613)$ mm	$d_2 = (16.4511 \pm 0.0061)$ mm

3.8 VALUATION OF MEASUREMENTS AND UNCERTAINTIES

Once a measurement value has been established and the uncertainty has been evaluated, a valuation may be carried out for the product or standard that has been calibrated: does it satisfy the specification, is it consistent with an earlier or other calibration, etc.? This section deals with the four situations: it can be checked whether a measured product meets its specification; whether two values and uncertainties are consistent; whether a result, in comparison with many other results, is consistent or not; and whether a measuring instrument, calibrated at many points within its measurement range, conforms to specifications.

3.8.1 Conformance or Nonconformance with Specifications

In the field of dimensional metrology, the specification is commonly a dimensional feature of a product as it is indicated on a technical drawing, with tolerance limits. Also, the tolerance may be given by a specification standard, for example, ISO 1502:1996 for screw calipers, or ISO 3650:1998 that gives the tolerances for gage blocks of several classes. The judgement by which it is decided whether a measurement result means a conformance or non-conformance with specification is called a *decision rule*. The decision rules for dimensional measurements are defined in ISO 14253-1:2018. In this section, we treat the default case of a measured length L_m of an object, with a lower specification limit (LSL) and an upper specification limit (USL), assuming that the uncertainty is smaller than the tolerance interval (USL – LSL >> $u(L_m)$) and that the probability distribution specifying the uncertainty is symmetric and Gaussian. This gives the situation as indicated in Figure 3.9 where the measured value L_m is within the tolerance limits set by LSL and USL, considering a measurement uncertainty $U(L_m) = k \cdot u(L_m)$, where the expanded measurement uncertainty is the standard uncertainty multiplied by the coverage factor $k = 1.65$, which corresponds to a coverage probability of approximately 90%. Here, a coverage probability of 90% is considered sufficient to provide a 95% probability of justly concluding conformity with specification if the measure L' (length with 90% confidence interval) is within the limits set by LSL and USL. This is called the conformance probability limit. Conformance with specifications can be concluded if L' is fully inside the interval set by LSL and

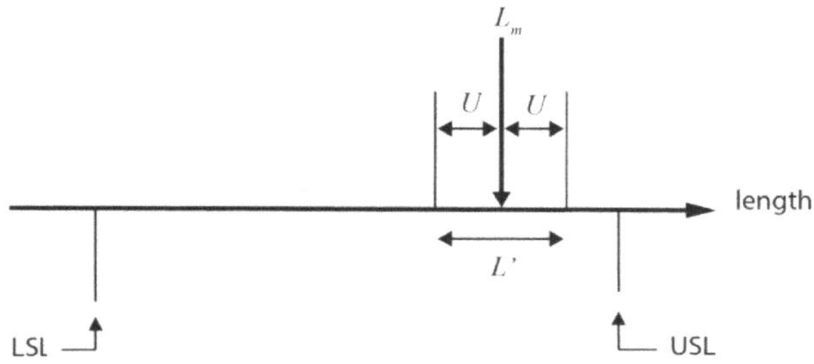

FIGURE 3.9 Schematic presentation of a measured value L_m with uncertainty U, giving an uncertainty interval L' that has to be considered in relation to LSL and USL.

ULS. Nonconformance with specifications can be concluded if L' is fully outside the interval set by LSL and USL. For the cases that L' overlaps with either LSL or USL, some decision has to be taken. The default decision rule is: 'the measurement uncertainty always counts against the party who is verifying conformity or nonconformity and therefore making the measurement'. This implies that a supplier taking the measurements must ensure that L' is fully inside the limits (as indicated in Figure 3.9), and that a customer who wants to claim that a product does not conform to its specification must ensure that L' is fully outside the limits. It is good practice that the decision rule is agreed, for example, in a contract, before a measurement is taken. Sometimes decision rules are implicitly or explicitly defined in (other) ISO standards or legislation. An example is taking a 'binary' (pass/fail) decision by requiring that the measurement value L_m is within the limits set by LSL and USL, so without taking the measurement uncertainty into account. However, in those cases, it is common to require that the measurement uncertainty is smaller than some limit, for example, smaller than 1/5th of the interval USL – LSL. This is known as 'the 20% rule'. Other decision rules may include putting $k = 2$ or even $k = 3$, or assuming different, even nonsymmetrical probability distribution functions. More information and considerations on this subject are given in ISO 14253-1:2018 and in ILAC Guide G8 (ILAC 2019).

As an example, the result of the thread measurement as given in Table 3.6 is taken: $d_2 = 16.4511$ mm and $u(d_2) = 3.2$. With $k = 1.6$, this gives the interval $L' = (16.4511 \pm 0.0051)$ mm. From ISO 1502:1996, it can be derived that LSL = 16.376 mm and USL = 16.581 mm. This means that L' is safely within the interval set by LSL and ULS and it can be stated that the pitch diameter d_2 of the gage is within the tolerance limits set by ISO 1502:1886.

3.8.2 Comparing Two Values, Both Given with Uncertainty

A measurement result x_1 with a measurement uncertainty $U(x_1)$ based on $k = 2$ can be compared with another result x_2 having a measurement uncertainty $U(x_2)$ based on $k = 2$, considering using the normalized error E_N that is defined as:

$$E_N = \frac{x_1 - x_2}{\sqrt{U^2\left(x_1\right) + U^2\left(x_2\right)}}. \tag{3.37}$$

It is common practice to consider the measurements as mutually consistent if $\left|E_N\right| < 1$. The background of this criterion is that, if $\left|E_N\right| > 1$, then the value of $y = x_1 - x_2$, considering equation (3.28) is significantly different from zero. Equation (3.37) can be used in several ways: to compare the results of two calibration laboratories, to compare the result of a calibration laboratory to a reference value, for example, measured at an NMI, to compare the performance of different calibration technicians, etc. When applying equation (3.37), care must be taken that the uncertainty evaluation does not contain terms that are the same for both measurements. For example, the calibration uncertainty of the same measuring instrument when the performance of different calibration technicians is compared, or the uncertainty in the pitch P when the pitch diameter of the same screw caliper is calibrated: in that case it is better to consider the simple pitch diameter. If $\left|E_N\right| > 1$, it is assumed that 'something is wrong', that is, one of the laboratories has underestimated its uncertainty by forgetting or underestimating a significant contribution to the uncertainty, making some mistake in aligning, reading, etc.

3.8.3 Comparing Multiple Values, All Given with Uncertainty

In comparisons between multiple laboratories, also called proficiency testing (see ISO/IEC 17043: 2023), it is common that one laboratory acts as the reference laboratory; this can be a NMI or a laboratory having a significantly lower uncertainty than most others. In this case, the E_N value calculated from equation (3.37) can be used for every individual result x_1 compared to the reference value x_2. When the results and uncertainties are obtained using comparable techniques and there is no reason to consider one laboratory as 'undoubted' reference laboratory, a mutual comparison can still be carried out. One possibility is to make a matrix of all mutual E_N values using equation (3.37). If one row (or column) has a significant number of $\left|E_N\right|$ values larger than 1, the participant corresponding with that row (or column) has likely a problem with its measuring method and/or uncertainty evaluation. Another analysis method is to calculate the weighed mean and using the weighed mean as a reference value; see also Section A.1.2. Taking the measured values of N participants identified by $I = 1\ldots N$ as x_i and the standard uncertainties as u_i gives for the reference value x_{ref} a weighed mean:

$$x_{ref} = \frac{\sum_{i=1}^{N} \frac{x_i}{u_i^2}}{\sum \frac{1}{u_i^2}}. \tag{3.38}$$

The standard uncertainty in this weighed mean can be derived using equation (3.28) and is given by (see Section A.1.1):

$$u_{\text{ref}} = \frac{1}{\sum_{i=1}^{N} \frac{1}{u_i^2}}. \tag{3.39}$$

The normalized error for every measurement result x_i is now given by:

$$E_{Ni} = \frac{x_i - x_{\text{ref}}}{2\sqrt{u_i^2 - u_{\text{ref}}^2}}. \tag{3.40}$$

The negative sign in the denominator of equation (3.40) ensures that every measurement is compared to the weighed mean of all the others using equation (3.37). If a measurement has an $|E_N|$ value larger than 1, and is the only or one of the few among all the results, this measurement can be excluded from the calculation, and equation (3.37) can be applied for this measurement with the reference measurement defined from equations (3.39) and (3.40). Methods exist to find a largest consistent subset of measurements that appear to be mutually consistent (Cox 2007). These analyses depend on traceable and independent measurements, and realistic uncertainty budgets. If one participant claims an unrealistically low uncertainty, the weighed mean will automatically be biased towards this measurement result. As this analysis is purely statistical, it cannot exclude the possibility that a few participants with $|E_N|$ values greater than 1 are in fact the only participants measuring correctly, while all the others have a similar uncorrected bias. In ISO 17025:2017 a distinction is made between intra-laboratory comparisons (within one laboratory) and interlaboratory comparisons (between different, independent laboratories). In intra-laboratory comparisons, correlations between measurements may occur, for example, the same standard is used for every measurement. In that case the common factors, in this case the uncertainty in the used standard, must be excluded from the uncertainty budget when judging if measurement results are mutually consistent.

3.8.4 Conformance to Specification of a Measuring Instrument: The Test Uncertainty

When calibrating a measuring instrument, for example, a micrometer using a number of gage blocks, or a CMM measuring a number of sizes using a step gage, the measurement uncertainty must include the instruments repeatability if the measurand is the deviation of the instrument at that specific point in its measurement range. However, if the measurand is the instruments specification, then the instruments repeatability is already expressed in the measured value that should be inside specification, and this repeatability should not be ‘double counted' by also incorporating this in the evaluation whether an instrument conforms to its specification. For this reason, the concept of 'test uncertainty' has emerged, that is mainly applied in acceptance tests of CMM's but is also applicable to other measuring instruments. The basics is that the instruments contributions to the uncertainty are not included in the uncertainty budget as these are already included in the measured values and their dispersion; the uncertainty in the used standards and calibration instruments (for example, a laser interferometer) and the environmental effects on these standards are the remaining uncertainty sources in this uncertainty budget. This has the advantage that the measuring instrument does not contribute to the uncertainty 'band' U indicated in Figure 3.9. A disadvantage is that interlaboratory comparisons may give inconsistent results when considering a certain point within a measurement volume; however, the conclusion conformance/nonconformance to specification must still be univocal between different laboratories. Details are given in ISO 14253-5:2015.

EXERCISES

1. Standard deviation in the mean
   Four measurements of a rod diameter are taken using a caliper. The measurements $x_1 \ldots x_4$ are used to estimate a mean rod diameter d:

$$d = \frac{x_1 + x_2 + x_3 + x_4}{4}. \quad (3.41)$$

   All four measurements have the same standard deviation s.
   a. Calculate the standard deviation in the mean diameter using the general law of uncertainty propagation, equation (3.23).
   b. Make an uncertainty budget similar to that in Table 3.5 with variables $x_1 \ldots x_4$ and calculate the uncertainty in d from there.
   c. Take the general case of the calculation of the mean of n measurements $y = \frac{\sum_{i=1}^{n} x_i}{n}$ and prove equation (3.4) using equation (3.23).

2. Mean and standard deviation
   The diameter of a cylinder axis is measured at several positions. The measuring instrument has no errors. The results are given in Table 3.9.

   Calculate:
   a. The standard deviation in the measurements.
   b. The mean diameter and the standard deviation in the mean diameter. Assume that these measurements are taken from a Gaussian distribution.

TABLE 3.9
Cylinder Diameter Measurements at Six Positions

Measurement No.	1	2	3	4	5	6
Diameter/mm	2.563	2.542	2.557	2.553	2.560	2.550

TABLE 3.10
Five Repeated Laser Interferometer Measurements

Measurement No.	1	2	3	4	5
Size/mm	10.0001	10.0001	10.0001	10.0002	10.0001

Calculate:

c. The diameter at any position with a probability of 95%.
d. The mean diameter with a probability of 95%.
e. The probability that the diameter at some position is smaller than 2.540 mm. Hint: use the MATLAB function *tcdf.*

3. Temperature effect on length measurement
A gage block with a nominal length of 10 mm is measured on a length measuring machine that is readout by a laser interferometer. The laser interferometer has no systematic uncertainty. The laser interferometer is not corrected for the linear expansion coefficient, so it measures the actual length. The repeated measurements are given in Table 3.10.
The room temperature is specified as $T = (21 \pm 1)$ °C and this is also the gage block temperature. The gage block length is defined as the length at 20 °C.
Assume a rectangular distribution for the temperature. The linear thermal expansion coefficient of the gage block is specified as $\alpha = (11.5 \pm 1.0) \cdot 10^{-6}\ \mathrm{K}^{-1}$. Assume a rectangular distribution for the linear thermal expansion coefficient.
a. Give the model function for the gage block length at 20 °C (see Section 2.2).
b. Give the standard uncertainty caused by the limited resolution of the laser interferometer.
c. Give the standard uncertainty caused by the fluctuation of the laser interferometer.
d. Give the standard uncertainty in the gage block length caused by the temperature deviation from 20 °C.
e. Give the standard uncertainty in the gage block length caused by the uncertainty in the expansion coefficient.

f. Calculate the gage block length at 20 °C and its uncertainty, based on $k = 2$, using an uncertainty budget and the partial derivatives of the model function.
g. Calculate the uncertainty based on $k = 2$ for the gage block length using an uncertainty budget with the direct calculation method.
h. Calculate the 95% confidence interval for the gage block length at 20 °C using a Monte-Carlo method.

4. Temperature effect on length measurement; same nominal temperature
A steel object is measured in comparison to a steel gage block with a nominal length of $L = 50$ mm. This situation is sketched in Figure 3.10. Before measuring, care is taken to ensure that the gage block and the object assume the ambient temperature of the measuring room. The remaining difference in temperature between the object and the gage block to be calibrated is estimated to be within ±0.05 K. Based on the calibration certificate and the manufacturer's data for the of the reference gage block to be calibrated the linear thermal expansion coefficient of the steel gage blocks is assumed to be within the interval $(11.5 \pm 1.0) \cdot 10^{-6}$ °C^{-1}. Combining the two rectangular distributions, the difference in linear thermal expansion coefficient is triangularly distributed within the limits $\pm 2 \cdot 10^{-6}$ °C^{-1}. The deviation of the mean temperature of measurement from the reference temperature $t_0 = 20$ °C is estimated to be within ±0.5 °C. The uncertainty in the measurement of the length difference is negligible.
a. Give the model function for the object length at 20 °C (see Section 3.5.1).
b. Calculate the uncertainty in the object length at 20 °C based on $k = 2$ using an uncertainty budget and the partial derivatives of the model function.

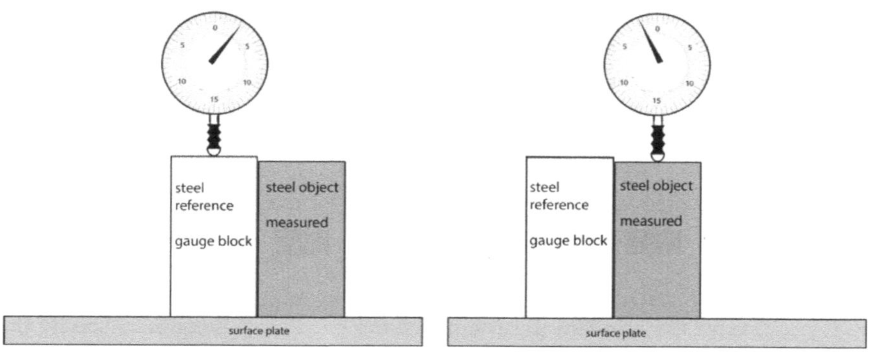

FIGURE 3.10 Measurement of a steel object in comparison with a steel gage block having a same nominal temperature and a same nominal coefficient of expansion.

Note that some uncertainty contributions are negligible because the differences in temperature and expansion coefficient are nominally zero. Therefore, second-order terms have to be taken into account in the evaluation of their uncertainty contribution resulting in the product of standard uncertainties associated with the factors of the product term $\Delta\alpha\left(\bar{T} - 20\,°C\right)$ in equation (2.9). This gives an additional term in the uncertainty budget of $u\left(\Delta\alpha\left(\bar{T} - 20\,°C\right)\right)\cdot L = 0.236\cdot10^{-6}\cdot L$. See also example S4 in EA4/02 (EA 2013).

c. Include this additional term in the uncertainty budget and calculate the uncertainty in the object length at 20 °C based on $k = 2$.
d. Calculate the 95% confidence interval based on the measurement of the length difference ΔL using a Monte-Carlo method and compare the result to the result obtained in b. and c.

5. Straightness reversal measurements
In Section 2.9, it is demonstrated how an absolute straightness profile can be obtained by measuring the profiles $z_A(x)$, $z_B(x)$ and $z_C(x)$ against each other in three combinations; by measurements denoted by $M_1(x)$, $M_2(x)$ and $M_3(x)$. For a given value of x, the standard uncertainty in $M_1(x)$, $M_2(x)$ and $M_3(\mathrm{x})$ has the same value $u(x)$. Calculate the standard uncertainty in the profile $z_A(x)$ at position x.
6. Effective degrees of freedom
Calculate the effective degrees of freedom for the uncertainty calculation presented in Table 3.6. Calculate the appropriate t-factor and the uncertainty interval with a probability of 95%. Compare with $k = 2$.
7. Gage block classification
The tolerance of a gage block of class 0 is ± (0.12 µm + $2\cdot10^{-6}\cdot l$), where l is the gauge block length in millimetres. In a calibration laboratory, the length of a gage block having a nominal length of 100 mm is measured l = (100.00025 ± 0.00010) mm, where the uncertainty is based on $k = 1.6$.
What does the calibration laboratory conclude about the class of the gage block?
   - May a reseller sell such a gage block as a class 0 gage block?
   - Can a buyer of such a gage block send it back to the seller, stating it is outside tolerance?
   - What should the measurement uncertainty be so that the result would not lead to any disputes?
8. Evaluation of comparison results
A reference ball diameter is measured by several laboratories. The results are given in Table 3.11.
   a. Lab 1 is considered as the reference laboratory. Calculate the E_N values of the other laboratories.
   b. It is decided that there is no specific reference laboratory. Calculate the weighed mean and the standard uncertainty in the weighed mean, and calculate the E_N values of all laboratories.

TABLE 3.11
Measured Ball Diameter and Their Standard Uncertainty Measured by Six Laboratories

Laboratory	Measured Diameter/mm	Standard Uncertainty/µm
Lab 1	25.40051	0.15
Lab 2	25.40070	1.50
Lab 3	25.39940	1.00
Lab 4	25.40025	0.20
Lab 5	25.39930	0.43
Lab 6	25.40010	0.55

c. Exclude the laboratory that probably has an issue and repeat the calculations as in b.
d. Take the weighed mean and its standard uncertainty as calculated in c. as reference and calculate (again) the E_N value of the laboratory that probably has an issue. Compare with the result found in b.

REFERENCES

Cox, M.G., 2007, The Evaluation of Key Comparison Data: Determining the Largest Consistent Subset. *Metrologia* **44(3):**187–200.

EA, 2013, EA-4/02. *Expression of the Uncertainty of Measurement in Calibration.* https://european-accreditation.org/wp-content/uploads/2018/10/EA-4-02.pdf accessed 2025-03-01.

EURAMET, 2012, Determination of Pitch Diameter of Parallel Thread Gauges by Mechanical Probing. www.euramet.org/Media/docs/Publications/calguides/EURAMET_cg-10__v_2.1_Determination_of_Pitch_Diameter.pdf accessed 2025-01-03.

Haitjema, H., 2021, Measurement Uncertainty. In: Leach, S. (ed.) *Basics of Precision Engineering.* Boca Raton, CRC Press.

ILAC, 2019, ILAC G8:09/2019. *Guidelines on Decision Rules and Statements of Conformity,* International Laboratory Accreditation Cooperation. https://ilac.org/publications-and-resources/ilac-guidance-series/ accessed 2025-01-03.

ISO 1502, 1996, General-Purpose Metric Screw Threads; Gauges and Gauging. Geneva, International Organization for standardization.

ISO 3650, 1998, Geometrical Product Specifications (GPS) – Length Standards – Gauge Blocks. Geneva, International Organization for Standardization.

ISO 14253-1 2017. Geometrical product specifications (GPS) – Inspection by measurement of workpieces and measuring equipment – Part 1: Decision rules for verifying conformity or nonconformity with specifications . Geneva, International Organization for Standardization.

ISO 14253-2, 2011, Geometrical Product Specifications (GPS) – Inspection by Measurement of Workpieces and Measuring Equipment – Part 2: Guidance for the Estimation of Uncertainty in GPS Measurement, in Calibration of Measuring Equipment and in Product Verification. Geneva, International Organization for Standardization.

ISO 14253-5, 1015, Geometrical Product Specifications (GPS) – Inspection by Measurement of Workpieces and Measuring Equipment – Part 5: Uncertainty in Verification Testing of Indicating Measuring Instruments. Geneva, International Organization for Standardization.

ISO 17043, 2023, Conformity Assessment. General Requirements for the Competence of Proficiency Testing Providers. Geneva, International Organization for Standardization.

ISO/IEC 17025, 2017, General Requirements for the Competence of Testing and Calibration Laboratories. Geneva, International Organization for Standardization.

JCGM 100, 2008, *GUM 1995 with Minor Corrections, Evaluation of Measurement Data — Guide to the Expression of Uncertainty in Measurement.* Paris, Bureau International de Poids et Mésures.

JCGM 101, 2008, *Evaluation of Measurement Data — Supplement 1 to the 'Guide to the Expression of Uncertainty in Measurement' —Propagation of Distributions Using a Monte Carlo Method.* Paris, Bureau International de Poids et Mésures.

Squires, G.L., 2008, *Practical Physics.* United Kingdom, Cambridge University.

Taylor, B.N. and Kuyatt, C.E., 1994, *Guidelines for Evaluating and Expressing the Uncertainty of NIST Measurement Results*, NIST Technical Note 1297, (NIST, Gaithersburg: USA).

4 Interferometry

4.1 INTRODUCTION

This section will introduce the aspects of optics that are required to understand interferometry. For a more thorough treatment of optics, the reader is referred to the literature (Born and Wolf 2013). For the treatment of light, electromagnetic waves of optical frequencies, usually called 'visible light' are considered. From Maxwell's equations, it follows that the electric field of a plane wave, with speed c, frequency f and wavelength λ, traveling in the z-direction, is given by:

$$\bar{E}(z,t) = \begin{pmatrix} E_x \\ E_y \end{pmatrix} e^{i(kz-\omega t)} , \tag{4.1}$$

where $\omega = 2\pi f = 2\pi c/\lambda$ is the circular frequency and, t is the time, i is the imaginary unit k is the circular wave number: $k = 2\pi/\lambda$. To simplify the mathematical description, it is assumed that $E_y = 0$, that is, the light is linearly polarized in the x-direction. The amplitude E_x can be obtained as a measurable quantity by the real part of equation (4.1). At the location $z = 0$, the electric field $E = E_x \cos\omega t$. This means that the electric field oscillates with a frequency f. For visible light, for example, green light (λ = 500 nm), this gives, with the speed of light defined as c = 299 792 458 m·s^{-1} (length definition, see Chapter 1), a frequency of $f = 6 \times 10^{14}$ Hz. As no electric circuit can directly follow such a high frequency, light properties are measured by averaging the cosine function over time. The intensity is given by half the square of the amplitude:

$$I(z) = \bar{E} \cdot \bar{E} = E_x^2 \cos^2(\omega t) = \frac{1}{2} E_x^2. \tag{4.2}$$

A distortion at $t = 0$ and $z = 0$, for example, a change in amplitude E_x in equation (4.1), will propagate as the same change at time t at location $z = \omega t/k = ct$, so the propagation velocity is indeed c.

DOI: 10.1201/9780367821746-4

In equation (4.1), the amplitudes E_x and E_y can both be complex. This general case is called elliptical polarization; the E-vector describes an ellipse in space. If E_x and E_y are both real, the light is called linearly polarized. Another special case is when $E_y = iE_x$; in this case, the vector describes a circle in space and this phenomenon is called circular polarization.

When light beams from different sources, or from the same source but via different paths, act on the same location, their electric fields can be added. This is called the principle of superposition and causes interference.

Interference is illustrated in Figure 4.1 where two light beams are mixed using a beam splitter. The light intensity on the screen is calculated from:

$$
\begin{aligned}
I(z) &= \left\langle \overline{E_1 + E_2} \cdot \overline{E_1 + E_2} \right\rangle = C\left\{E_1 \cos\left(k_1 z_1 - \omega_1 t\right) + E_2 \cos\left(k_2 z_2 - \omega_2 t\right)\right\}^2 \\
&= C\left\{E_1 \cos^2\left(k_1 z - \omega_1 t\right) + E_2 \cos^2\left(k_2 z - \omega_2 t\right) + 2E_1 E_2 \cos\left(k_1 z_1 - \omega_1 t\right)\cos\left(k_2 z_2 - \omega_2 t\right)\right\} \\
&= C\left\{I_1 + I_2 + 2\sqrt{I_1 I_2}\cos\left(\left(k_1 z_1 - k_2 z_2\right) + \left(\omega_2 - \omega_1\right)t\right) + \cos\left(\left(k_1 z_1 + k_2 z_2\right) + \left(\omega_2 + \omega_1\right)t\right)\right\} \\
&= C\left\{I_1 + I_2 + 2\sqrt{I_1 I_2}\cos\left(\left(k_1 z_1 - K_2 z_2\right) + \left(\omega_2 - \omega_1\right)t\right)\right\},
\end{aligned}
\tag{4.3}
$$

where C is a constant, and it is considered that the term $\left(\omega_2 + \omega_1\right)$ represents a frequency too large to be measured making the second cosine term vanish. Visible, stable interference can appear when the wavelengths are the same and there is a determined phase relationship between the superimposed waves. If the wavelengths are not the same, the intensity varies with the frequency difference between the two beams.

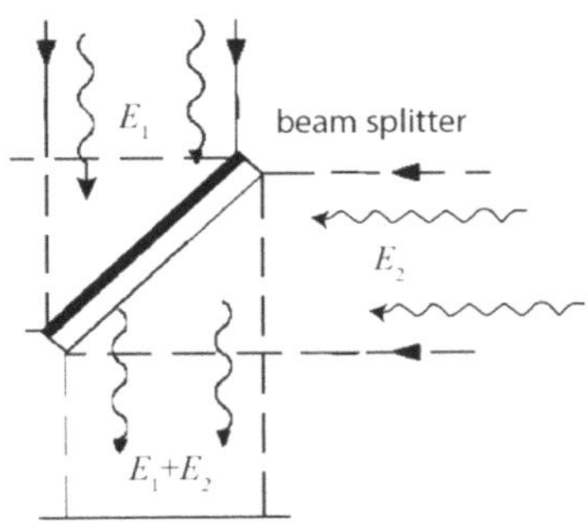

FIGURE 4.1 Schematic of the superposition of two monochromatic waves with amplitude E_1 and E_2.

4.2 INTERFEROMETRY MODEL FOR A MICHELSON INTERFEROMETER

In Figure 4.2, the principle of the Michelson interferometer is sketched; with the lens collimating the light, this is known as a Twyman-Green interferometer. If the light source emits a single wavelength λ, this means that the two beams have the same angular frequency ω when the mirrors do not move. The optical path length is $2z_1$ for one beam and $2(z_1+\Delta L)$ for the other beam. The factor of two is because the light travels this length twice. Because of this factor of two, interference phenomena mostly repeat at distances and displacements corresponding to $\lambda/2$. For this case, equation (4.3) reduces to:

$$\begin{aligned} I(''L) &= C\left\{I_1 + I_2 + 2\sqrt{I_1 I_2}\cos(2k\Delta L)\right\} \\ &= C(I_1 + I_2)\left\{1 + \frac{2\sqrt{I_1 I_2}}{I_1 + I_2}\cos\left(\frac{4\pi\Delta L}{\lambda}\right)\right\}. \end{aligned} \tag{4.4}$$

The intensity is maximal if $\Delta L = 0, \pm\frac{\lambda}{2}, \pm\lambda, \ldots.$ and minimal for $\Delta L = \pm\frac{\lambda}{4}, \pm\frac{3\lambda}{4}, \ldots$

When mirror M_2 is moved, the intensity shows a sinusoidal form as a function of ΔL with a wavelength of $\frac{\lambda}{2}$, mean intensity of $C(I_1+I_2)$ and a modulation amplitude $\frac{2\sqrt{I_1 I_2}}{I_1 + I_2}$. The visibility of the interference was defined by Michelson as:

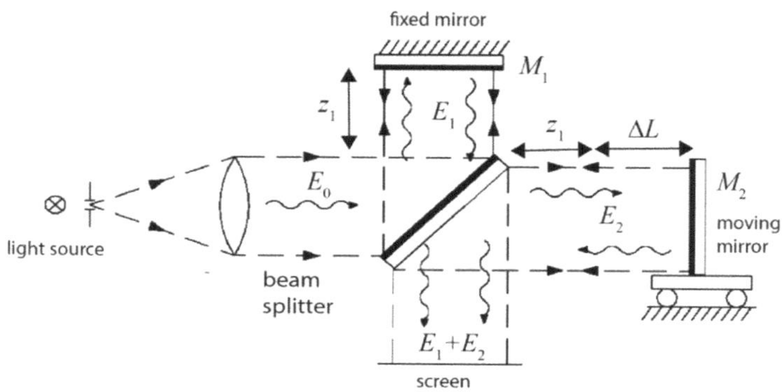

FIGURE 4.2 Schematic of a Michelson/Twyman Green interferometer.

$$V = \frac{I_{max} - I_{min}}{I_{max} + I_{min}}. \tag{4.5}$$

The visibility defines a relative contrast. For the signal, or pattern for a slightly tilted mirror, given in equation (4.4), this gives

$$V = \frac{2\sqrt{I_1 I_2}}{I_1 + I_2}. \tag{4.6}$$

The visibility becomes maximal when $I_1 = I_2$, in which case the minimum intensity is zero. For $I_1 = I_2$, equation (4.4) reduces to

$$I(\Delta L) = 2CI\{1 + \cos(2k\Delta L)\}. \tag{4.7}$$

In general, a light source will not emit a single wavelength, but a wavelength distribution $I(k)$. The intensity at displacement ΔL is now obtained by integration over the light source intensity:

$$\begin{aligned} I(\Delta L) &= \int_{k=0}^{k=\infty} 2CI(k)\{1 + \cos(2k\Delta L)\}dk \\ &= \int_{k=0}^{k=\infty} 2CI(k)dk + \int_{k=0}^{k=\infty} 2CI(k)\cos(2k\Delta L)dk \end{aligned} \tag{4.8}$$

In equation (4.8), the second term shows that the variation in intensity as a function of the displacement is the inverse cosine transform of the light source spectrum $I(k)$. As some phase effects will be present in the mirrors and the beam splitter, it can be stated more generally that the variation in intensity as a function of the displacement is the inverse Fourier transform of the light source spectrum $I(k)$. For a broad-band light source, the cosine functions in equation (4.8) will overlap and extinguish each other more and more for larger pathlengths ΔL. This is illustrated in Figure 4.3 for a broadband 'white light' source and a light source limited by a green filter with a central wavelength of 0.5 µm and a spectral bandwidth of 25 nm. Depending on the application, it can be advantageous to have a broadband light source, a limited bandwidth or a bandwidth as small as possible. If the localization of the interference is required, a broad bandwidth can be advantageous; this is used in white-light interferometry, also called coherence scanning interferometry (CSI). A limited bandwidth can be useful if the interference is used over a limited range, for example, some millimeters, but interference over larger path lengths can be problematic, for example, in interference microscopes where interference

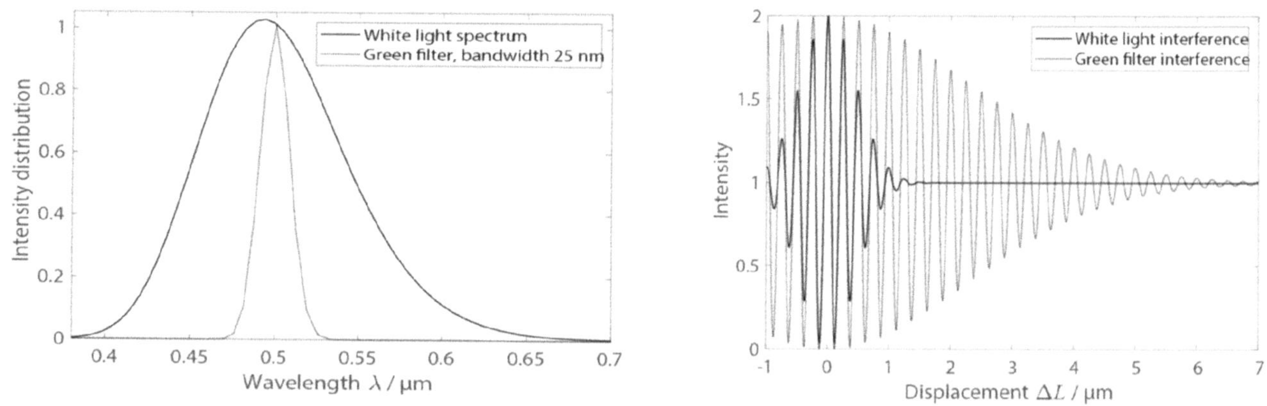

FIGURE 4.3 Spectral distribution of two light sources and the correspondence interference as a function of the displacement ΔL.

of optical elements outside the region of interest can be disruptive. The smallest possible bandwidth is useful for displacement measurements over macroscopic lengths from a few meters to several kilometers.

The distance ΔL over which the visibility drops by 50% is called the coherence length; this is related to the spectral bandwidth. For small bandwidths, the relationship is:

$$L_c = \frac{\lambda_0}{\Delta\lambda}\lambda_0 = \frac{k_0}{\Delta k}\lambda_0 = Q\lambda_0, \tag{4.9}$$

where λ_0 is the central wavelength, k_0 is the central circular wavenumber, $\Delta\lambda$ and Δk are the respective bandwidths and Q is the quality factor which determines over how many wavelengths the interference is easily detectable. Table 4.1 lists a few typical light sources with their quality factors and typical coherence lengths. In the early 20th century, the cadmium spectral lamp was used for interference over macroscopic distances. Michelson's determination of the cadmium lamp wavelength related to the meter standard was a breakthrough toward a meter definition based on physical constants; the definition of the ångström in 1907 was a first step in this process. The orange-red line of the ^{86}Kr spectral lamp was used as the meter definition from 1963 until 1983. This definition was possible as, with some effort, interference over a meter length difference was possible and a length up to one meter could be measured using interferometry. Since then, stabilized lasers have been employed that enable interference over large distances; commercially used laser interferometer systems can measure displacements well up to 50 m. Nowadays, light emitting diodes (LEDs) are used as broadband light sources; laser diodes are used as light sources with a smaller bandwidth.

TABLE 4.1
Quality Factor and Coherence Length of Some Light Sources

Light Source	Q	L_c	λ_0/nm	Color
Light bulb	1	0.5 µm	525	'white'
Green LED	10	5.5 µm	550	green
Green laser diode	100	55 µm	550	green
He spectral lamp	34000	20 mm	587	orange
Cd spectral lamp	$2.5\cdot10^5$	160 mm	644	red
Ångström definition 1907-1960				
Krypton spectral lamp	$1.5\cdot10^6$	0.8 m	565	orange
Meter definition 1960-1983				
Stabilized He-Ne laser	$8\cdot10^7$	50 m	632.8	red

4.3 WAVELENGTH, FREQUENCY AND REFRACTIVE INDEX

Interferometric measurements are commonly carried out in ambient air, see also Section 2.3. Under normal conditions, the refractive index of air is typically $n_{air} \approx 1.000\,273$. The wavelength in air of a light source is related to the frequency, the vacuum wavelength and the speed of light through:

$$\lambda_a = n_a \cdot \lambda_v = n_a \cdot \frac{c}{f}, \tag{4.10}$$

where λ_a is the wavelength in air, n_a is the refractive index of the air, λ_v is the wavelength in vacuum, f is the frequency and c is the speed of light (see Section 2.1).

The refractive index can be determined by measuring the optical path length difference between a distance measured in air compared to the same length measured in vacuum; an instrument performing this operation is called a refractometer. Refractometers are only used when relative uncertainties in the 10^{-8} range are needed. In due course, it has been shown that uncertainties up to $5 \cdot 10^{-8}$ can be achieved by carefully measuring the ambient conditions. For the refractive index of normal ambient air, empirical equations have been derived that give the refractive index as a function of the air temperature, atmospheric pressure and humidity. The most extensive evaluations have been made by Edlén (Edlen 1966), Ciddor (Ciddor 1996) and Bönsch (Bönsch and Potulski 1998). The calculation using these equations, with some corrections, is available on the internet (Stone and Zimmerman 2001). Here we give the simplified equations after Muijlwijk (Muijlwijk 1988). The refractive index as a function of wavelength λ, expressed in µm (throughout) at standard pressure (1000 hPa) and air temperature ($T_a = 20$ °C) is defined as:

$$(n-1)_\lambda = \left(26445.5 + 148.7 \cdot \lambda^{-2} + 1.205 \cdot \lambda^{-4} + 0.02712 \cdot \lambda^{-6}\right) \cdot 10^{-8}. \tag{4.11}$$

This equation is corrected for a deviating air density that depends on the temperature and pressure:

$$(n-1)_{t,p} = \frac{(n-1)_\lambda \cdot p \cdot 10^{-3}}{\left(1 + 0.0034198 \cdot \left(T_a - 20\,°C\right)\right)}. \tag{4.12}$$

An additional correction is made for the relative humidity to obtain a value for n_a

$$n_a = 1 + (n-1)_{t,p} - 0.87 \cdot RH \cdot \left(1 + 0.06 \cdot \left(T_a - 20\,°C\right)\left(1 - 0.008 \cdot \lambda^{-2}\right)\right) \cdot 10^{-8}. \tag{4.13}$$

In these equations, λ is the vacuum wavelength in micrometers, p is the atmospheric pressure in hecto-pascals, T_a is the air temperature in degrees Celsius and RH is

TABLE 4.2
Sensitivity of the Air Refractive Index for the Ambient Conditions and Their Effect on a Measured Length

Condition	Sensitivity of *n*	Uncertainty in a Measured Length *L*	Uncertainty Needed for an Uncertainty in Length $u(L) < 5 \cdot 10^{-7} \cdot L$
T_a	$\frac{\partial n}{\partial T_a} = -0.95 \cdot 10^{-6}\ °C^{-1}$	$u(L) = 0.96 \cdot 10^{-6} \cdot L \cdot u(T_a)$	$u(T_a) < 0.48$ °C
P	$\frac{\partial n}{\partial p} = 0.27 \cdot 10^{-6}$ hPa^{-1}	$u(L) = 0.27 \cdot 10^{-6} \cdot L \cdot u(p)$	$u(p) < 1.4$ hPa
H	$\frac{\partial n}{\partial \mathrm{RH}} = 0.91 \cdot 10^{-8}$ %RH^{-1}	$u(L) = 0.91 \cdot 10^{-8} \cdot L \cdot u(H)$	$u(H) < 45$ %RH

the relative humidity in percent. These equations give the refractive index with a relative uncertainty of about $2 \cdot 10^{-8}$ for λ between 0.45 µm and 0.65 µm, p between 900 hPa and 1100 hPa, and T_a between 18 °C and 22 °C. Outside this range, the Ciddor equations can be considered. Once the refractive index can be adequately calculated, it is important to consider the propagation of the uncertainties in the measured atmospheric conditions; usually the used wavelength is adequately known. Table 4.1 gives the effect of the uncertainty in the atmospheric conditions in the uncertainty of the air refractive index n_a. This value is also the relative uncertainty in any measured length using this wavelength. Table 4.2 indicates that the relative humidity is the least critical condition. Measurement of the air temperature is not trivial because of possible self-heat of electronic sensors, the need of measuring the temperature close to the measurement path, heat radiation of the environment, etc. A failing temperature measurement because of a broken sensor can easily be noticed as humans can feel the temperature within a few degrees Celsius. A failing pressure measurement because of a broken sensor is more difficult to notice.

4.4 DISPLACEMENT INTERFEROMETRY

An important type of commercially available interferometers for application in dimensional metrology are the displacement measuring interferometers. These are commonly referred to as 'laser interferometer', 'calibrator', 'laser calibration system', 'optical laser encoder', 'laser interferometry system', 'displacement measuring interferometer system' or 'laser transducer system'. The common aspect is that displacements are measured by interferometry. The optical scheme is based on the Michelson interferometer as shown in Figure 4.2.

As a light source, a stabilized laser is used and the mirrors are replaced by retroreflectors, also called corner cubes. These optical elements have the property that light is reflected in the same direction as from where it is received. This omits the need of a careful alignment of flat mirrors, as is needed by the set up shown in Figure 4.2.

4.4.1 Homodyne Laser Interferometers

Homodyne laser interferometers are based on a stabilized laser of which one single wavelength is used as the light source for the interferometer operation. The beam is either circularly polarized or linearly polarized with the polarization direction oriented under 45° so that the beam is split into two beams that are linearly polarized in orthogonal directions of about equal intensity in the polarizing beam splitter, see Figure 4.4. When the beams are recombined, interference takes place after they pass a linear polarizer. Half of the beam is first fed through a quarter-wave ($\lambda/4$) plate; this gives a phase shift of 90° of one beam relative to the other. The reference photodiode signal I_{ref} can be subtracted from the interference signal I_0 giving (see equation (4.7)):

$$S_A = I_0 - I_{\text{ref}} = I_{\text{ref}}\left(1+\cos\frac{4\pi\Delta L}{\lambda}\right) - I_{\text{ref}} = I_{\text{ref}}\cos\frac{4\pi\Delta L}{\lambda}. \tag{4.14}$$

The waveplate introduces and additional phase shift of $\pi/4$ between the two beams. Subtracting the reference signal from I_{90} gives:

$$S_B = I_{90} - I_{\text{ref}} = I_{\text{ref}}\left(1+\cos\left(\frac{4\pi\Delta L}{\lambda}-\frac{\pi}{4}\right)\right) - I_{\text{ref}} = I_{\text{ref}}\sin\frac{4\pi\Delta L}{\lambda}. \tag{4.15}$$

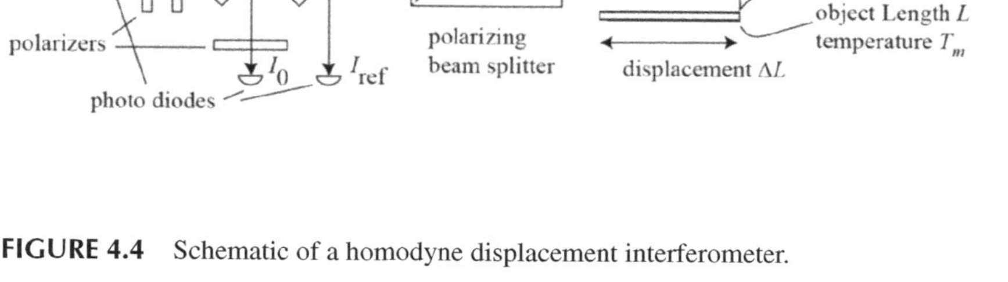

FIGURE 4.4 Schematic of a homodyne displacement interferometer.

The displacement can be derived from equations (4.14) and (4.15) as:

$$\Delta L = \frac{\lambda}{4\pi}\tan^{-1}\left(\frac{S_B}{S_A}\right) = \frac{\lambda_a}{4\pi}\phi, \tag{4.16}$$

where ϕ is the phase angle. Obviously, the displacement calculated from equation (4.16) is periodic with a period of $\lambda/2$. Within one period, deviations may occur because of un-equal amplification, non-perfect polarization splitting, beam alignment errors, etc. These may lead to errors in the nanometer-range that are called nonlinearity errors or interpolation errors. These errors can be modeled and compensated to some extent; this is commonly called the Heydemann correction (Heydemann 1981) and is treated in Section A.5.5. Using transistor-transistor logic (TTL), electronic circuits can be designed that add or subtract a 'count', that is, a distance of $\lambda/8$, every time either S_B or S_A crosses zero, where the sign of the count (plus or minus) depends on the sign of the other signal when S_B or S_A crosses zero. This is illustrated in Figure 4.5 that gives signals S_A and S_B, the calculated displacement according to equation (4.16) combined with the counting at the zero crossings of the signals for the typical wavelength of a red He-Ne laser (λ = 632.8 nm) and a displacement of 1 μm. The measured displacement ΔL is given by:

$$\Delta L = \frac{(N+\phi/(2\pi))\lambda_a}{2} = \frac{(N+\phi/(2\pi))\lambda_v}{2n_a(T_a, p, RH)}, \tag{4.17}$$

where N is the number of 'counts', that is, the number of times that the phase ϕ in equation (4.16) has completed a full circle. If this displacement is intended to measure a material size, for example, a length bar or a scale, this size must

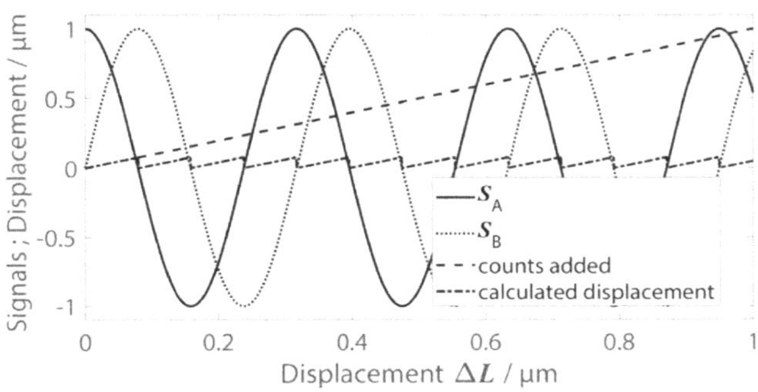

FIGURE 4.5 Signals S_A and S_B and the calculated displacement without and with added counts of $\lambda/8$.

be corrected to 20 °C. Combining equations (4.17) and equation (2.6) gives the general system equation for a laser interferometer measurement:

$$L_{20} = \frac{\left(N + \frac{\phi}{2\pi}\right)\lambda_v}{2n_a\left(T_a, p, RH\right)}\left(1 + \alpha\left(T_m - 20\,^\circ C\right)\right). \tag{4.18}$$

Equation (4.18) contains all elements for which a laser interferometer measurement can be sensitive. In practical measurements, it may be necessary to take Abbe- and cosine errors (see Section 2.4) into account as well. Commercial systems commonly have measures to prevent the system from 'missing counts' (N), for example, by putting thresholds on the intensities measured. As not all systems use the correct equations for the refractive index calculation, this may also cause deviations.

4.4.2 Heterodyne Laser Interferometers

In heterodyne laser interferometers, two slightly different wavelengths are used that are both linearly polarized orthogonal to each other. The basic set-up is indicated in Figure 4.6.

The frequency difference can be generated by using the Zeeman effect. Due to this effect, two frequencies are generated by applying an axial magnetic field to the laser tube. With this effect, the laser frequency may be stabilized as well by equaling both intensities. The frequency split is limited to a maximum of about 4 MHz. Another way to generate a frequency shift is with an acousto-optic modulator; for example, a Bragg cell driven by a quartz oscillator. This enables a frequency split of 20 MHz or more.

The light emitted by the laser source is split, one part passes a combining polarizer and falls onto a detector with a band-pass filter. The resulting signal is an alternating signal with a beat frequency equal to the split frequency in the laser. This signal forms the reference measurement I_{ref}, given by:

$$I_{ref} = C\cos\left(2\pi(f_2 - f_1)t + \left(\phi_{02} - \phi_{01}\right)\right), \tag{4.19}$$

where C is a constant, f_1 and f_2 are the frequencies and ϕ_{01} and ϕ_{02} are the respective phases.

The other part of the light enters the interferometer optics (Figure 4.6), consisting of a polarizing beam splitter and two corner cubes reflectors. In the polarizing beam splitter, the two frequencies are split according to their polarization. Frequency f_1 is reflected by the polarizing beam splitter and enters the reference arm, then is reflected by the reference corner cube and is again reflected by the polarizing beam splitter. Frequency f_2 is transmitted by the polarizing beam splitter and enters the measurement arm, then is reflected by the moving retro-reflector and is

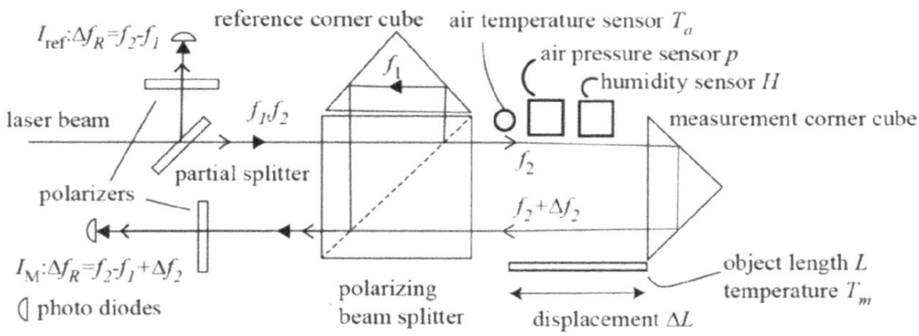

FIGURE 4.6 Schematic of a heterodyne laser interferometer system.

again transmitted by the beam splitter. Ideally, both frequencies emerge from the polarizing beam splitter in their unchanged polarization orthogonal to each other. To enable interference, the beams are transmitted through a polarizer at 45° with their polarization axes. After the polarizer, the interference signal is detected by the detector giving the measurement signal:

$$I_M = C\cos\left(2\pi(f_2 - f_1)t + \left(\phi_{02} - \phi_{01}\right) + \left(\phi_{\mathrm{m}} - \phi_{\mathrm{r}}\right)\right), \tag{4.20}$$

where $(\phi_{\mathrm{m}} - \phi_{\mathrm{r}})$ is the difference in phase between the signal in the measurement arm and the reference arm. This term consists of a constant term in the reference arm ϕ_{r} and a phase in the measurement arm ϕ_{m} that may change from its original value ϕ_{mo} as a result of the moving retro-reflector:

$$\phi_m - \phi_{m0} = \Delta\phi. \tag{4.21}$$

In a practical application, the measurement retro-reflector is displaced in order to measure a displacement or a size. As a retro-reflector moves in the measurement arm over a distance ΔL with velocity v, during this displacement, a Doppler shift is generated for frequency f_2:

$$\Delta f = \frac{2vn_a f_2}{c} \tag{4.22}$$

with v the velocity of the moving retro-reflector, n_a the refractive index of the air through which the light travels and c the speed of light in a vacuum. From equation (4.22), it follows that the maximum traveling speed of the target is limited by the finite frequency shift between the two frequencies in the laser source: if Δf becomes larger than the basic frequency difference between f_1 and f_2, the meaning of the measured frequency difference becomes ambiguous. The effective phase change in the interference signal resulting from the Doppler shift is calculated from:

$$\Delta\phi = \int_{t_1}^{t_2} 2\pi\Delta f dt = \int_{t_1}^{t_2} 2\pi \frac{2vn_a f_2}{c} dt = \frac{4\pi n_a f_2}{c} \int_{0}^{L/v} v dt = \frac{4\pi n_a f_2}{c} \Delta L \quad (4.23)$$

where ΔL is the displacement of the retro-reflector. This means that by measuring the phase change between the measurement signal and the reference signal, and measuring the frequency difference as a number of light–dark transitions (counts) N, the displacement of the corner cube is determined by:

$$\Delta L = \frac{\left(N + \Delta\phi/(2\pi)\right)\lambda_{2,a}}{2} = \frac{\left(N + \Delta\phi/(2\pi)\right)\lambda_{2,v}}{2n_a\left(T_a, p, RH\right)}. \quad (4.24)$$

Note that equation (4.24) is similar in principle to equation (4.17); this means that equation (4.18) applies to a heterodyne laser interferometer system as well, and the uncertainties propagate similarly. Normally the frequency difference between the two beams is too small to make a useful distinction of whether f_1 or f_2 is used for the measured displacement. For example, a 4 MHz difference corresponds to a relative difference in measured length of about $1 \cdot 10^{-8}$. In industrial measurement practice, such as the calibration of machine tools or the measurement of reference specimens, a user will not notice differences between a homodyne or a heterodyne laser interferometer system. Even the same beam splitters and corner cubes can be used. As a heterodyne interferometer operates at high frequencies, a high signal-to-noise ratio may be maintained when the laser beam is split, for example, to measure several displacements and angles simultaneously, as is needed in complicated positioning systems such as those used in wafer steppers.

4.5 SURFACE INTERFEROMETRY

4.5.1 Twyman-Green Interferometer with Monochromatic Light

The interferometer as sketched in Figure 4.2 can be used as a surface interferometer, this is an interferometer that measures the topography of a surface relative to a reference surface; this reference surface is commonly a flat mirror. Figure 4.2 is modified in Figure 4.7 as a Twyman-Green interferometer for surface measurements. The measuring mirror is replaced by an object with a reflecting surface characterized by a topography $z(x,y)$, this means that one height z can be attributed to any (x,y) coordinate on the objects surface. This is also called a 2.5D representation. Normally the surface deviations are relatively small, in the micrometer-region or less, so that the reflection direction can be approximated as perpendicular. In general, the object will have a small tilt around the x- and y-axis,

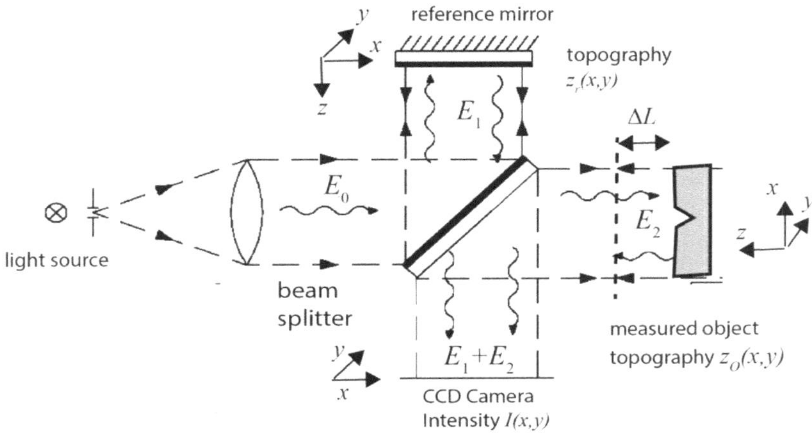

FIGURE 4.7 Schematic for interferometric surface measurements using a Twyman-Green interferometer.

given by θ_x and θ_y, and an offset ΔL from the point where both light paths are equal. For the intensity on the detector, equation (4.7) can be rewritten as:

$$I(x,y) = 2CI\left\{1 + V(x,y)\cos\left(\frac{4\pi}{\lambda}\left(\Delta L + z_0(x,y) + x\theta_y + y\theta_x - z_r(x,y)\right)\right)\right\}. \quad (4.25)$$

For a perfectly flat surface and reference surface $z_0(x,y) = z_r(x,y) = 0$, the image will give straight light-dark lines with spatial wavelength $\lambda_x = \frac{\lambda}{2\theta_y}$ and $\lambda_y = \frac{\lambda}{2\theta_x}$, respectively. Adjusting the tilt may make the straight interference lines disappear and the interference pattern resulting from the object $z_0(x,y)$ remains. This is illustrated in Figure 4.8 that shows the interference pattern of a flat with a groove with a depth of 4λ with different tilts. The interference lines can be considered as contour lines of equal height. A quick look on a single image may give a semiquantitative evaluation that can be rather accurate. For example, if equidistant straight lines are observed, then the surface will be flat within $\lambda/2$. Concentric circles indicate sphericity, and so on.

For a quantitative evaluation, the most common evaluation is that the phase is shifted in a few discrete steps by displacing either the object, the reference surface or the optical system relative to the object. This changes the ΔL term in equation (4.25). As an illustration, we give the equations where the object is moved in four steps of $\lambda/8$, omitting the tilt, and for a perfectly flat reference surface $z_r(x,y) = 0$ and a perfect visibility $V(x,y) = 1$. The four intensities can be written as following:

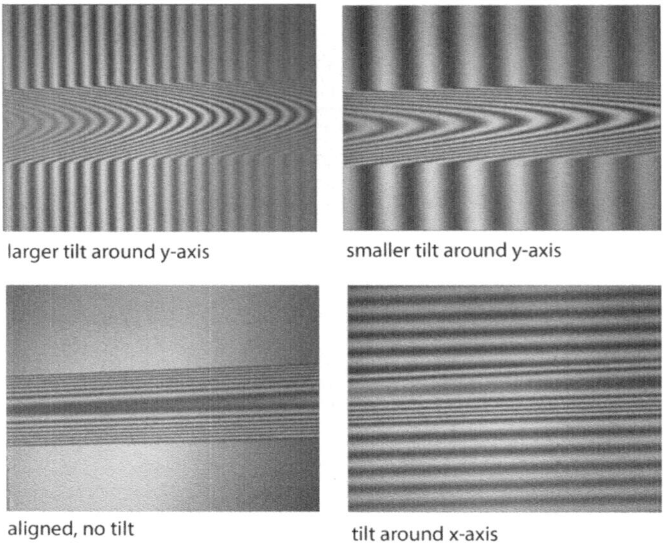

FIGURE 4.8 Images of interference patters, also called interferograms, of a flat surface with a horizontal central groove, with different tilts.

$$\begin{aligned}
I_A(x,y) &= 2CI\left\{1+\cos\left(\frac{4\pi}{\lambda}\left(\Delta L + z_O(x,y)\right)\right)\right\},\\
I_B(x,y) &= 2CI\left\{1+\cos\left(\frac{4\pi}{\lambda}\left(\Delta L + \frac{\lambda}{8} + z_O(x,y)\right)\right)\right\}\\
&= 2CI\left\{1-\sin\left(\frac{4\pi}{\lambda}\left(\Delta L + z_O(x,y)\right)\right)\right\},\\
I_C(x,y) &= 2CI\left\{1+\cos\left(\frac{4\pi}{\lambda}\left(\Delta L + \frac{\lambda}{4} + z_O(x,y)\right)\right)\right\}\\
&= 2CI\left\{1-\cos\left(\frac{4\pi}{\lambda}\left(\Delta L + z_O(x,y)\right)\right)\right\},\\
I_D(x,y) &= 2CI\left\{1+\cos\left(\frac{4\pi}{\lambda}\left(\Delta L + \frac{3\lambda}{8} + z_O(x,y)\right)\right)\right\}\\
&= 2CI\left\{1+\sin\left(\frac{4\pi}{\lambda}\left(\Delta L + z_O(x,y)\right)\right)\right\}.
\end{aligned} \tag{4.26}$$

From equations (4.26), the topography can be derived from the phase:

$$z(x,y)=\frac{\lambda}{4\pi}\tan^{-1}\left(\frac{I_D(x,y)-I_B(x,y)}{I_A(x,y)-I_C(x,y)}\right)+N(x,y)\frac{\lambda}{2}+\Delta L, \quad (4.27)$$

where N is an integer number. If the surface is continuous, the number N can be obtained by a process called unwrapping; when there is a discontinuity because the arctan-term in equation (4.27) changes from 2π to 0 when progressing along a line, N can be increased by 1 to keep a continuous surface. This unwrapping appears smoothly along one line; however, in 2D this is far from trivial, especially when the measurements are noisy or if the surface contains discontinuities. In the literature, several unwrapping strategies are discussed (Malacara 1998). The interference intensity can be obtained from

$$4CI(x,y)V(x,y)=\sqrt{\left(I_D(x,y)-I_B(x,y)\right)^2+\left(I_A(x,y)-I_C(x,y)\right)^2}. \quad (4.28)$$

Equation (4.28) can be used to put a threshold on the measurements that can be considered valid before applying equation (4.27). This may prevent phase-jumps due to an incorrect calculation of $N(x,y)$. Note that equations (4.27) and (4.27) in fact determine numerically the phase and amplitude of the first Fourier component with a period of $\lambda/2$. Instead of the four steps described in equation (4.26), several other strategies to obtain a surface map are possible that are more or less sensitive to noise and calibration and/or nonlinearities in the applied displacement. These can be found in the extensive literature (Malacara 1998; Gasvik 2002).

The evaluation according to equations (4.26) and (4.27) are illustrated in Figure 4.9 that gives as an example for a 0.55 μm deep groove and a wavelength of λ = 0.633 μm, the four images according to equation (4.26), the calculated phase according to equation (4.27), the unwrapped phase and the levelling of the topography relative to the least-squares plane.

4.5.2 Fizeau Interferometer

The interferometer discussed in Section 4.5.1 has the disadvantage that the path lengths inside the interferometer can be rather long, which may give rise to noise and fluctuations due to air turbulence inside the optical paths. For this reason, for surface interferometry on larger (50–1000 mm diameter), nominally flat, surfaces, the Fizeau configuration is more common. The optical scheme of a typical Fizeau interferometer is shown in Figure 4.10. A collimated beam passes through the reference mirror that is semi-transparent. The front side of this reference mirror acts as the reference mirror. The backside of this mirror has either an antireflection coating or it has a small wedge, so reflections of the back side do not interfere with the front side or the object. The object is put opposite to the reference mirror. Now

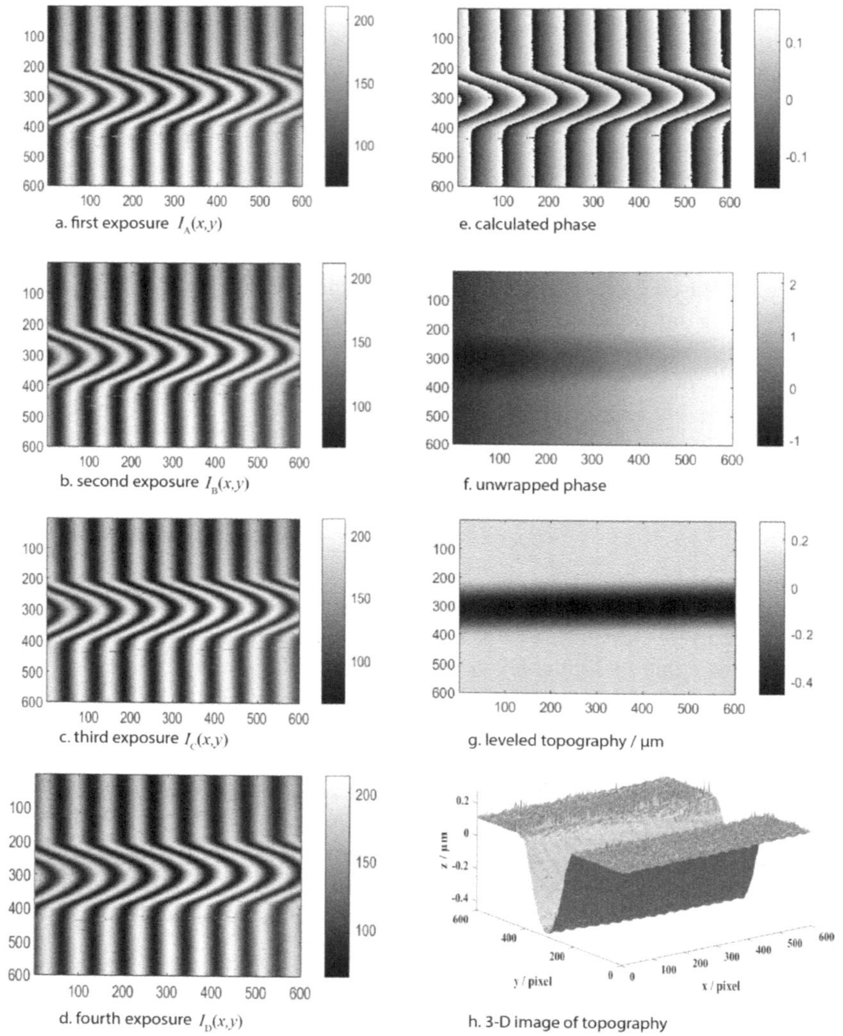

FIGURE 4.9 Four interferograms and their evaluation into a height map. On the left-hand side the interferogram images in steps of $\lambda/8$ (equation (4.27)). At the right-hand side, the phase evaluation according to equation (4.27) with $\lambda = 0.633$ μm, the unwrapping, the leveling and the scaling.

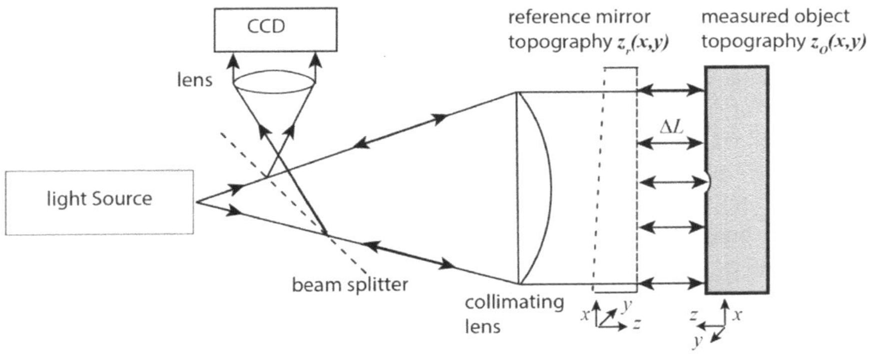

FIGURE 4.10 Optical configuration of a Fizeau interferometer.

interference occurs between the beam reflected from the object and the reference surface. Similar to equation (4.25), the interference intensity is given by:

$$I(x,y) = 2CI\left\{1 + V(x,y)\cos\left(\frac{4\pi}{\lambda}\left(\Delta L + z_O(x,y) + x\theta_y + y\theta_x + z_r(x,y)\right)\right)\right\}. \quad (4.29)$$

An important difference to equation (4.25) is that in equation (4.29), the topographies of the reference surface and the measured object, considered in their respective workpiece coordinates, are added, while in equation (4.25), these are subtracted. The reflectance of both the object and the reference surface can still be matched for optimal visibility; however, the reflectance of the reference surface should not be too high to avoid multiple reflections. Interference fringes can be observed directly to make a semiquantitative evaluation of the surface, or the object or the reference surface can be moved in a few steps and equations (4.26) and (4.27) can be used to obtain a complete surface topography $z_O(x,y)$. With some modifications, spheres and parabolas can also be measured; see the literature (Malacara 1991) for further details.

4.5.3 White-Light Interferometry: Twyman-Green Interferometer with Broadband Light

The configuration and evaluations shown in Sections 4.5.1 and 4.5.2 have the advantage of a rather straightforward evaluation and a traceability to the meter definition through the wavelength used. For surface interferometry, often an unstabilized laser provides enough accuracy for the wavelength, as the measured surface deviations are usually small. The need for a continuous surface and the difficulties of unwrapping in the case of surfaces with discontinuities or steps makes these interferometers less appropriate for more general use. For this reason, the monochromatic light source in Figure 4.7 may be replaced with a broadband light source that enables the determination of the absolute topography of an object without the need for unwrapping. As, in this case, the wavelength λ can no longer be the measure for the displacement, the displacement ΔL must be determined in an independent way, for example, by an LVDT, a capacitive displacement sensor or by a laser interferometer as described in Section 4.4. As a pathlength difference $\Delta L = 0$ must be established for every object coordinate $z_O(x,y)$, the Fizeau configuration cannot be used here and a Michelson interferometer configuration, or some variation of it, must be used. The measurement procedure is that the object in the configuration as shown in Figure 4.7 is displaced through the virtual plane where $\Delta L = 0$, while the displacement d_z of the object is measured and recorded independently. Then, for every recorded pixel of the object image, there is a signal similar to the signals depicted in Figure 4.3. From this signal, the displacement for which $\Delta L = 0$ must be determined. This corresponds to an absolute coordinate $z_0(x,y)$ on the surface for which no unwrapping is needed. This is illustrated in Figure 4.11 for a tilted surface that contains a groove, in fact the same surface that is used in Figure 4.9. The light source is a broadband source with a central wavelength of $\lambda_c = 500$ nm and a spectral bandwidth of about 100 nm. Figure 4.11 shows at the left-hand side one of the 144 images that were recorded when the surface was displaced in steps of 1/8th of the central wavelength. The location on the surface where $\Delta L = 0$ can be identified as the locations where the interference intensity is highest. Figure 4.11 gives at the right-hand side the intensity as a function of the object displacement d_z in terms of the recorded gray levels at the x-location given by pixel number 600, and the y locations given by $y = 240$, $y = 495$ and $y = 720$. Functions as shown in Figure 4.11b are called correlograms because the topography is calculated from their relative shift, that is, their correlation. In general, there are two basic ways of calculation: the envelope calculation that calculates the effective intensity and looks at what location the intensity is maximal, and the phase calculation, that establishes the location where the phase is exactly zero near the intensity maximum. To determine the relative shift between the correlograms, manufacturers use their own, partly proprietary, methods; see, for example, Larkin (Larkin 1996) and de Groot (de Groot 1995). As an example, we give a straightforward method here. Suppose the steps by which the object is displaced correspond to approximately $\lambda_c/4$; then the phase $\phi(d_i, x, y)$ and the modulation envelope intensity $E(d_i, x, y)$ can be calculated at any position d_i by considering the signals, up to two displacement steps forward and backward, with symmetric versions of equations (4.27) and (4.28):

$$\phi(d_i, x, y) = \tan^{-1} \frac{I(d_{i+1}, x, y) - I(d_{i-1}, x, y)}{\dfrac{I(d_{i+2}, x, y) + I(d_{i-2}, x, y)}{2} - I(d_i, x, y)}, \tag{4.30}$$

with $d_{i+k} = d_i + k\lambda_c / 4$, and

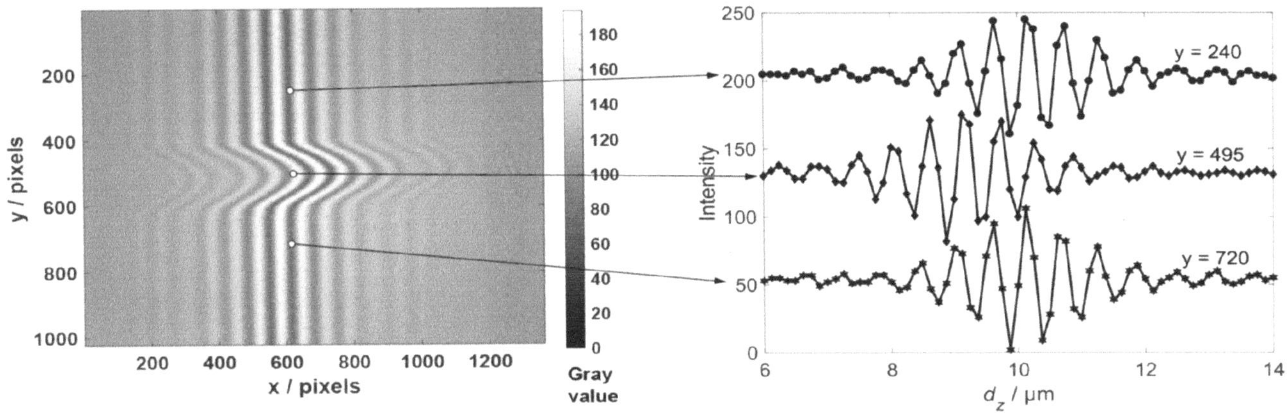

a. Image at $d_z = 10$ µm

b. Intensities at $x = 600$; shifted

FIGURE 4.11 a. White light interferogram at $d_z = 10$ µm. b. Intensity as a function of d_z for $x = 600$ at different y-locations.

$$E(d_i,x,y) = \sqrt{\left(I(d_{i+1},x,y)-I(d_{i-1},x,y)\right)^2+\left(\frac{I(d_{i+2},x,y)+I(d_{i-2},x,y)}{2}-I(d_i,x,y)\right)^2}. \quad (4.31)$$

These functions are illustrated in Figure 4.12 for pixel intensity $I(d,600,720)$ as given in Figure 4.11b. The scan position for which the modulation envelope intensity reaches its maximum is a measure of the surface height z_0. This can be refined by fitting this envelope function to a model, or by taking the center of gravity of the envelope intensity over a limited range from the maximum $d_{\max} = \max(E(d_z))$, for example:

$$z_O(\text{envelope}) = \frac{\sum_{d_{\max}-2\,\mu m}^{d_{\max}+2\,\mu m} d_z E^2(d_z)}{\sum_{d_{\max}-2\,\mu m}^{d_{\max}+2\,\mu m} E^2(d_z)}. \quad (4.32)$$

Commonly a height resolution in the nanometer range can be achieved by applying equation (4.32). A further refinement is possible by searching for the zero-crossing of the phase that is closest the z_0 value as calculated in equation (4.32). This can be done by either linear interpolation of by fitting a straight line through the calculated – and unwrapped – phase and take the intercept with the $\phi = 0$ axis. This

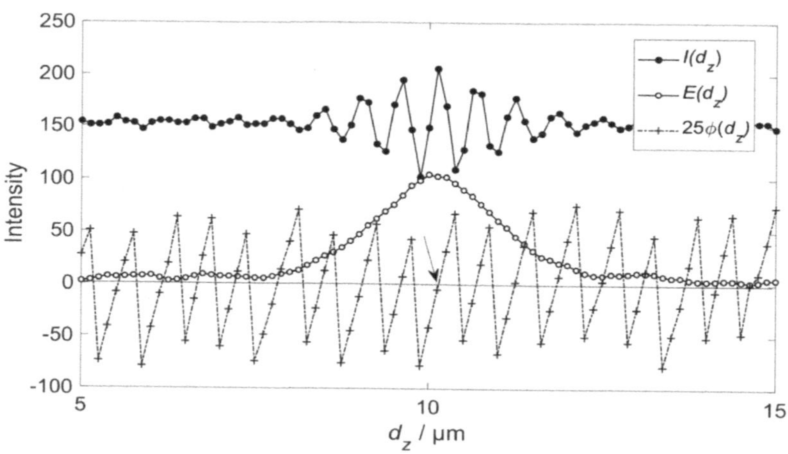

FIGURE 4.12 Pixel intensity, envelope intensity and the calculated phase, enlarged for better visibility, as a function of the displacement d_z for pixel (600,720) as displayed in Figure 4.11.

is called the phase evaluation, and with this commonly sub-nanometer resolution can be achieved. However, there is the risk of 'phase jumps' in the case of noise signals when the 'wrong' zero crossing is calculated and a jump or peak of $\lambda/2$ in the surface is observed. Calculating z_0 for every pixel (x,y) gives the 2.5D topography like in Figure 4.9f. This can be further evaluated by leveling the surface and/or displaying it as a quasi 3D image as illustrated in Figure 4.9g and h.

4.6 GAGE BLOCK INTERFEROMETRY

Gage blocks are characterized by having two parallel, optically flat, measuring faces. This enables the measurement of their length by interferometry, especially when one of the faces is wrung to a base plate. The definition of a gage block length is based on the fact that the most accurate length measurement can be made in a gage block interferometer. In this instrument, the gage block length is determined in terms of a calibrated wavelength and thus this gives a connection and traceability between the optical frequencies and wavelengths on one side, and material workpieces and objects that have an actual length on the other side. Therefore, a gage block interferometer and its operation are considered a cornerstone of dimensional metrology.

Gage block interferometers can be designed as a Twyman-Green or a Fizeau configuration, where the former is more common. For the characteristics discussed in this section, either configuration can be considered, however, it is illustrated by a variation of the Twyman-Green interferometer that enables the selection of various wavelength from a spectral source: the Kösters interferometer, see Figure 4.13.

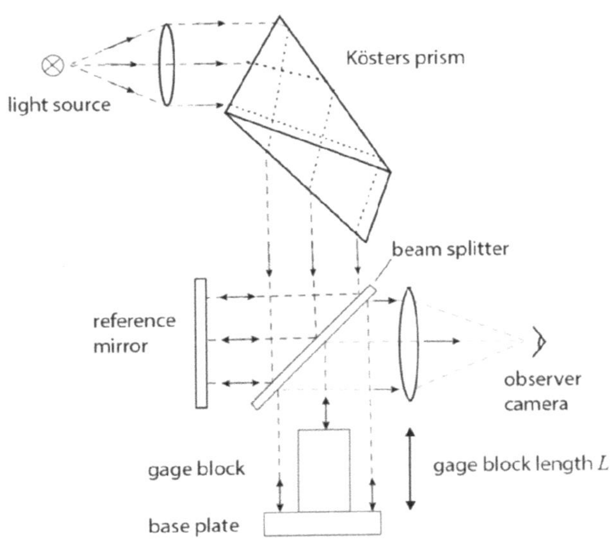

FIGURE 4.13 Schema of a gage block interferometer containing a gage block.

From a multiwavelength spectral source, one wavelength is selected by rotating the Kösters-prism. The base-plate gage block assembly interferes with the reference surface. When the assembly is slightly tilted, straight parallel interference lines appear as with a flat surface, both for the baseplate surface and the gage block surface. If the gage block length equals an integer number of half wavelength, these interference lines are not interrupted. However, when this is not the case, the interference lines from the gage blocks are shifted relative to the lines coming from the baseplate. This relative shift is called the fraction *f* and is related to the difference in phase. Figure 4.14 gives both the ideal situation and an image of an actual gage block measurement. Assuming a perfectly flat reference mirror and base plate, the possible lengths the gage block can have is determined by an unknown number of half wavelengths *N* plus a fraction of half a wavelength as following:

$$L = \frac{\lambda_v}{2n(\lambda)}\left\{N + \frac{\left[\phi_{\text{gage block}} - \phi_{\text{base plate}}\right]}{2\pi}\right\} = \frac{\lambda_v}{2n(\lambda)}\{N + f\}. \quad (4.33)$$

If the gage block length is already known within half a wavelength, that is, within ±0.15 μm, the unknown number *N* can be estimated and the actual gage block length can be determined from the fraction evaluation of a single measurement. In the traditional gage block interferometers, this is commonly done by selecting several wavelength from spectral lamps with a series of spectral lines with known and well-defined wavelengths, such as cadmium, mercury or krypton; these wavelengths are even defined in the framework of the meter definition (BIPM 2024). A requirement is that the coherence length must be larger than the length difference to be measured. For gage block lengths less than 20 mm, a helium spectral lamp

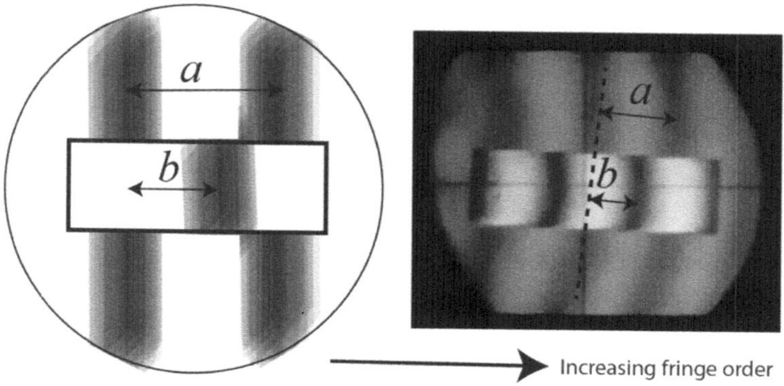

FIGURE 4.14 Theoretical and actual interference pattern and the evaluation of the fraction $f = b/a$.

can be used. In modern interferometers, stabilized lasers are used that enable the measurement of gage blocks up to 1 m length. For multiple wavelengths, equation (4.33) must be valid for every wavelength λ_i and measured fraction f_i used, giving:

$$L = \frac{\lambda_1}{2n(\lambda_1)}\{N_1 + f_1\} = \frac{\lambda_2}{2n(\lambda_2)}\{N_2 + f_2\} = \frac{\lambda_3}{2n(\lambda_3)}\{N_3 + f_3\} = \ldots \tag{4.34}$$

As the measured fractions will never exactly coincide with the theoretical fractions for a given length, the problem to be solved is actually a least-squares problem, where the function *Q(L)* is to be minimized for the *K* wavelengths and corresponding fractions measured (see also Section 10.5.3):

$$Q(L) = \sum_{i=1}^{K}\left(L - \frac{\lambda_i}{2n(\lambda_i, T_a, p, R)}(N_i + f_i)\right)^2. \tag{4.35}$$

Equation (4.35) can be used explicitly, for example, by varying the length over 2 μm from its nominal length with steps of 10 nm or less and looking for the length for which *Q(L)* is minimal, or the method of exact fractions can be used that implicitly uses equation (4.35). This procedure is as follows:

- one takes the longest wavelength, most toward 'red' (λ_1); this requires the fewest calculations.
- starting from the nominal length of the gage block (and possibly the fraction), calculate the nominal integer number of wavelengths $N_{1,\mathrm{nom}}$:

$$N_{1,\mathrm{nom}} = \mathrm{round}\left(\frac{2n(\lambda_1)}{L_n \lambda_{1,v}}\right). \tag{4.36}$$

This is the number of half-wavelengths that fit into the nominal length L_n. The actual length will be close to this value, for example, within ±0.6 μm, so $N_j = N_{1,\mathrm{nom}} \pm 2$. From this, five potential lengths belong to the measurement of the fraction of the longest wavelength:

$$L_j = \frac{\lambda_1}{2n(\lambda_1)}(N_{1n} + f_1 + j) \tag{4.37}$$

with $j = -2,-1,0,1,2$.

For these possible lengths, the theoretical fractions of the other wavelengths f_{ti} are calculated:

TABLE 4.3
Calculation of a Gage Block Length of Nominally 5 mm Length by the Exact Fractions Method

$\lambda_i/n/\mu m$	f_i	N_{1n}	$L_{N1n-2}+f_1$ /µm	$L_{N1n-1}+f_1$ /µm	$L_{N1n}+f_1$ /µm	$L_{N1n+1}+f_1$ /µm	$L_{N1n+2}+f_1$ /µm
		14 974	4 999.35	4 999.69	5 000.02	5 000.35	5 000.69
0.667 818	0.2		0.20	0.20	0.20	0.20	0.20
0.501 570	0.7	f_{t2j}	0.81	0.15	0.48	0.81	0.13
0.471 317	0.8	f_{t3j}	0.39	0.83	0.22	0.64	0.06
Standard Deviation	$\sqrt{\sum_{i=2}^{K}\frac{\left(f_{tij}-f_i\right)^2}{K-1}}$		0.21	0.28	0.31	**0.10**	0.25

$$f_{tij} = \mathrm{rem}\left(\frac{2L_j n(\lambda_i)}{\lambda_i}\right). \tag{4.38}$$

The length at which the difference between the theoretical and measured fractions is smallest is taken as the absolute length. When determining these differences, the continuous nature of the fractions should be noted: the difference between fractions of, for example, 0.10 and 0.90 is 0.20, not 0.80. From this, the number of wavelengths N_i of the other wavelengths follow, and together with the measured fraction, also the measured lengths at the other wavelengths. As the final value for L, the average of these values is taken. An example for a determination of a gage block with a nominal length of 5 mm is shown in Table 4.3. Here it can be seen that at the length L =5 000.35 µm, the measured fractions best match the theoretical ones. The deviations from this length measured with the other two wavelengths follows from the deviation of the fraction:

$\Delta f_2 = 0.7 - 0.81 = -0.11$; $\Delta L_2 = -0.027$ µm. $\Delta f_3 = 0.8 - 0.64$; $\Delta L_3 = +0.037$ µm. Together with the 0 deviation of the first wavelength, this gives an additional average correction of +3 nm, so the final length becomes L = 5 000.353 mm = 5 mm + 0.353 µm. This length is to be corrected for the gage block temperature T_m to obtain the gage block length at 20 °C; this means that for the used wavelengths the system equation takes the form:

$$L_{20} = \frac{(N+f)\lambda_v}{2n_a\left(\lambda_v, T_a, p, RH\right)}\left(1+\alpha\left(T_m - 20\ ^\circ\mathrm{C}\right)\right). \tag{4.39}$$

Note that equation (4.39) is essentially the same as equation (4.18); this implies that an uncertainty evaluation contains similar elements and considerations regarding atmospheric conditions and material temperature. In addition, corrections must be made for a potential difference in phase change between the base plate and the gage block (see Section 4.8), and for the fact that the light source is not a point source but has a final size; this is called the aperture correction (Haitjema 2014). The fractions can be estimated visually by eye, by a phase stepping method as given in Section 4.5.1, or directly from a recorded image as in Figure 4.14. Visually, a repeatability of 10% in the fraction determination is possible, this corresponds to a repeatability of $\lambda/20 \approx 0.025$ μm in length. If the repeatability in the fraction determination is worse than about 10%, the minima in equation (4.35) cannot be distinguished clearly, and the length becomes indeterminate, see also Section A.5.3. This means that this measurement is either very accurate, or useless. The smallest final uncertainties achieved in interferometric gage block calibrations are around 0.02 μm, even with advanced fringe evaluation methods (Prieto et al. 2022).

4.7 MICROSCOPIC SURFACE INTERFEROMETRY

A major part of surface interferometry takes place in the microscopic domain, where surfaces lateral ranging from a few millimeters squared to about 50 μm × 50 μm are studied. These surfaces are studied with microscopes where the interferometer is integrated in to the objective. From a metrology perspective, the most important characteristic of a microscope objective is its numerical aperture (NA) that defines the convergence of the light beam and characteristics such as the lateral resolution and focal depth. Figure 4.15 depicts the definition of the numerical aperture.

Microscopic-interferometry objectives are based on the Michelson interferometer principle where three types are most common:

1. Michelson-type objective
   The microscope objective based on a cube beam splitter in the converging imaging beam path is denoted as a Michelson-type interference objective. Because of the additional space the cubic beam splitter takes, and the optical

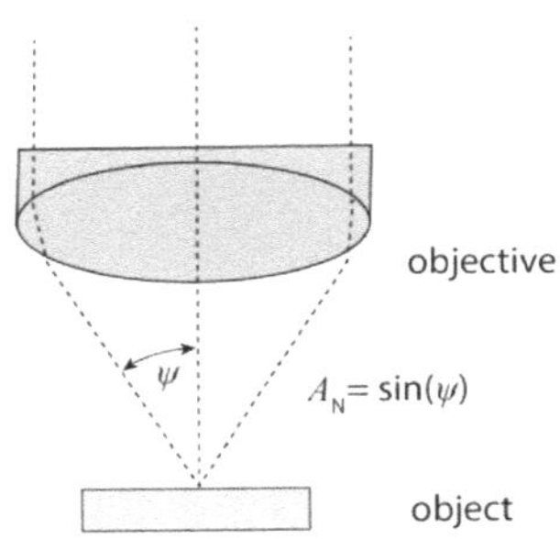

FIGURE 4.15 Definition of numerical aperture (NA), denoted A_N.

aberrations caused by a converging beams that passes a cubic object, this type is limited to low NA and a relatively large field of view (FOV).

2. Mirau-type objective
   In the Mirau-type objective, the beam splitter consist of a semitransparent flat that partially reflects the light 180° back into the objective where it is focused on to a small reference mirror in the beam path. In this type, the optical beam is less distorted and the interference optics takes less space, therefore, they can be used for larger magnifications and higher NA values.
3. Linnik-type objective
   In this type of objective, the beam is split before it enters the objective part, and the reference beam is directed to an objective reference mirror assembly. The advantage is that there are no obstructions and distortions in the measurement beam. The obvious disadvantage is that two – preferably highly similar – objectives are needed and for white-light interferometry, there are additional alignment difficulties as the focus and white-light interference point must coincide for both objectives. The optical schemes for the three interference objective types are given in Figure 4.16.

Microscope systems equipped with one of the three types of interference objective can measure surface topography using the principle of phase shifting interferometry or white-light interferometry, using the methods as described in Sections 4.5.1 and 4.5.3. For phase-shifting interferometry, it must be noted that, because of the converging optical beam, the wavelength cannot automatically be used as the length reference. For this case, the wavelength must be multiplied by a so-called obliquity factor that may range from close to 1 for low NA objectives, up to about 1.10 for a 50× Mirau objective with NA = 0.55 (de Groot 2017). Characteristics for phase-shifting interference microscopes are given in ISO 25178-603:2025. Surface topography measurement systems based on white-light interferometry are classified as coherence scanning interferometry instruments in ISO 25178-604:2025.

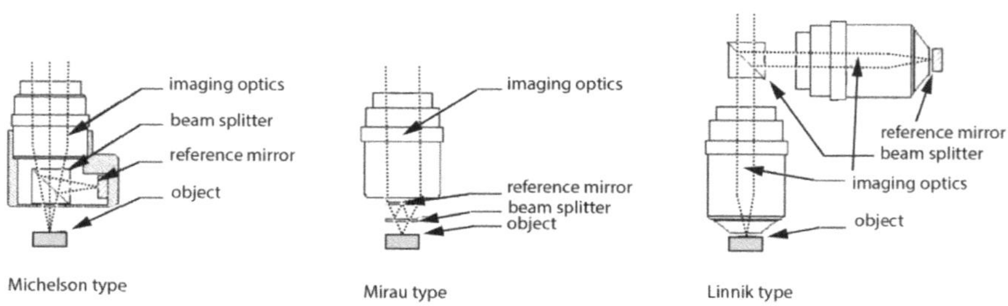

FIGURE 4.16 Interference microscope objective types.

4.8 THE ELECTROMAGNETIC SURFACE

In the preceding sections, it was assumed that the surface to be measured was homogeneous, meaning that the optical properties are identical over the surface, and free of coatings or optical effects. However, when measuring a surface optically, the obtained surface topography may differ from the surface that would be measured by a small mechanical probe. This is due to the fact that electromagnetic radiation may not be fully reflected at the material/air interface, but in fact it may penetrate somewhat into the surface when reflected, or a phase shift is generated by a transparent coating, affecting the effective height surface height that is detected. For a homogeneous surface, the reflectance properties are determined by the complex refractive index $\tilde{n} = n - \text{j}k$; where j is the imaginary unit and k accounts for absorption in the material. Especially for metals the value of k may be considerable, causing a phase shift at the air–material interface. For a dielectric surface ($k = 0$), there is a π phase change on reflection (at normal incidence), this means a π phase difference between the incident and reflected beams. For materials with free electrons at their surfaces (for example, metals and semiconductors), there will be a ($\pi - \delta$) phase change on reflection, where δ is given by:

$$\delta = \tan^{-1}\frac{2k}{1 - n^2 - k^2} \tag{4.40}$$

for normal incidence. This may be a constant value over the surface and not affect the measured topography, but if there are different materials inside the region measured by interferometry, deviations may arise. For example, for a chrome step on a glass substrate, the difference in phase change on reflection gives rise to an error in the measured height of approximately 20 nm at a wavelength 633 nm. A related effect is the so-called wringing correction when measuring a steel gage block wrung on a transparent optical flat in a gage block interferometer (see Section 4.6); then the correction may be approximately 40 nm. This may make the topography deduced from an optical measurement, the 'electromagnetic surface', deviate from the mechanical surface as defined in ISO 14406:2010 (see also Section 9.4.3) when measured using an optical interferometer or another optical surface topography measuring instrument. For steel gage blocks wrung on glass, Doiron and Beers noted correction values between 25 and 55 nm (Doiron and Beers 2014). Some typical values of phase shift effects in white-light interferometry are given by Harasaki (Harasaki et al. 2001).

EXERCISES

1. Coherence length
   Calculate the coherence length of the following light sources:
   a. A LED with a wavelength of 600 nm and a bandwidth of 1 nm.
   b. A monomode red He-Ne laser (λ = 632.8 nm) with a bandwidth of 10 kHz.

   c. A multimode red He-Ne laser with several lines within the relative frequency range of $1.5 \cdot 10^{-6}$.
2. Refractive index
   A laser interferometer system is corrected for atmospheric conditions at sea level. Then it is used in mountainous areas, where the atmospheric pressure is 100 hPa lower at 1000 m altitude. Calculate the deviation of the laser interferometer. Does it indicate too much (too long) or too little (too short)?
3. Gage block interferometer
   A gage block of 12 mm ± 0.5 μm is measured in a gage block interferometer using a spectral lamp with wavelengths in air (λ_a) of exactly 600, 500 and 400 nm.
   a. You may use two of these wavelengths for the length calibration. Give the advantages and disadvantages of each choice.
   b. A measurement is made with these three wavelengths. The measured fractions are 33%, 40% and 50%, respectively. Calculate the length of the gage block.
4. Interference pattern
   An interferogram, recorded with a Fizeau interferometer, is shown in Figure 4.17 at full size (1:1). The center of the object is elevated relative to the edges. The wavelength used is $\lambda = 633$ nm.

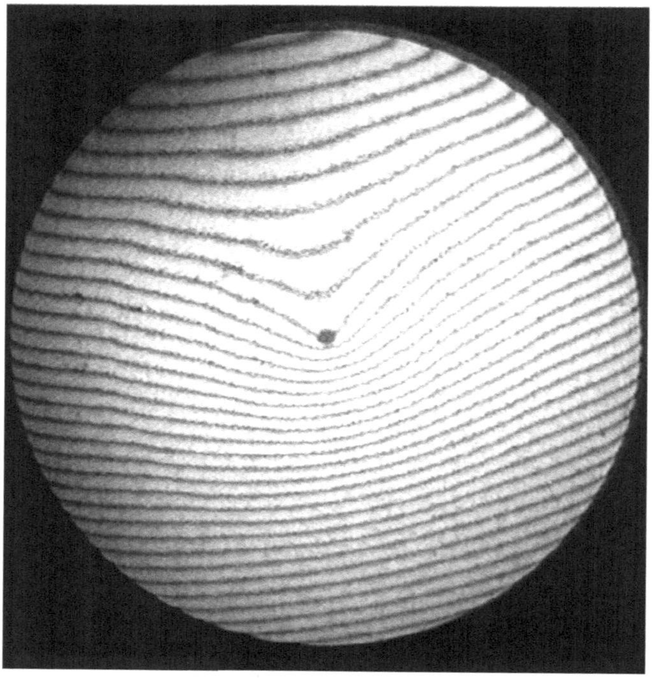

FIGURE 4.17 Fringe pattern taken with a Fizeau interferometer with a light source with a wavelength $\lambda = 633$ nm.

a. Determine the tilt, expressed in arcseconds, of the object with respect to the reference plane in both the x- and y-directions.
b. Using a ruler on a line through the center, determine the straightness deviation of the object.

5. Fringe-phase evaluation
In a Fizeau interferometer, four phase steps of $\lambda/8$ are taken and four images are recorded.
For one pixel in the images, the measured intensities (gray-values) are $I_1 = 19$, $I_2 = 105$, $I_3 = 190$ and $I_4 = 91$. The wavelength $\lambda = 633$ nm.
a. Calculate the height, apart from an integer number of half wavelengths.
b. Calculate the uncertainty in the height, assuming an independent standard uncertainty of seven gray-values in each of the four intensities $I_1 .. I_4$.

6. Gage block interferometer
In a gage block interferometer, a gage block is measured with a length of 100 mm ± 0.5 µm. The gage block is wrung on to an optical flat of the same material as the gage block. The interferogram is taken with a spectral source having two wavelengths: $\lambda_a = 0.5$ µm (green) and $\lambda_a = 0.4$ µm (blue) in the air in which the measurement takes place. Both interferograms are displayed in Figure 4.18. The optical flat-gage block assembly is slightly rotated so that the left side is slightly higher with respect to the reference plane inside the interferometer. This means that the order of interference increases toward the right.
a. Determine the gage block length with a resolution of 0.01 µm.
b. Evaluate the uncertainty in this measurement, based on these two interferograms only; omitting all other influencing quantities.
c. Calculate the angle (in arcseconds) of the optical flat-gage block assembly relative to the reference plane. The lateral sizes of the gage block are 9 mm × 35 mm.
The gage block has a linear expansion coefficient of $\alpha = 10 \cdot 10^{-6}$ K^{-1}.

Interference pattern for green light $\lambda = 0.5$ µm

Interference pattern for blue light $\lambda = 0.4$ µm

FIGURE 4.18 Interference pattern of a gage block wrung on a platen for different wavelength.

TABLE 4.4
Variation of the Recorded Intensity $I(d_z)$ (Gray-Value of the Image) as a Function of the Displacement d_z

d_z/μm	0.125	0.25	0.375	0.5	0.625	0.75	0.875	1	1.125	1.25	1.375	1.5
$I(d_z)$	104	115	139	135	107	94	127	155	127	76	101	156
d_z/μm	1.625	1.75	1.875	2	2.125	2.25	2.38	2.5	2.625	2.75	2.875	3
$I(d_z)$	154	84	88	139	156	115	90	117	142	127	108	107

d. Give an uncertainty budget for the gage block length with as contributions:
   - The fraction evaluation as calculated in b.
   - A standard uncertainty in the air temperature of 0.1 °C:

$$T_{\text{air}} = (20.00 \pm 0.10)\ °\text{C}$$

   - A standard uncertainty in the atmospheric pressure of 0.4 hPa.

$$P_{\text{air}} = (1013.25 \pm 0.40)\ \text{hPa}$$

   - A standard uncertainty in the gage block temperature of 0.01 °C.

$$T_m = (20.000 \pm 0.010)\ °\text{C}$$

Calculate the total uncertainty in the gage block length based on $k = 2$.

7. White-light correlogram analysis
   A measurement using white-light interferometry creates an image stack. For one pixel out of this stack, the variation of the recorded intensity (gray value of the image) as a function of the displacement d_z is given in Table 4.4.
   a. Calculate the envelope intensity as a function of the height.
   b. Give the height where the envelope intensity is maximal.
   c. Calculate the height as the center of gravity of the square of the envelope intensity.
   d. Calculate the phase in a region ±0.5 μm from the center of gravity.
   e. Determine the location where the phase is zero near the center of gravity. Use the 'atan2' function for this calculation.

   These calculations are best performed using a spreadsheet.

REFERENCES

BIPM, 2024, Recommended Values of Standard Frequencies www.bipm.org/en/publications/mises-en-pratique/standard-frequencies-metre; accessed 10-11-2024, see document on spectral lamps.

Bönsch, G. and Potulski, E., 1998, Measurement of the Refractive Index of Air and Comparison with Modified Edlén's Formulae. *Metrol.* **35 (2):**133–139.

Born, M. and Wolf, E., 2013, *Principles of Optics.* Cambridge, Cambridge University Press

Ciddor, P.E., 1996, Refractive Index of Air: New Equations for the Visible and Near Infrared *Appl. Opt* **35(9)**:1566–1573.

De Groot, P.J., 2017, Interference Microscopy for Surface Structure Analysis. In: Yoshizawa, T. (ed)., 2nd ed., *Handbook of Optical Metrology: Principles and Applications.* Boca Raton, CRC Press.

De Groot, P. and Deck L., 1995, Surface Profiling by Analysis of White-Light Interferograms in the Spatial Frequency Domain. *J. Mod. Opt.* **42(2):**389–401.

Doiron, T. and Beers, J.S., 2014, *The Gage Block Handbook.* NIST monograph 180. , Gaithersburg, USA, NIST.

Edlén, B., 1966, The Refractive Index of Air. *Metrol.* **2(2)**:71–80.

Gasvik, K.J., 2002, *Optical Metrology.* New Jersey, Wiley.

Haitjema, H., 2014, Length Traceability Using Interferometry. In: Leach, R.L., *Fundamental Principles of Engineering Nanometrology,* 2nd ed. Amsterdam, Elsevier.

Harasaki, A, Schmit, J. and Wyant, J.C., 2001, Offset of Coherent Envelope Position due to Phase Change on Reflection. *Appl. Opt.* **40:**2102–2106.

Heydemann, P.L.M., 1981, Determination and Correction of Quadrature Fringe Measurement Errors in Interferometers. *Appl. Opt.* **20**:3382–3384.

ISO 14406, 2010, Geometrical Product Specifications (GPS) – Extraction. Geneva, International Organization for Standardization.

ISO 25178-603, 2025, Geometrical Product Specifications (GPS) – Surface Texture: Areal. Part 603: Design and Characteristics on Non-Contact (Phase-Shifting Interferometry) Instruments. Geneva, International Organization for Standardization.

ISO 25178-604, 2025, Geometrical Product Specifications (GPS) – Surface Texture: Areal. Part 604: Nominal Characteristics of Non-Contact (Coherence Scanning Interferometry) Instruments. Geneva, International Organization for Standardization.

Larkin, K.G., 1996, Efficient Nonlinear Algorithm for Envelope Detection in White Light Interferometry. *J. Opt. Soc, Am. A: Optics and Image Science, and Vision* **13(4):**832–843.

Malacara, D., 1991, *Optical Shop Testing.* New Jersey, Wiley.

Malacara, D., Servin, M., Malacara, Z., 1998, *Interferogram Analysis for Optical Testing.* New York, Marcel Dekker.

Muijlwijk, R., 1988, Update of the Edlén Formulae for the Refractive Index of Air. *Metrol.* **25(3)**189.

Prieto, E., Rodriguez, J., Bartl, G., Pirée, H., Meli, F., Picotto, G.B., Franco, M. and Balling, P., 2022, Measurement of Short Gauge Blocks by Interferometry. *Metrol.* **59(1A)**.

Stone, J. and Zimmerman, J. 2001, Index of Refraction of Air. http://emtoolbox.nist.gov/Wavelength/Documentation.asp accessed 2024-05-17.

5 Sensor Principles

5.1 MECHANICAL SENSORS

Common dial indicators and comparators are based on principles of mechanical transmission of displacement. Figure 5.1 shows three different types of indicators: the lever indicator, dial indicator and comparator. The difference in designation between the dial gage and comparator refers to the different mechanical drive principles and the associated accuracy and measuring ranges. The general principle of a lever indicator is a measuring pin provided with a rack that drives a pinion (Figure 5.2a). The resulting rotation is increased by further gears until a dial can be driven. The latter usually has one revolution per millimeter and a subdivision of 0.01 mm. With a larger range, a small dial indicates the number of millimeters. The measuring range is usually 10 mm. In order to prevent play in the gears, one gear wheel is connected to a spiral spring which ensures that the gear wheels are always pushed to one side. However, the measuring force due to friction is considerably different for the inward and outward movement of the measuring pin, as a result of which hysteresis may occur. The uncertainty of a measurement with lever indicators is of the order of 0.01 mm. In a dial indicator, use is made of a mechanism other than a rack. With this principle, the measuring pin presses against the short end of a lever (Figure 5.2b).The long end of the lever is provided with toothing, which drives a pointer via a number of gear wheels. The measuring range of a dial indicator is generally small, usually 0.1 mm, with a resolution of 0.001 mm. Application fields for dial indicators are, therefore, accurate measurements in a small measuring range, in particular for comparison measurements. Like the lever indicators, the dial indicators suffer from hysteresis in the measuring force due to the friction. A special version of the comparator, based on Johansson's torsion spring principle, overcomes this problem (Figure 5.2c). With this principle, the gears have been replaced by elastic elements. The measuring pin applies a force via a lever to a twisted leaf spring, to which a dial is attached. Because of the force, the spring twists to a greater or lesser extent and the dial rotates. With this principle, there is no play and hysteresis either. This allows resolutions of 1 μm to 0.1 μm to be achieved with a measuring range of 50 μm. Applications are very accurate

DOI: 10.1201/9780367821746-5

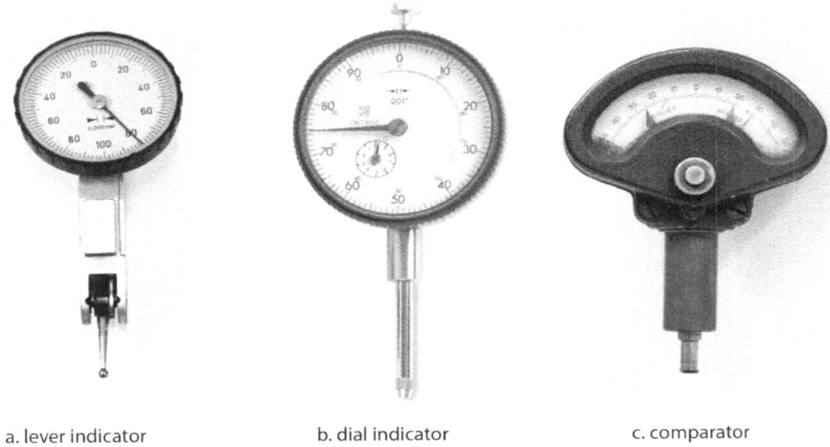

FIGURE 5.1 Three embodiments of indicators: from left to right: a. lever indicator, b. dial indicator and c. comparator.

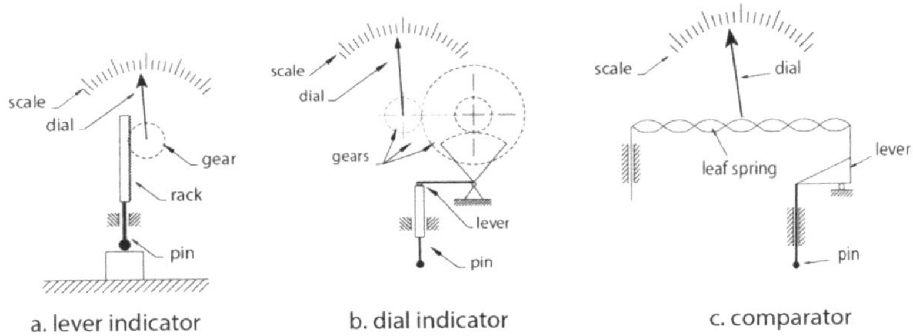

FIGURE 5.2 Schematics of the mechanisms of indicators: from left to right: a.: rack and pinion, b.: gears and c.: elastic elements.

displacement measurements, such as comparison measurements with gage blocks. However, the readout is visual: a scale reading has to be taken visually. For this reason, in due course, linear variable differential transformers (LVDTs), see Section 5.3, have mainly taken over the application of these analog devices.

5.2 SENSORS BASED ON RESISTANCE

Displacement transducers based on the varying resistance operating principle rely on the relationship between the length and the resistance of a wire. For this, it can be written:

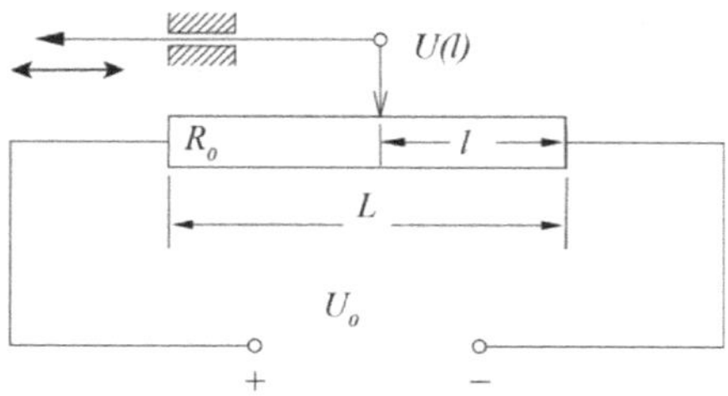

FIGURE 5.3 Schematic display of a displacement transducer based on a resistance sensor.

$$R = \frac{l \cdot \rho}{A} \tag{5.1}$$

where R is the resistance, l is the length, ρ is the specific resistance and A is the cross-sectional area. Figure 5.3 shows a displacement sensor diagrammatically. The resistor consists of a number of wire turns homogeneously distributed over the length l. The displacement sensor bridges, depending on the position, a certain fraction of the resistance. In this way, the position is converted into a voltage U. For $U(l)$, it can be written:

$$U(l) = \frac{U_0}{L} l \tag{5.2}$$

where L is the length of the resistor, l is the relative position, U_0 is the supply voltage and $U(l)$ is the output voltage.

5.3 SENSORS BASED ON INDUCTANCE: LVDT

The inductive measuring principle is based on the change of the inductance in a coil. For the inductance L_e in a coil,

$$L_e = \frac{\mu_r \mu_0 \cdot n^2 \cdot A}{L}, \tag{5.3}$$

which gives for the induced voltage U_i when an AC current $i(t)$ is applied:

$$U_i = -L_e \frac{di}{dt}, \tag{5.4}$$

where L_e is the self-induction in $V \cdot s \cdot A^{-1}$, μ_r is the relative permeability, μ_0 is the absolute permeability in vacuum $\mu_0 = 4\pi \cdot 10^{-7}\ V \cdot s \cdot A^{-1} m^{-1}$, n is the number of windings, A is the cross-sectional area, L is the length of the coil and i is the current. In Figure 5.4, an inductive displacement sensor is shown schematically. By

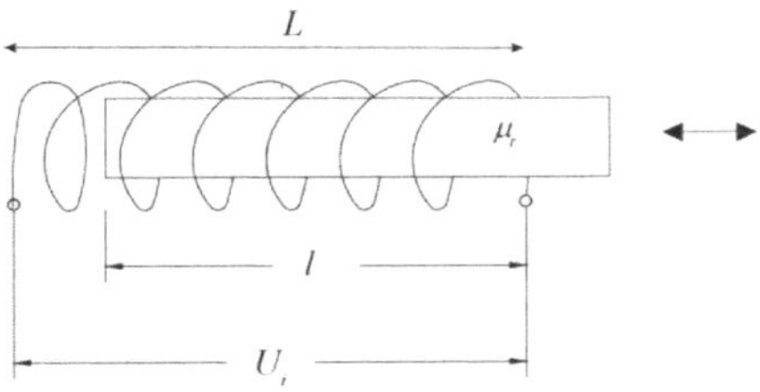

FIGURE 5.4 Schematic view of an inductive displacement transducer.

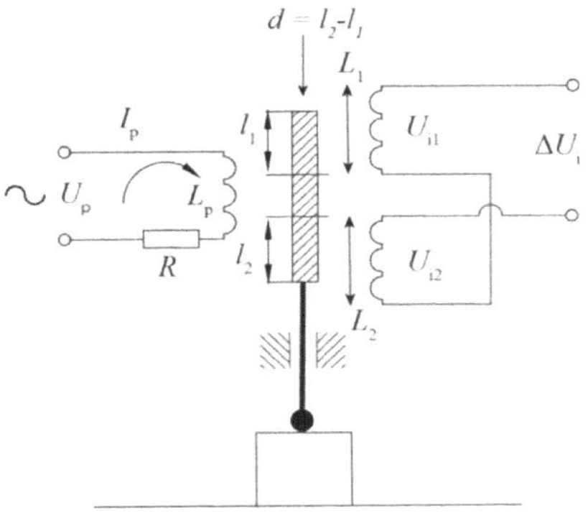

FIGURE 5.5 Principle of an linear variable differential transformer (LVDT) based on three coils.

a change of the effective μ, for example, by displacement of the core in the coil, the inductance and the voltage change as a function of the length l:

$$L_e = f(l) \approx \frac{\mu_r \mu_0 \cdot n^2 \cdot A}{l} + \frac{\mu_0 \cdot n^2 \cdot A}{L - l}. \tag{5.5}$$

LVDTs come in versions with two and three coils. In Figure 5.5, a schematic of an LVDT is given in which three coils are used. In this system, one of the coils, the primary coil, is fed with an alternating voltage. In the two secondary coils, a voltage is induced, due to the mutual inductance between each of the two secondary coils and the primary coil. In order to obtain a good coupling between the coils, in practice, the primary coil is not placed next to the two other coils, as the diagram suggests, but is wound around it. With a displacement of the kernel, the length l_1 decreases and the length l_2 increases, creating a difference between U_{i1} and U_{i2} which, after rectification, acts as a measure of the displacement d:

$$\Delta U_i = U_{i2} - U_{i1} = f(l_2 - l_1) = f(d). \tag{5.6}$$

This voltage ΔU_i will be approximately proportional to the movement of the core. When the bridge is in equilibrium, $l_1 = l_2$ and $\Delta U = 0$. This means that around this point the voltage can be considerably amplified. This not only increases the sensitivity but also it limits the measuring range. With such a system, a resolution of 1 nm can be achieved when the gain is increased such that the measuring range is only a few micrometers. The measuring range can be up to 100 mm but 1 mm is more common. In each case, an AC voltage measuring bridge with a data processing system is required for the measuring systems described above. Sometimes several sensors are connected alternately to the measuring bridge via a switch (multiplexer). In this way, several sensors can be used in a relatively inexpensive manner. This is particularly useful because different applications are based on difference measurements, in which at least two sensors are used. The resolution that is feasible for inductive sensors is 10 nm. With special precautions and further reinforcement, 1 nm is possible. The absolute accuracy depends on the embodiment and the electronics, and is often of the order of 1% of the measured displacement. LVDTs can be used separately (see Figure 5.5b) or can be part of a measurement system such as a CMM-probe, a gage block comparator, a probe of a roughness or roundness tester, and so on. More details about the various types of LVDT and their electronic circuits can be found elsewhere (Doebelin 2003).

5.4 SENSOR BASED ON CAPACITANCE: CAPACITIVE SENSORS

Capacitive sensors are based on the change of the capacity of a capacitor, which changes depending on a displacement. The capacity of a flat capacitor is given by:

$$C = \frac{\varepsilon_r \cdot \varepsilon_0 \cdot A}{d}, \tag{5.7}$$

where C is the capacitance, A is the effective common surface area of the capacitor plates, d is the distance between the plates, ε_0 is the dielectric constant of vacuum: $\varepsilon_0 = 8{,}854\ 187\ 817 \times 10^{-12}\ \mathrm{F\ m^{-1}}$ and ε_r is the relative dielectric constant of the medium between the plates. Capacitive displacement sensors are based on the variation of the capacitance on displacement, due to variation of the plate distance, the surface area of the capacitor or the medium in the capacitor. Figure 5.6 shows the

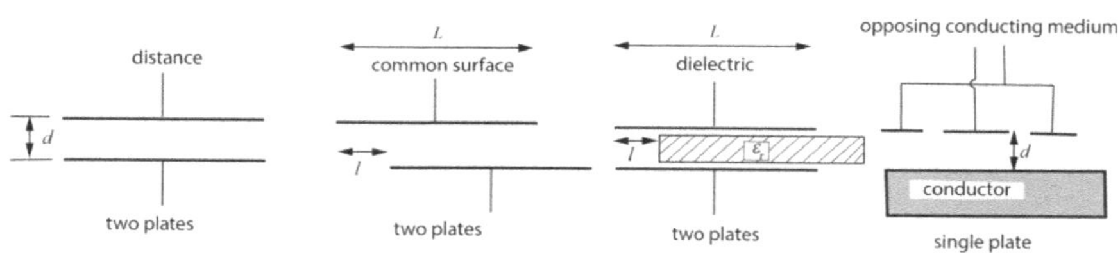

FIGURE 5.6 Possibilities to use capacitors to detect a distance or displacement.

different possibilities of variation. These will be briefly discussed below. Variation of the distance between the plates gives a reciprocal relationship between capacity and displacement as indicated in Equation (5.7).This means that some linearization will be needed in the electronics. Variation of the 'common' length and thus the surface gives a linear relationship between capacitance C and the displacement l:

$$C = \frac{\varepsilon_r \cdot \varepsilon_0 \cdot A}{d} \cdot \frac{l}{L}. \tag{5.8}$$

A third possibility is the variation of the dielectric medium by sliding a plate between the capacitor plates with dielectric constant ε_r. The total capacitance is the sum of the partial capacitances with and without the dielectric medium:

$$C = \frac{\varepsilon_0 \cdot A}{d} \cdot \frac{l}{L} + \frac{\varepsilon_r \cdot \varepsilon_0 \cdot A}{d} \cdot \frac{L-l}{L} = \frac{\varepsilon_0 \cdot A}{d \cdot L} \cdot \left(l\left(1-\varepsilon_r\right) + \varepsilon_r L \right). \tag{5.9}$$

This gives a linear dependence of the capacitance with the displacement. The configuration using a single plate uses the capacitance between two conductors positioned in a plane: the capacitance between these conductors depends on the distance d of an opposing conducting medium. Such capacitive single-plate sensors measure against all kinds of conductive surfaces and are easier to handle mechanically than two-plate assemblies. The electronics is commonly designed such that there is a basic working distance (offset) that gives a zero signal. This is accomplished by a Wheatstone bridge that contains a reference capacitor that corresponds to the capacitance at the nominal working distance. This electronic scheme is given in Figure 5.7 (see also Figure 1.10). In Figure 5.7, the reference capacitor C_r can be taken equal to the capacitance C at a nominal distance d. The variable resistance R and the fixed resistors R_0 can be chosen to achieve an appropriate amplification. A special case of sensor is an electronic level where a rotation of the device generates a displacement of a moveable central electrode relative to two fixed electrodes. This is schematically depicted in Figure 5.8. For C_1 and C_2 it can be written:

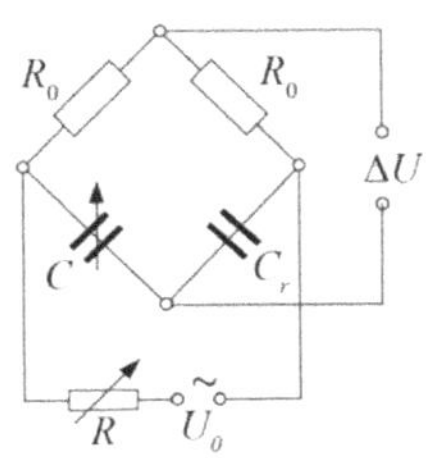

FIGURE 5.7 Electrical scheme for a capacitive probe.

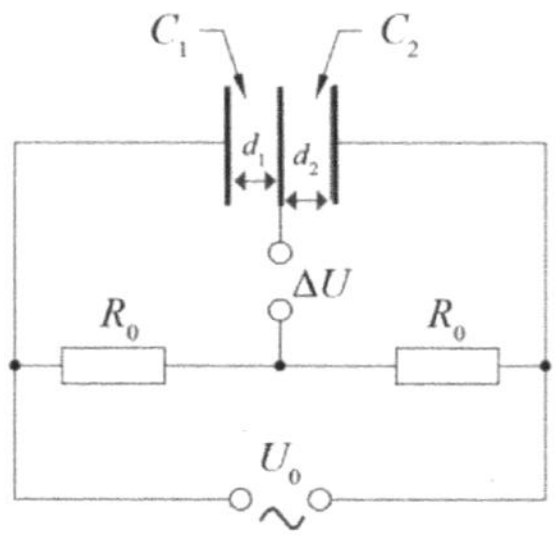

FIGURE 5.8 Electrical scheme for a capacitive three-plate probe as it is used in electronic levels.

$$C_1 = \frac{\varepsilon_r \cdot \varepsilon_0 \cdot A}{d_1} = \frac{\varepsilon_r \cdot \varepsilon_0 \cdot A}{d_0 + \Delta d} \text{ and } C_2 = \frac{\varepsilon_r \cdot \varepsilon_0 \cdot A}{d_2} = \frac{\varepsilon_r \cdot \varepsilon_0 \cdot A}{d_0 - \Delta d}, \tag{5.10}$$

where d_0 is the mean of d_1 and d_2, and Δd is the displacement from the central position.

For the voltage difference, ΔU, it can be written:

$$\Delta U \sim U \frac{C_2 - C_1}{C_1 + C_2}. \tag{5.11}$$

Substitution of equation (5.10) in to equation (5.11) gives for $\Delta d << d$:

$$\frac{\Delta U}{U} \sim \frac{\Delta d}{d}. \tag{5.12}$$

For small displacements Δd, the output voltage difference ΔU is proportional to the displacement; when applied to an electronic level, this implies that the output is proportional to the (small) angle.

5.5 OPTO-ELECTRONIC DEVICES

Opto-electronic devices include sensors that supply an electrical current as a result of an incident light beam. A number of these are used in measuring technology, often as a sensor part in a measuring instrument. The most important opto-electronic sensors will be discussed here.

5.5.1 Position-Sensitive Detector

A position sensing detector (PSD), also called a lateral photodiode, consists of a light-sensitive silicon surface on a support. If there is a light spot on this surface

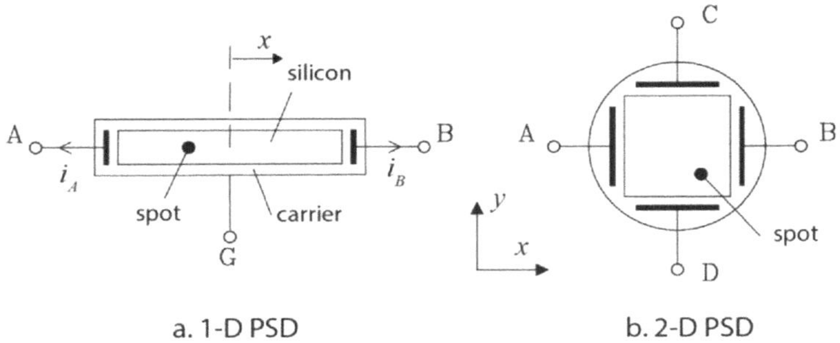

FIGURE 5.9 Schematic of the working principle of 1D (a) and 2D (b) PSDs.

(see Figure 5.9), a current will flow. Because the light intensity of the spot can vary due to environmental influences, the voltage difference is divided by the total voltage, so that for the position x holds:

$$x \approx \frac{V_A - V_B}{V_A + V_B}, \tag{5.13}$$

where V_A and V_B are the voltages at point A and B, respectively, in Figure 5.9a. In the case where terminal G is grounded, the sensor is used in the so-called photovoltaic mode. If a supply voltage (5–15 V) is applied at point G, the sensor operates in photoconductive mode, which is particularly suitable for fast measurements. A laser diode or a helium-neon (He-Ne) laser are generally used for the light spot. In addition to the 1D version of the PSD discussed here, 2D PSDs are also available (Figure 5.9b). Their operation is similar to the 1D PSD type. PSDs are available in many embodiments, in terms of dimensions and accuracy. Typical dimensions of the sensitive surface are several millimeters. Resolutions smaller than 1 μm are feasible. The accuracy depends, among other things, on the linearity of the sensor.

5.5.2 CCD and CMOS Sensors

The operation of charge coupled devices (CCD) will be briefly discussed here. A CCD sensor consists of a large number of CCD cells similar to photodiodes and placed in a matrix on a support (IC). Figure 5.10 shows a row with CCD cells schematically. Due to incident light, a charge is generated in a CCD cell, which is proportional to the amount of light that falls on the cell. The distribution of the charge over the cells thus provides an image of the light distribution over the CCD sensor whereby, for example, the position of a light spot can be determined. A challenge with a CCD sensor, however, is the rapid reading of the charge from the cells. For a rapid serial reading, the cells are connected to a so-called shift register, which transports the charge packets from the cells to a readout integrated into the IC. In order to get a correct picture of the exposure situation of the CCD sensor at a certain

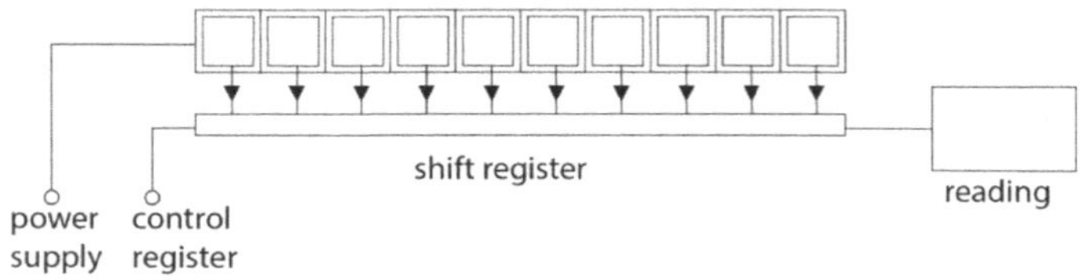

FIGURE 5.10 Schematic of a 1D CCD sensor.

moment, the transfer time of the charges to the readout circuit must be taken into account. Application of CCD sensors is not limited to the position determination of light spots, such as in PSDs, to measure displacements. Because the exposure can be determined over a large number of cells, a CCD sensor can be used as a camera. This application is particularly interesting when coupling the CCD sensor to a computer, which can analyze the acquired image, for example, with regard to the dimensioning of the object of which the image is included. Some applications of CCD cameras in dimensional metrology are the use in measuring microscopes ('optical CMMs', see Section 8.2.1), surface topography measuring systems (see Section 7.9.3), Fizeau interferometers (see Section 4.5.2), X-ray computed tomography systems (see Section 8.5), fringe projection systems (see Section 8.4), photogrammetry and vision systems in general. Conventional CCD sensors contain several million pixels with pixel sizes in the order of 2–10 µm. The resolution and accuracy depends on the optics used to display an object on the CCD sensor and the interpolation between the pixels. For some years now, complementary metal-oxide semiconductors (CMOS) have been taking over the role of CCDs. One of the biggest differences between CCD and CMOS sensors is that each CMOS pixel has an amplifier. Apart from this, each column of pixels has its own analog-to-digital converter (ADC), and some CMOS configurations may even have an ADC for each pixel. Once a pixel detects the light, the ADC amplifies and processes it. More ADCs make it possible for CMOS to process several pixels simultaneously.

5.5.3 Linear Encoders

Optical linear encoders are a means to read a scale/ruler electronically and automatically. Optical linear encoders are extensively used because of their noncontact, high-resolution measurement characteristics and because their output signal is easily converted into an electronic position or displacement signal. The range of applications of optical encoders spans from measurement of angular or linear motion in low-cost consumer products (inkjet printers) to advanced industrial systems (machine tools, robots and stage positioning in lithographic projection tools). An optical encoder contains a detection system that can detect light emanating from a patterned movable member. The patterned member is usually referred to as the 'optical scale'. The scale typically contains one or more sets of parallel lines with a predetermined constant pitch: that is the distance between

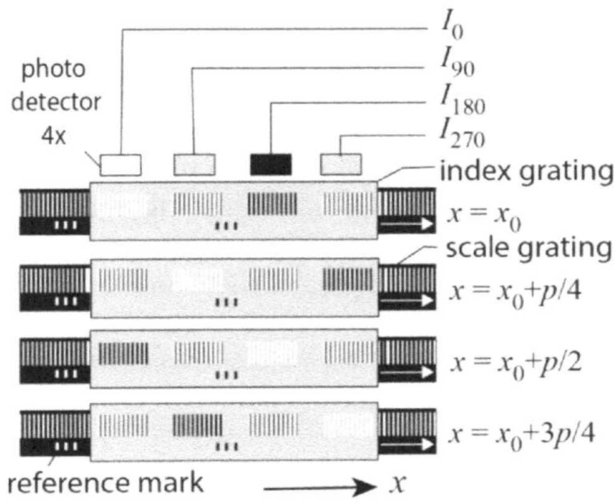

FIGURE 5.11 Schematic of a linear encoder consisting of a scale with pitch p and four masks with the same pitch but mutually shifted by 90°.

the lines. The detected light is converted into an electrical signal and may be processed as digital square waves or as analog signals, where analog signals allow for resolution enhancement by means of interpolation. To obtain various phase signals in parallel from the same scale, the reticle mask is equipped with multiple grating areas that have predefined offsets. In practice, often four-phase quadrature detectors signals are generated. An example is given in Figure 5.11 that depicts the scale being moved over four masks that have the same pitch as the scale, but are mutually shifted. This shift makes the detected intensities through the masks vary periodically. For accurate detection using phase quadrature interpolation, the optical system should be designed to generate a sinusoidal signal. The captured diffraction orders of the imaging system, the spectral width, the size and shape of the source and the distance between the gratings affect the sharpness of the imaged grating. With a proper encoder design, a grating can be imaged as a sinusoidal image such that, effectively, a sinusoidal modulation of the detector signal is obtained (Durschmid 1993). In that case, the signals can be written as:

$$
\begin{aligned}
I_o(x) &= C\left\{1 + \cos\left(\frac{2\pi}{p}(x - x_0)\right)\right\}, \\
I_{90}(x) &= C\left\{1 + \cos\left(\frac{2\pi}{p}\left(x + \frac{p}{4} - x_0\right)\right)\right\} = C\left\{1 - \sin\left(\frac{2\pi}{p}(x - x_0)\right)\right\}, \qquad (5.14) \\
I_{180}(x) &= C\left\{1 + \cos\left(\frac{2\pi}{p}\left(x + \frac{p}{2} - x_0\right)\right)\right\} = C\left\{1 - \cos\left(\frac{2\pi}{p}(x - x_0)\right)\right\},
\end{aligned}
$$

$$I_{270}(x) = C\left\{1 + \cos\left(\frac{2\pi}{p}\left(x + \frac{3p}{4} - x_0\right)\right)\right\} = C\left\{1 + \sin\left(\frac{2\pi}{p}(x - x_0)\right)\right\},$$

where C is a constant, x_0 is the initial position, x is the displacement of the scale and p is the pitch of the scale. Note the similarity to equations (4.26) to (4.27) in Section 4.5.1, where the half-wavelength $\lambda/2$ is replaced by the pitch p that depends on the manufacturing accuracy of the scale. By taking the difference of signals that have an opposite phase: $(I_0 - I_{180})$ and $(I_{270} - I_{90})$, and dividing these, the constant C can be eliminated and the sensitivity for transmission losses of the electrical signal, causing offset variations, can be reduced, giving

$$x = x_0 + \frac{p}{2\pi}\tan^{-1}\left(\frac{I_{270} - I_{90}}{I_0 - I_{180}}\right). \tag{5.15}$$

The signals can be detected and counted with photodiodes and an electronic circuit. The number of light–dark transitions is a measure of the displacement. Typical encoders have a pitch of 20 µm or 40 µm and are typically interpolated with a resolution up to 0.1 µm or smaller.

5.5.4 Interferential Encoders

For a pitch $p < 5$ µm, diffraction effects become significant and the principle of using a mask can no longer be used. Instead, these diffraction effects can be exploited in interferential encoders, also known as 'grating interferometers'. As an illustration, in Figure 5.12, the interaction between a periodic grating structure and a coherent light source for a transmissive grating is depicted. The illuminated grating may be considered as an array of coherent point sources, each emitting a spherical wave. In specific propagation directions, the waves from all point sources interfere constructively and will propagate as a plane wavefront. For constructive interference, the optical path length from an input wavefront to an output wavefront must differ by integer multiples of unity for adjacent slits of the scale grating. This integer multiple m is called the order of diffraction. Between the entrance angle θ_i and the output angle θ_m of a diffracted beam of order m, wavelength λ and grating pitch p the relationship is:

$$\sin\theta_i + \sin\theta_m = \frac{m\lambda}{p}. \tag{5.16}$$

If the input beam is perpendicular to the grating, equation (5.16) simplifies to:

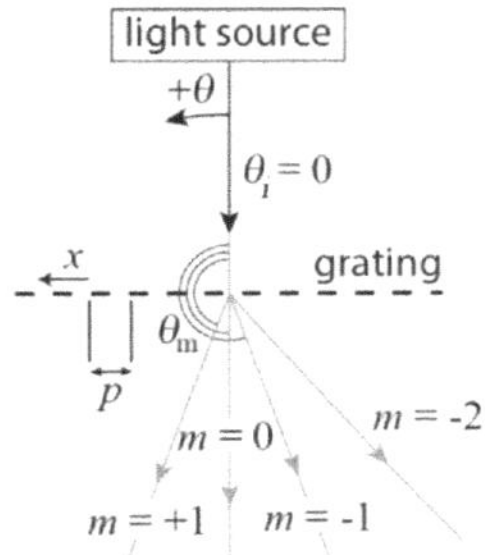

FIGURE 5.12 Transmissive diffraction grating with angles according to the grating equation (5.17).

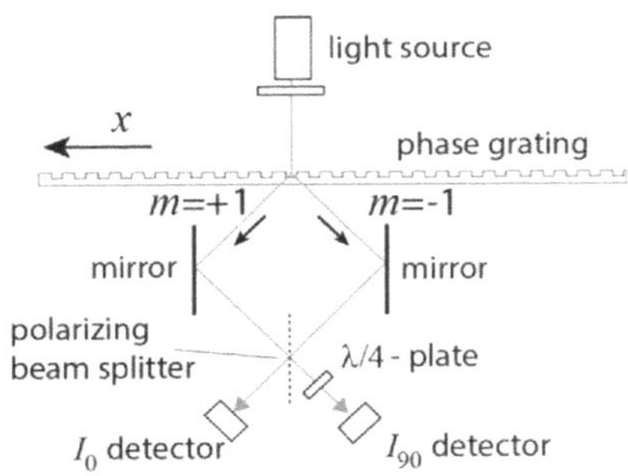

FIGURE 5.13 Schematic presentation of a basic transmissive diffraction encoder.

$$\sin\theta_m = \frac{m\lambda}{p}. \tag{5.17}$$

In most designs, a phase grating is used instead of a partially transmissive grating, so that the zeroth order is suppressed. In Figure 5.13, a basic interferential grating encoder is shown. A coherent beam is diffracted into a $m = +1$ and $m = -1$ diffraction order by a transmissive-phase grating, forming the two interferometer arms with opposite phase offsets as a function of grating displacement. When the two interferometer arms are combined, a sinusoidal interference signal as a function of grating displacement is obtained with a period of half the grating pitch. The calculation of the displacement x may be carried out in a similar manner to that for a homodyne laser interferometer (Section 4.4.1). Details about more elaborate designs and the ways of phase detection can be found in the literature (Cosijns et al. 2018).

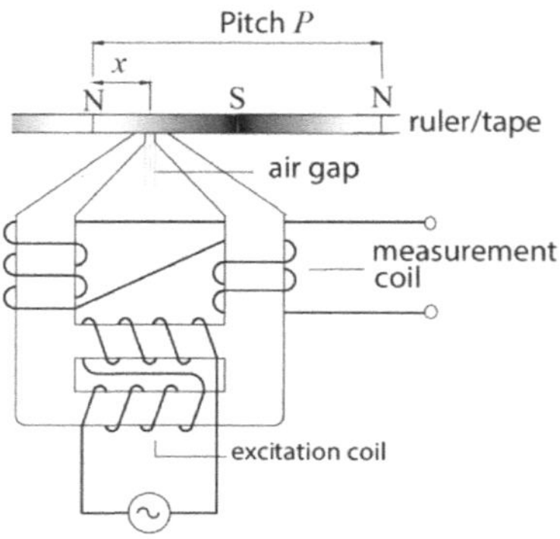

FIGURE 5.14 Schematic of a magnetic tape encoder.

5.6 MAGNETIC-INDUCTIVE ENCODERS

The magnetic-inductive principle is based on a ruler with magnetic graduation of exchanging magnetic field directions, denoted by N (north) and S (south), resulting in a sinusoidal magnetic pattern with a constant pitch *P*. Figure 5.14 shows the principle of this system. A sensor head can be moved along the ruler, which can count the changes of the magnetic field. Since the magnetic field has a known (sinusoidal) course within a pitch, the position x within the pitch can also be determined accurately by means of interpolation. In practice, two sensor heads are employed that are shifted by $P/4$, and enable phase-quadrature detection similar to that in interferometry and optical encoders. The pitch *P* is typically 400 μm while interpolation is possible up to 5 nm resolution. Systems based on this principle may be less sensitive to harsh environments characterized by the presence of condensation or oil compared to optical encoder systems.

5.7 CONFOCAL POINT SENSOR

A confocal sensor point sensor is an optical sensor that can measure the distance to a point (in fact a small area) on a surface. A simplified schematic is given in Figure 5.15. The principle is that a signal is generated when the point on the surface is out of focus. This signal can be used directly as a measure for a measured displacement relative to the focus position, or the error signal is used in a feedback loop to position the lens such that the point in focus again, where a displacement sensor measures the lens position. A combination of the two may be that, while the surface is scanned in the x-direction, the low-frequency part of the signal is used as a feedback to the lens to keep it nominally in focus, while the high-frequency part is used to measure small surface features. This principle is used in CD- and DVD players. Such a sensor can be used for contactless surface topography measurements when integrated in a surface topography measuring system. The spot size determines the lateral resolution and provides the focal shift parameter for

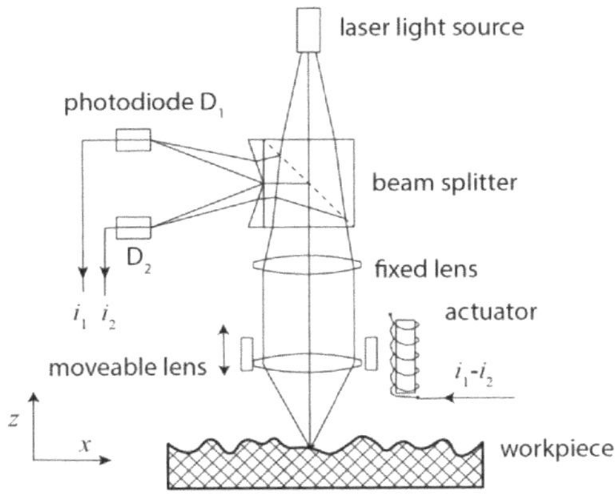

FIGURE 5.15 Schematic diagram of a point autofocus probe.

surface measurement. When a parallel beam with a uniform intensity distribution enters the objective, the spot size W_{su} at the focal plane is given by:

$$W_{su} = 1.22 \cdot \frac{\lambda}{A_N}, \tag{5.18}$$

where λ is the wavelength of the light and A_N is the numerical aperture (NA) of the objective; see Section 7.9.3.3. However, in the case of a Gaussian intensity distribution, as with a laser, the spot size W_{sg} can be expressed as the diameter where the intensity is reduced to $1/e^2 = 14\%$ of the maximum intensity:

$$W_{sg} = 0.64 \cdot \frac{\lambda}{A_N}. \tag{5.19}$$

5.8 CONFOCAL CHROMATIC PROBE

The basic design of a chromatic confocal probe resembles that of the confocal point sensor, but where normally the objective is designed as an ensemble of lenses that minimizes chromatic aberration (i.e., the effect that the focal length of a lens depends on the wavelength due to the dispersion of common optical materials); in this design, the chromatic aberration is intentionally used to discriminate the height of the surface the probe is pointing to. Figure 5.16 gives the schematic diagram. Within a limited range, either the 'blue' side of the spectrum is in focus, giving a closer focus distance, or the 'red' side of the spectrum is in focus giving a longer focus distance. The focus point is imaged on a diaphragm before a spectrometer. The intensity of the color that is in focus will be maximal at that

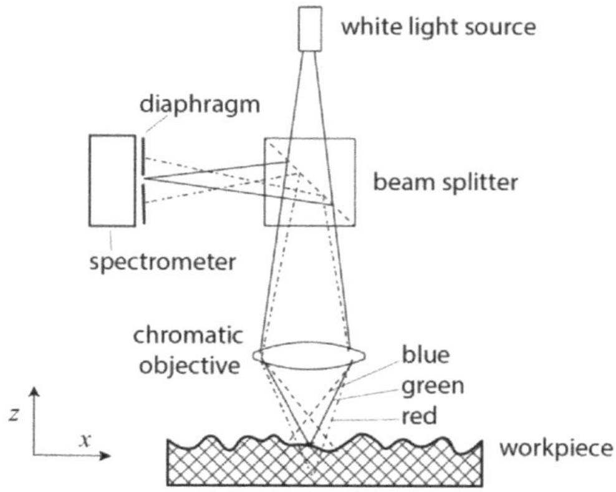

FIGURE 5.16 Schematic diagram of a confocal chromatic probe.

point. Inside the spectrometer the light is dispersed by either a prism or a grating and imaged on a CCD-array (see Section 5.5.2), see Figure 5.10. The location of the maximum intensity on this array corresponds to the detected color and, through some calculations and calibration, this can be related to the height of the measured surface. The advantage of this assembly is that it does not contain any moving parts; the evaluation of the CCD-array directly corresponds to the height of the measured surface. The measurement range depends on the lens material and has its limitations. A typical value is a measurement range of 0.4 mm at a focal length (working distance) of 10 mm. The limit of the lateral resolution may correspond to that of the confocal point sensor (see Section 5.7) with potentially some additional effect of the colors that are not in focus that broaden the focal point.

5.9 LASER LINE SCANNER

An laser line scanner (LLS) projects one or more planes of laser light onto the workpiece surface and tracks the intersection thereof with a sensor. This allows the determination of surface coordinates by triangulation, as illustrated in Figure 5.17. A laser diode projects a line of light on the workpiece. Simultaneously, a camera consisting of a lens and a CCD sensor detects the image. From the image on the CCD sensor, the intersection of the plane generated by the diode laser and the workpiece surface can be determined as a profile of (y,z) coordinates. As a result of their noncontact character, LLSs have a high measuring speed and a high density of sampling points. The accuracy of the results of optical scanners is not only dependent on the positioning and orientation with respect to the workpiece surface, but the accuracy is also dependent on the characteristics of the surface such as color, roughness and reflection and external factors such as vibrations, temperature and ambient light. A typical specification is a field of view (y-range × z-range) of

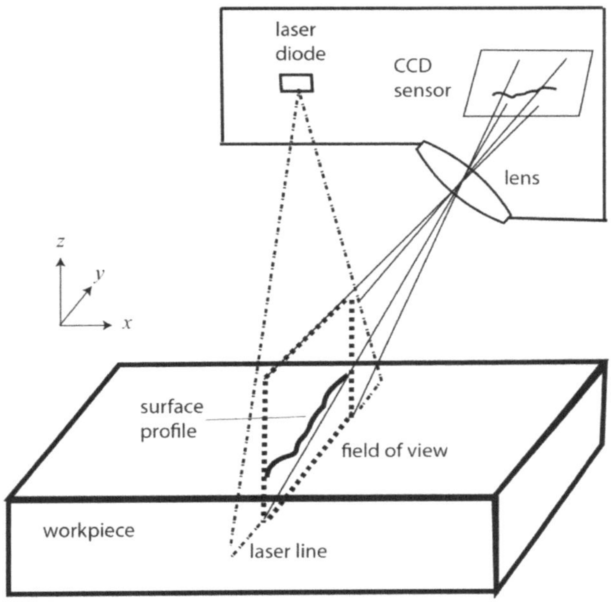

FIGURE 5.17 Measurement principle of a laser line scanner (LLS).

100 × 60 mm, a standoff distance of 100 mm, data collection rate 400 000 points/s and a resolution about 40 μm.

EXERCISES

1. Capacitive transducer
   A (square) capacitive transducer can be used to detect a displacement by either varying the axial distance or by applying a lateral displacement. Compare the sensitivities of both these principles choosing a feasible geometry.
2. Inductance by a coil
   Figure 5.18 gives the cross-section of a cylindrical coil C. The coil has a length l_s where in the cylinder, a Kernel K is moved such that this is inside the cylinder over a length l_k.
   The self-induction coefficient L_e of this coil can be expressed in l_k, l_c, the section A, the number of windings n, the magnetic permeability in vacuum μ_0, the relative permeability of air μ_l and of this kernel μ_k. The windings are homogeneously distributed over l_c against each other and have a diameter d. Give the expression for L_e in terms of μ_0, μ_k, n, A, l_c, l_k and d.

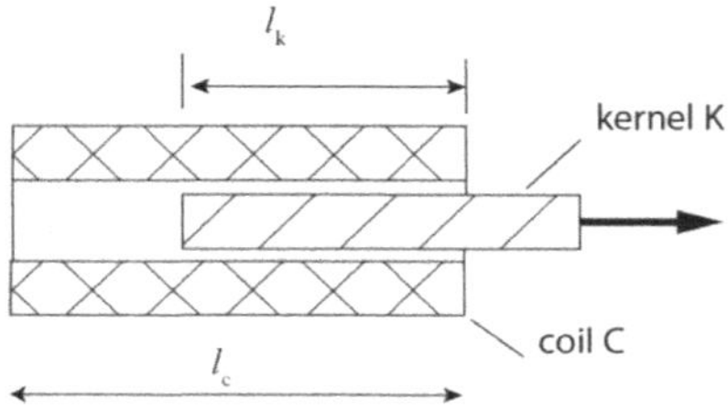

FIGURE 5.18 Cross-section of a cylindrical coil.

3. Interferential encoder
   An interferential encoder is designed, the light source is a laser diode with a wavelength $\lambda = 500$ nm. Give the minimal and maximal pitch of the grating that can be usefully applied.
4. Laser interferometer versus linear encoder
   A stabilized laser with a nominal wavelength $\lambda = 633$ nm is available as a light source for designing a positioning system for displacements up to 300 mm. Consider employing this laser for a displacement interferometer system or for an interferential encoder. Discuss advantages/disadvantages of each system regarding laser stability, air refractive index, material temperature, grating pitch, etc.

REFERENCES

Cosijns, S.J.A.G., Jansen, M.J. and Haitjema, H., 2018, Advanced Optical Incremental Sensors: Encoders and Interferometers. In: Yurish, S.Y. ed. *Smart Sensors and MEMS: Intelligent Sensing Devices and Microsystems for Industrial Applications,* 2nd Edition. Amsterdam, Elsevier.

Doebelin, E.O., 2003, *Measurement Systems, Application and Design.* New York: Mc Graw Hill.

Durschmid, F., 1993, Untersuchungen zur Messsignalerzeugung und Messsignalvera rbeitung inkrementaler Langenmesssysteme unter besonderer Berucksichtigung der analyse und Reduzierungsystematischer und zufalliger Messabweichungen, Ph.D. diss, Technischen Universität Illmenau.

6 Material Dimensional and Angle Measures

6.1 1D LENGTH STANDARDS

For standards that determine a size it may be useful to distinguish between two-sided and one-sided sizes. Two-sided sizes can be inside- or outside measures. This is illustrated for a plug with a cylindrical hole in Figure 6.1. The hole must be probed from the inside, the plug must be probed from the outside. The probe taking these measurements may need corrections for probe size, indentation, probe bending, etc., and these corrections may be opposite for the two cases.

This need for correction is illustrated in Figure 6.2 for mechanical measurements, where a probe is used that contacts the surface with a measurement force and is read out by a scale. The scale reading must be corrected for the probe size, indentation and bending of the shaft. These corrections may be opposite for inside- and outside measurements. When the measurement is one-sided, meaning that the probe comes from the same direction for both measurements that determine a size, these corrections cancel.

For optical measurements, a similar situation exists when lateral distances are read-out by an optical system, for example a profile projector, a vision system or in an XCT measurement. A light-dark transition will never be perfectly sharp, and a measurement system must decide a threshold for defining and edge, that may cause a systematic error that is opposite for inside- and outside measurements. This is illustrated in Figure 6.3. One-sided optical measurements are far less sensitive to the threshold definition compared to two-sided optical measurements.

6.1.1 Gage Blocks

A gage block is a material measure of rectangular section, made of wear-resistant material, with one pair of planar, mutually parallel measuring faces, which can be wrung to the measuring faces of other gage blocks to make composite assemblies, or to similarly finished surfaces of auxiliary plates (often called platens), for length measurements. Gage blocks thus define a length that can be between 0.1 mm and 1 m. The most commonly used gage blocks have a standardized cross-section of

DOI: 10.1201/9780367821746-6

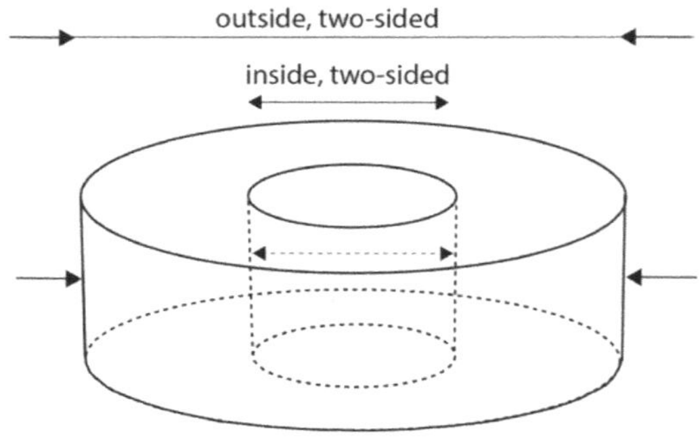

FIGURE 6.1 Definition of inside- and outside sizes of a cylindrical plug and hole.

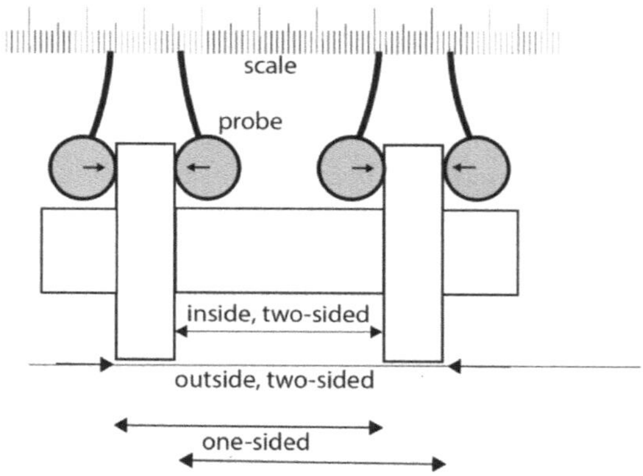

FIGURE 6.2 Illustration of one-sided and two-sided sizes measured with a probing system.

(9 mm by 35 mm) with a nominal length l_n > 10 mm and (9 mm by 30 mm) for 0.5 mm < l_n <10 mm. For a nominal length < 6 mm, the value is engraved in one of the measuring faces; for a nominal length > 6 mm, the nominal length is indicated on a side. We speak of short (l_n < 100 mm) or long gage blocks (l_n > 100 mm). In addition, there are also square and cylindrical versions. Cylindrical long reference specimen with parallel end faces are called 'length bars', regular parallel gage blocks are just called 'gage blocks', 'slip gauges' or 'gauge blocks'. Short gage blocks are calibrated in a vertical position, and long gage blocks are calibrated in a horizontal position. The flatness of the measuring surfaces is such (< 0.1 µm) that gage blocks can be placed on top of each other with this adding little extra uncertainty in length. This process is called wringing. The various surface forces makes these surfaces stick together. The central length l_c of a gage block is defined as the length from the center to the ground surface on which it is (or

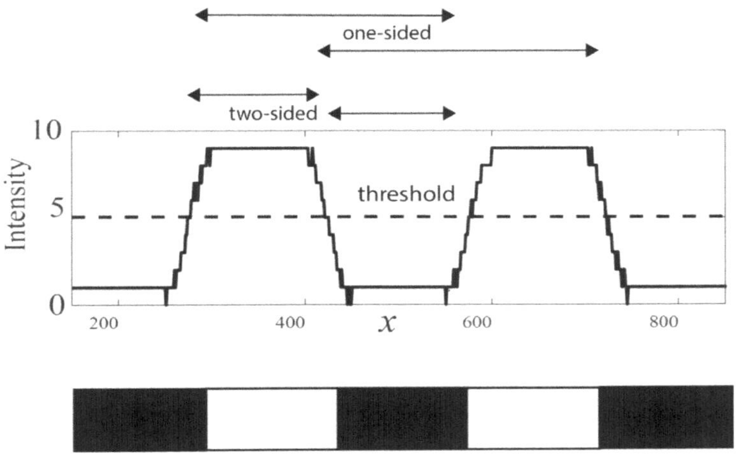

FIGURE 6.3 Illustration of the measurement of one-and two-sided sizes of a pattern by a vision system.

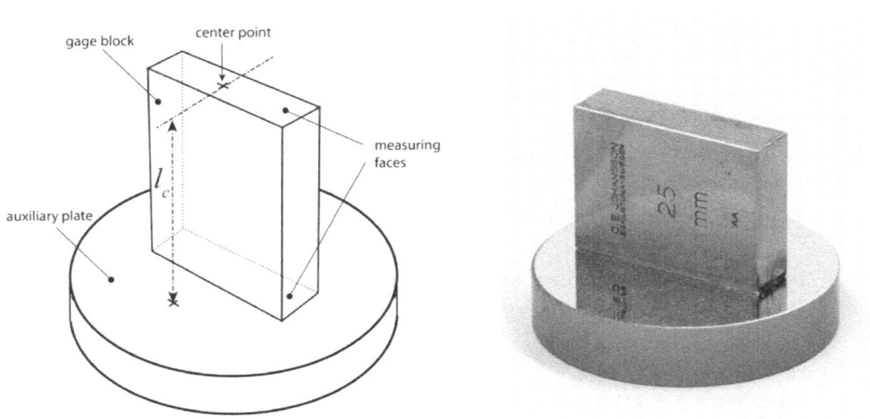

FIGURE 6.4 Left: definition of the central length l_c of a gage block. Right: gage block wrung on an auxiliary plate.

would have been) wrung, of the same material and with the same surface condition as the gage block. This enables a gage block to be measured by interferometry, thus establishing the connection of a light wavelength with a physical object; see Section 4.5.4. This is illustrated in Figure 6.4. The gage block calibration by interferometry is essentially a one-sided measurement; however, with some small corrections, this size is generally accepted as the two-sided size. Before assembly, the measuring surfaces of the gage blocks must be cleaned and the gage blocks must be stacked by sliding the surfaces over each other. They must then remain attached. This stitching is called wringing. The more gage blocks are stacked in this way, the larger the total uncertainty becomes. However, in theory this only depends on the uncertainty of the individual gage blocks. ISO 3650:1998 defines the metric, rectangular type gage blocks, the British standard BS 4311-1:2007 defines rectangular gage blocks of imperial (inch) size, and the USA standard ASME B89.1.9:2002 defines both metric and imperial system gage blocks, and both rectangular and square cross-sections. Gage blocks are divided into different classes. For each class, the permitted dimensional and form deviation is specified in detail in the standards and can be approximately specified by a constant factor plus a length-dependent factor that is proportional to the nominal length l_n . The classes as defined in ISO 3650:1998 are shown in Table 3.1. The K (Calibration) class has the strictest requirements for parallelism, so that it can be calibrated most accurately. In a gage block interferometer, these gage blocks can be calibrated according to the definition with an uncertainty of typically 0.02 µm. Gage blocks of other classes can then be calibrated with a small loss of accuracy in a gage block comparator with the calibrated class K gage blocks as a standard. ISO 3650:1998 gives an overview of the terms and requirements with regard to gage blocks, such as tolerances for the squareness of the side-faces, the hardness, etc. Gage blocks are available in different types of materials, where in particular the wear resistance, corrosion sensitivity and the coefficient of expansion are important. Materials are:

- hardened steel
  These gage blocks are the most commonly used. The linear thermal expansion coefficient is commonly $\alpha = (11.5 \pm 1.0)\cdot 10^{-6}/\text{K}$. This is nominally the same as most steel objects, so the expansion is naturally compensated if the temperature deviates from 20 °C for both the gage block assembly and the object to which it is compared (see Section 2.2). Disadvantages are the susceptibility to corrosion, so they must be kept greased when not in use, and they are scratch-sensitive.
- ceramics
  Gage blocks based on ceramics are commonly made of zirconium-oxide (ZrO_2). Advantages are that these are lighter than steel and, more importantly, corrosion-free. Disadvantages are an expansion coefficient that is slightly lower than steel (typically $\alpha = (9.3 \pm 0.5)\cdot 10^{-6}/\text{K}$) and a poor thermal conduction properties, which means that it may take more time for a temperature change to settle.
- hard metal
  Gage blocks based of hard metal may be based on chromium oxide (CrO_2; $\alpha = 4.4\cdot 10^{-6}/\text{K}$) or tungsten carbide (WoC; $\alpha = 7.6\cdot 10^{-6}/\text{K}$). Advantages are a better scratch resistance compared to steel, less sensitivity to corrosion, and they wring relatively easily. Disadvantages are the coefficient of expansion that is different from steel, they are relatively expensive and they are not available in lengths > 100 mm.

Less common are gage blocks made from quartz, aluminum, and gage blocks made from materials (also ceramics) having an ultra-low expansion coefficient. When m gage blocks are stacked, the total nominal length l_{nt} is given by:

$$l_{nt} = l_{n1} + l_{n2} + \ldots + l_{nm.} \tag{6.1}$$

The tolerance of every individual gage block can be specified as (see Table 6.1):

$$dl = \left| l_c - l_n \right| \leq A + B{\cdot}l_n. \tag{6.2}$$

When stacking and assessing the uncertainty, it is commonly assumed that the length deviations are correlated: the largest uncertainties of one manufacturer producing one gage block set may be systematic and equal for every gage block. Therefore, the uncertainties are added linearly. The limit for the total deviation then becomes:

$$dl_t = dl_{n1} + dl_{n2} + \ldots + dl_{nm} = m{\cdot}A + B{\cdot}l_{nt}. \tag{6.3}$$

This is typically a 'worst case' uncertainty evaluation for which a rectangular distribution can be assumed; see Chapter 3. Gage blocks are available in sets from 2 mm or 3 mm to 100 mm, from which any measure can be composed in steps of 1 μm. For example:

- a forty-six piece set consists of nine gage blocks ranging from 1.001 mm to 1.009 mm; nine gage blocks ranging from 1.01 mm to 1.09 mm; nine gage blocks ranging from 1.1 mm to 1.9 mm, nine gage blocks ranging from 1 mm to 9 mm and ten gage blocks ranging from 10 mm to 1.9 mm

TABLE 6.1
Gage Block Classes and the Corresponding Major Tolerances according to ISO 3650:1998

Class	Length Tolerance for Nominal length l_n $A + B \cdot l_n$	Parallelism Tolerance for Nominal Length l_n $A + B \cdot l_n$
K	0.20 μm + $4{\cdot}10^{-6}{\cdot}\ l_n$	0.05 μm + $2{\cdot}10^{-7}{\cdot}\ l_n$
0	0.12 μm + $2{\cdot}10^{-6}{\cdot}\ l_n$	0.10 μm + $3{\cdot}10^{-7}{\cdot}\ l_n$
1	0.20 μm + $4{\cdot}10^{-6}{\cdot}\ l_n$	0.16 μm + $5{\cdot}10^{-7}{\cdot}\ l_n$
2	0.45 μm + $8{\cdot}10^{-6}{\cdot}\ l_n$	0.30 μm + $7{\cdot}10^{-7}{\cdot}\ l_n$

FIGURE 6.5 Picture of a 112-piece ceramic gage block set.

- an eighty-seven set consists of nine gage blocks ranging from 1.001 mm to 1.009 mm; forty-nine gage blocks ranging from 1.01 mm to 1.49 mm; nineteen gage blocks ranging from 0.5 mm to 9.5 mm and ten gage blocks ranging from 10 mm to 100 mm
- a 112-piece set consists of one 1.0005 mm gage block, nine gage blocks ranging from 1.001 mm to 1.009 mm; forty-nine gage blocks ranging from 1.01 mm to 1.49 mm; forty-nine gage blocks ranging from 0.5 mm to 24.5 mm and four gage blocks ranging from 25 mm to 100 mm

Figure 6.5 shows a 112-piece ceramic gage block set. The advantage of larger sets is that less gage blocks need to be stacked to achieve a certain length and smaller sizes may be composed with high accuracy. A usual set of long gage blocks contains gage blocks of 125 mm, 150 mm, 175 mm, 200 mm, 250 mm, 300 mm, 400 mm and 500 mm. A special set for the calibration of micrometers contains gage blocks of 2.5 mm, 5.1 mm, 7.7 mm, 10.3 mm, 12.9 mm, 15.0 mm, 17.6 mm, 20.2 mm, 22.8 mm and 25.0 mm. This set is composed in such a way that periodic deviations in the screw spindle are detected. Also, there is a special set for calibrating a gage block comparator. Analogously, there are imperial (inch) sets, including several sets of inch gage blocks for inch micrometers. The gage dimensions determine an outside size. Using gage block accessories, a set of extended gage blocks can also be used to determine an internal dimension or to define a distance between two points. Gage blocks are considered as the basics of dimensional metrology where the size of objects is concerned. Typical is an advantageous price/quality ratio: for a few tens of euros one obtains a reference size with sub-micrometer accuracy and two surfaces of optical quality.

6.1.2 Plug Gages

A cylindrical standard for determining an outside measure is called a pen caliper, plug gage or setting pin. When these are used for determining the diameter of a hole,

FIGURE 6.6 Left: a nominal 50 mm plug gage. Right: a nominal 40 mm ring gage.

they are also referred to as pin gages. Very thin pins are flexible and are then called measuring wires. Three measuring wires of nominally equal diameter are used in the measurement of pitch diameter of screw threads, see Chapter 3. The tolerance on the size is a few micrometers. Because the size cannot be determined directly by interferometry (without mechanical probing), a pen caliper cannot be calibrated much better than with an uncertainty of 0.1 μm. Nevertheless, a calibrated pen caliper enables a useful comparative measure when calibrating a cylindrical outer dimension because the shape and alignment of the probe is less critical than when comparing a cylindrical surface to a gage block (with parallel faces). The form accuracy (roundness, cylindricity) is important for accurate measurement of the size. A typical plug gage is depicted in Figure 6.6

6.1.3 Ring Gages, Setting Rings

Analogously to the pin gage for a cylindrical outer dimension, the ring gage forms a cylindrical inner diameter. Setting rings are used especially for the calibration/setting of two-point or three-point hole (inside diameter) measuring instruments. Calibration of ring calipers is somewhat more difficult than pen calipers because it is necessary to measure from the inside while the reference measure will ultimately be a gage block (outer size); the best reachable uncertainty is around 0.2 μm. Furthermore, the considerations as mentioned for the plug gage apply. A typical ring gage is depicted in Figure 6.6

6.1.4 Step Gage

A step gage is a standard with a number of measurement planes which often consist of gage blocks that protrude from the base, which define a number of internal and external measures. This standard can be used for setting and calibrating calipers, micrometers and height gages. In addition, it can be used for the verification and calibration of measuring axes of 3D coordinate measuring machines. Figure 6.7 shows a typical embodiment.

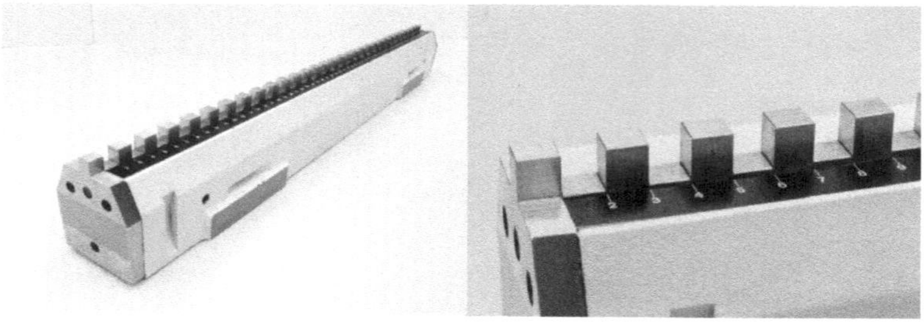

FIGURE 6.7 Embodiment of a step gage with 10 mm steps (left) and detail of a step gage (right).

6.1.5 Ruler, Line Standard

Both traditionally (until 1960) defined as the international primary standard as well common in daily use is the ruler, tape measure or, when used as standard, the line standard. The line standard consists of a number of parallel lines on a carrier of usually metal, plastic or glass. The size is commonly defined as the distance between the center lines of two stripes. For use in a workshop or measuring room there are steel rulers according to DIN 866:2006, accurate to 0.05 mm. The scale lines and the Vernier of a caliper are also line standards. A version in glass of 1 mm divided into fifty or 100 lines is used in optical microscopy and may be called a stage ruler or a calibration slide. Equidistant lines applied to glass or steel can be supplied with a reading system and are called linear encoders; see Section 5.5.3.

6.1.6 Ball, Ball Bar

(Measuring) balls are used as diameter standards. The most common probes of linear height gages and of 3D coordinate measuring machines have a spherical shape. This shape has been chosen because, when a ball is probed with a spherical probe, always a spherical form is detected: the diameter of this measured sphere is the sum of the ball and the probe diameter. This is illustrated in Figure 6.8. By using a reference ball with a calibrated diameter, the (effective) diameter of a probe can be determined on a coordinate measuring machine, which can be taken into account for all subsequent measurements. A ball bar consists of two balls at the ends of a rod. Figure 6.9 shows a typical example. The length can be defined as the distance between the end points or as the distance between the centers of the balls. When calibrating the axis of a coordinate measuring machine, the length between the centers does not depend on the diameter of the probe ball. All the objects mentioned establish a distance in one dimension, but in order to calibrate and use them with minimal uncertainty, requirements are also imposed on the form accuracy in the other two dimensions: parallelism for the gage blocks, line

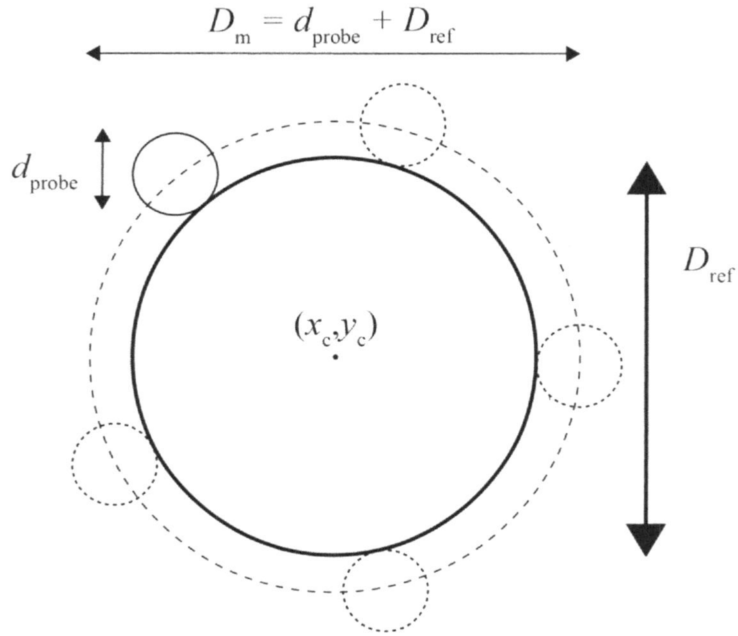

FIGURE 6.8 2D illustration of a sphere with diameter D_{ref} probed with a spherical probe with diameter d_{probe}. The CMM measures effectively a sphere with diameter $D_m = D_{ref} + d_{probe}$, with the same center coordinates.

FIGURE 6.9 Typical embodiment of a ball bar.

quality for line stripe sizes, sphericity for balls, roundness and cylindricity for ring gages, etc.

6.2 ANGULAR STANDARDS

6.2.1 Angle Gage Blocks

In a similar manner as parallel gage blocks are available as a reference for length measurements, angle gage blocks are available as a reference for angle measurements. As with the parallel gage blocks, the surfaces have to meet stringent accuracy requirements so that they can be wrung together. The basic accuracy is in the accuracy of the angle between the two end faces: typically 1", 2" or 5". The angular gage blocks are supplied in sets, with which different angles can be assembled. Such a set is constructed in a way in which a large number of angles can be assembled with a small number of angular gage blocks, where an angle gage block can be used both to enlarge an angle or to decrease an angle (see Figure 6.10).

When assembling angular gage blocks, it is important that the faces are perpendicular to the same plane. If this is not the case, an angular deviation $\Delta\varphi$ appears as a result of this misalignment, as shown in Figure 6.11 for two angle gage blocks, and given by

$$\Delta\varphi = \frac{\alpha^2}{2}\sin(\varphi_1)\cos(2\varphi_2). \tag{6.4}$$

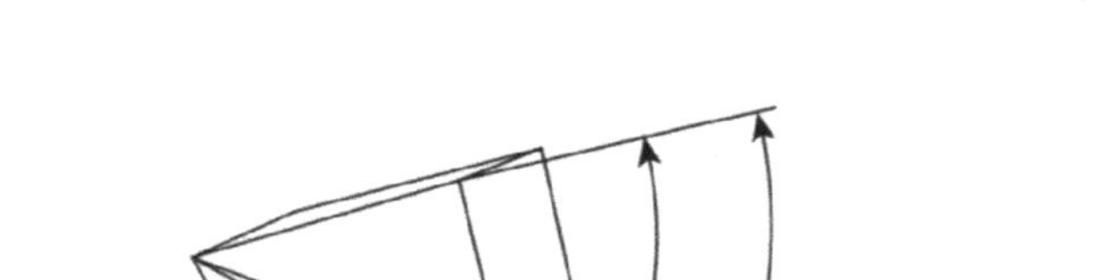

FIGURE 6.10 Illustration of the adding and subtracting of angles using angle gage blocks.

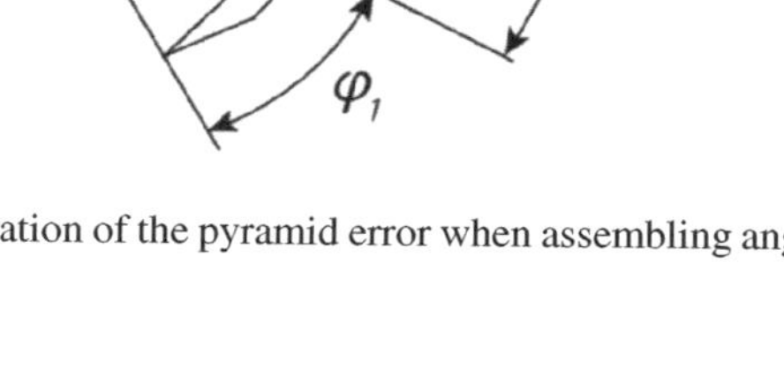

FIGURE 6.11 Illustration of the pyramid error when assembling angular gage blocks with a slight misalignment.

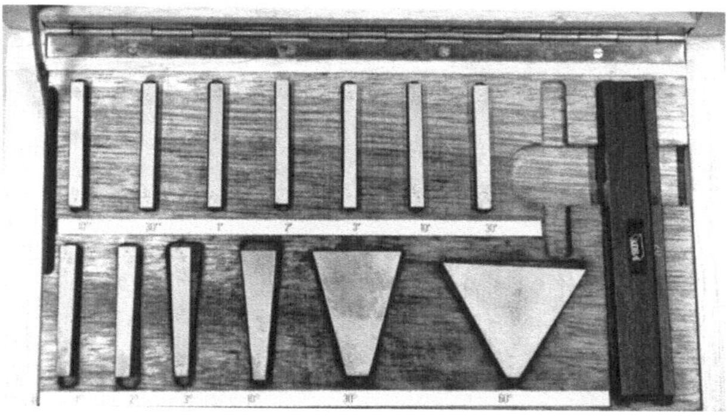

FIGURE 6.12 Thirteen-piece angle gage block set.

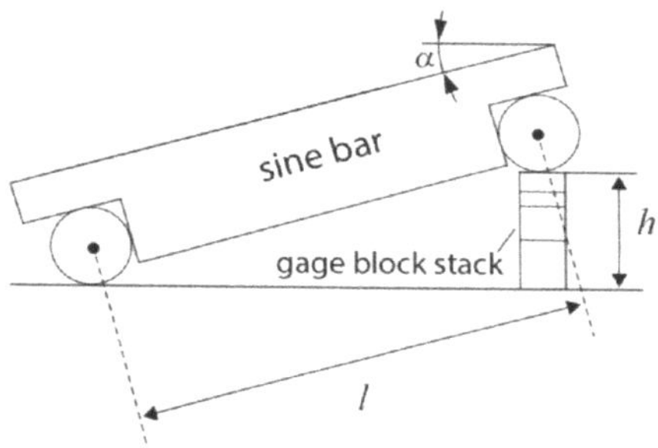

FIGURE 6.13 Sine bar of length l with an angle α generated by a gage block stack of height h.

This error is called the pyramid error. Figure 6.12 shows a thirteen-piece set of angular gage blocks, consisting of blocks that embody angles of nominally 10″, 30″, 1’, 2’, 10’, 30’, 1°, 2°, 3°,10°, 30° and 60°. This set enables the user to assemble any angle with a resolution of 10″.

6.2.2 The Sine Bar

The sine bar, sometimes called a sine ruler, is a measuring device for the accurate setting and measurement of angles up to about 40°. The principle is simple but unsurpassed in its accuracy and is indicated in Figure 6.13. A beam is mounted on two identical cylindrical rolls, which are positioned parallel with a known fixed distance from each other. The angle can be set by placing gage blocks under one of the rolls, or by displacing or moving them in another way.

The sine bar is essentially characterized by the distance between the rollers l. If one of the rollers is displaced over a height h, the generated angle α relative to $h = 0$ for a sine bar without any form errors is:

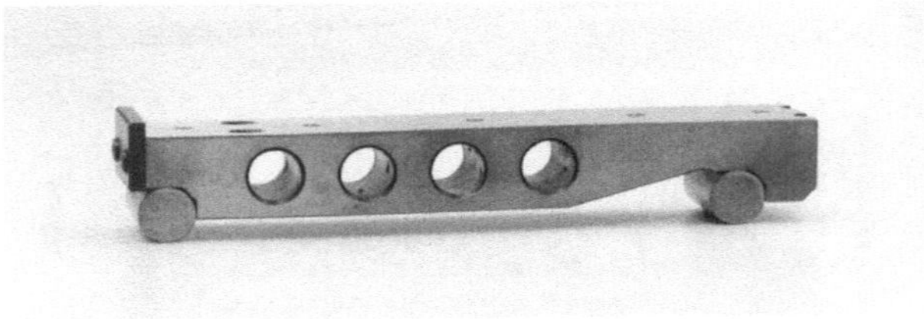

FIGURE 6.14 Embodiment of a sine bar with $l = 200$ mm.

$$\sin\alpha = \frac{h}{l} \text{ or } \alpha = \sin^{-1}\frac{h}{l}. \tag{6.5}$$

The uncertainty in the angle measurement depends on the uncertainty in the height h, the characteristic length l and the deviations of the ideal form: the parallelism, cylindricity and difference in diameter of both rollers. The propagation of uncertainties of h and l to the angle α can be derived using equations (6.5) and (3.28), giving:

$$\frac{u(\alpha)}{\tan\alpha} = \sqrt{\left(\frac{u(l)}{l}\right)^2 + \left(\frac{u(h)}{h}\right)^2}, \tag{6.6}$$

where h is the height, assembled using for example gage blocks, l is the distance between the rolls and $u(\alpha)$, $u(l)$ and $u(h)$ are the standard uncertainties in α, l and h respectively. The dimensions of the sine bar, and the accuracy requirements are laid down in, for example, the standard DIN 2273:1979. If only small angles (< 1°) are concerned, the accuracy can be improved by making l large. With such an arrangement (electronic) levels and autocollimators can be calibrated. Other applications of the sine bar include: highly precise adjustment of angle (uncertainty < 0,1″) and measuring the angles of objects. Figure 6.14 shows an embodiment of a sine bar.

6.2.3 Polygon

A polygon is an angular standard in the form of a regular polygon, usually made of glass, steel or hard metal, with reflective side surfaces: the polygonal surfaces. The number of faces can range from three under a polygonal angle of 120° to seventy-two under a polygonal angle of 5°. Figure 6.15 shows a photograph of a 12-sided polygon. Combined with an autocollimator, a polygon is ideally suited for the calibration of indexing tables and angle sensors. A feature of polygons is that the sum of all polygon angles is 360°. This also means that the average angle of an n-sided polygon is exactly 360°/n. A twelve-sided polygon is thus an intrinsic

FIGURE 6.15 Photograph of a 12-sided polygon.

standard of an angle of 30° and multiples thereof, i.e. twelve different orientations of the polygon. The feature that the sum of all polygon angles is 360° is also used in the calibration of all individual angles of a polygon. This is treated further in Chapter 7, exercise 1 and in Section 10.5.3.

6.2.4 Square, Squareness Standard

Squareness standards, or 90° standards, are important references for alignment of the motion axes of machine tools and measuring instruments. They are often made from steel, granite or ceramics, and are available in square, rectangular or triangular shape and in different sizes.

Squareness standards may range from simple steel squares that can be found in many toolboxes to large granite squares up to meters in size. Figure 6.16 shows an image of a square. Although it is an angle standard, deviations are mostly expressed in micrometers as a deviation from the ideal form. For example, the tolerance t according to DIN 875:2005 class 00 is:

$$t = 2\,\mu\text{m} + 1 \cdot 10^{-5} \cdot L \tag{6.7}$$

where L is the length of the considered side and the other side acts as the reference. Another embodiment is a cylindrical square, where the squareness of the cylinder faces is defined relative to the ground face. As the tooling is similar, the quality of the tooling commonly allows the faces of 90° standards to also be used as straightness measurement standards.

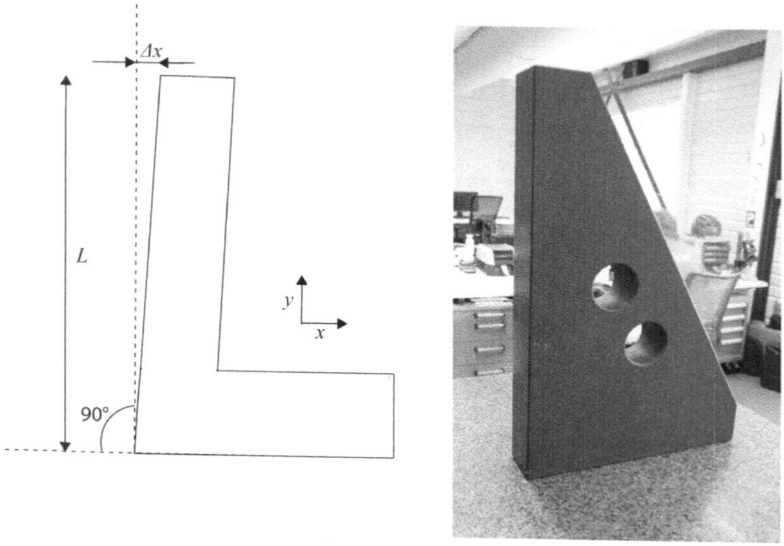

FIGURE 6.16 Schematic of a square (left) with form deviation Δx and picture of an embodiment in granite (right) with $L = 600$ mm.

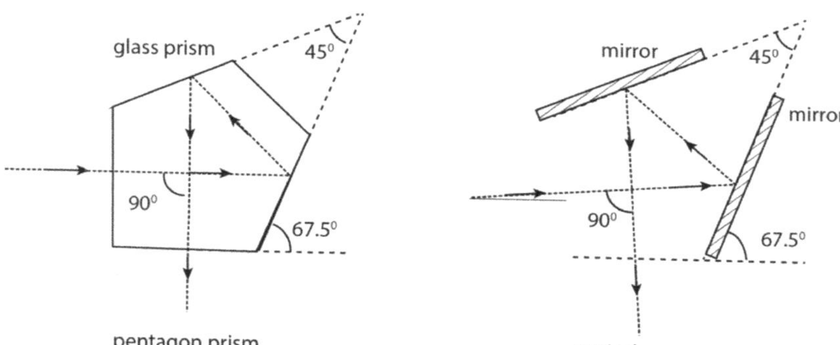

FIGURE 6.17 Schematics of optical squareness standards. Left-hand side: pentagon prism. Right-hand side: optical square.

6.2.5 Optical Squareness Standards: Pentagon Prism, Optical Square

As optical squareness standards, the pentagon prism, also called pentaprism, and the optical square are in use. The pentaprism is a 5-sided glass body where two faces are coated so that light is reflected; these two faces have a mutual angle of 45°, see Figure 6.17. This results in the property that it reflects a light beam at nominally 90°, independent of small angular deviations of the incoming beam or small angular deviations of a linear movement of the pentaprism. This is also achieved by mounting two mirrors at 45°, see the right-hand side of Figure 6.17. This assembly is commonly called an optical square, however pentaprisms may also be called optical squares. These optical squareness references are used combined with an autocollimator or a laser interferometer system with straightness optics; where a straightness measurement can be extended to a squareness measurement. The deviation of the true 90° angle is commonly within a few arcseconds. This can be corrected for, while the property of the reflection of a constant angle is maintained.

6.3 FORM STANDARDS

6.3.1 Straightness Standard: The Straight-Edge

As a reference for straightness, straight-edges are available in steel, granite and ceramics. The straightness is measured at a line on the center of the surface in the direction perpendicular to the surface. A typical straight-edge is depicted in Figure 6.18. Requirements for the straightness, in fact the flatness, of the faces are laid down in DIN 874:2003. For example, the tolerance t for the flatness according to DIN 874:2003 class 00 is:

$$t = 1\ \mu\text{m} + 6.7 \cdot 10^{-6} \cdot L. \tag{6.8}$$

If a straight-edge is two-sided it may also have tolerances for parallelism. A stretched taut wire may also be used as a straightness reference and is mostly used for lengths over several meters, where the deviation from straightness in the vertical direction increases with the length, due to sag of the taut wire.

6.3.2 Flatness Standards

An undisturbed liquid surface can be considered flat because the gravity vector is parallel over relatively large areas. The earth's curvature causes a flatness deviation of 0.1 µm over a length of 2 m. This deviation scales quadratic with the maximum length involved. Such surfaces have been used as the reference in Fizeau interferometers, but vibrations and complex operation have impeded their

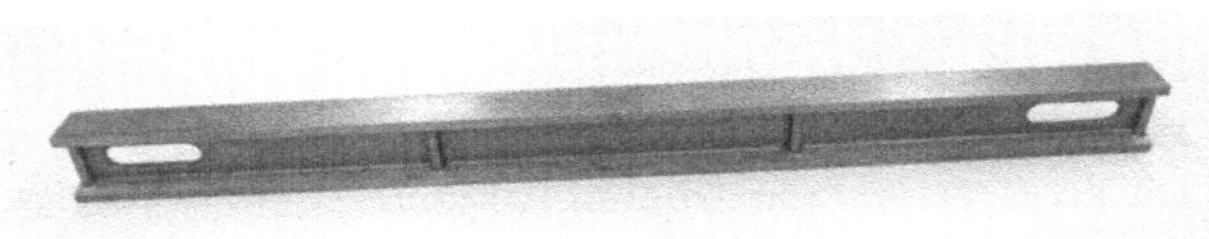

FIGURE 6.18 Photograph of a straight-edge with a length of 1 m (L=1 m).

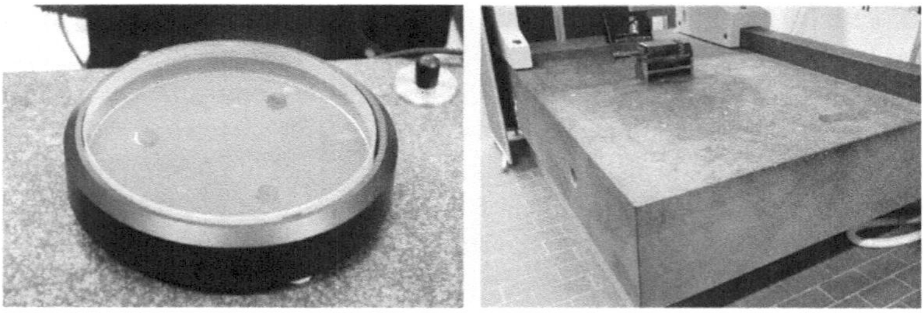

FIGURE 6.19 Photographs of reference flats: Left: 150 mm diameter optical flat. Right: 2 m × 3 m flatness base of a CMM.

use. As solid embodiments of flatness references are in use: atomically clean monocrystalline surfaces for measurements at the sub-micrometer levels, optical flats for sizes in the 10 mm to 300 mm ranges and granite surface plates for ranges up to several meters. Optical flats are often specified as being flat within a fraction of a wavelength, for example $\lambda/10$. This definition is based on the common calibration using a Fizeau interferometer where the used wavelength may be that of a 'red' He-Ne laser: $\lambda = 632.8$ nm; see Section 4.5. Granite surface plates are, for example, specified according to grades 00 to 3 in DIN 876:1984, where the tolerance for the flatness according to DIN 876:1984 class 00 is:

$$t = 2\ \mu\text{m} + 2 \cdot 10^{-6} \cdot L \tag{6.9}$$

where L is the size of the longest side of the plate. Cast iron and ceramic surface plates are also available. The advantage of granite relative to steel is that it is much lighter, easy to process and corrosion free. Therefore, granite is often used as a base for larger metrological instruments and machine tools, such as CMMs, wafer steppers, telescopes, etc. Figure 6.19 depicts as examples a 150 mm diameter optical flat and a 2 m x 3 m granite base of a CMM.

6.3.3 Roundness Standards

Roundness standards are available as glass hemispheres, that are mainly used for the calibration of roundness and cylindricity measuring machines. Typical roundness deviations of these standards are in the 5 nm to 20 nm range. A special artefact for the calibration of the amplification of roundness testers is the so-called flick-standard: if a small part of a cylinder is flattened it is recorded as a groove in a roundness measurement, and the depth of the groove can be used as a calibration value for the probe used. Figure 6.20 depicts a roundness standard and a flick standard.

FIGURE 6.20 Photographs of roundness standards: Left: 40 mm diameter roundness standard (hemisphere), right: 30 mm diameter flick standard with 294 μm 'flick'.

6.4 SMALL SCALE AND SURFACE TEXTURE ARTIFACTS

For surface texture measurements on a small scale a plethora of standards have been developed that are summarized in the literature (Carmignato et al. 2020), in ISO 5436-1:2000 and ISO 25178-70:2014. As the reference areas can be small, several kinds of reference specimen can be assembled on a single base, as described in the literature (Leach et al. 2014; Eifler et al. 2020). In this section, most commonly used are treated.

6.4.1 Groove Depth and Step Height Standards

Step height and groove depth standards are available for the surface height, commonly denoted as the z-direction, of surface topography measuring instruments such as stylus instruments, coherence scanning interferometers, etc. Typical heights may range from the nanometers to the millimeter range. Relevant for these measurements is the definition of the step height so that this can be reproduced by the calibration laboratory as well as by the user in same manner. Figure 6.21 gives the definition of a groove depth (or step height) as it is defined in ISO 5436-1:2000.

In Figure 6.21 the dotted reference line is given by the equation:

$$z(x) = \alpha x + \beta + d\delta. \tag{6.10}$$

For a groove width w, a region of $3w$ is considered where the groove edges are omitted over a length of $w/3$ at each side. The areas A, B and C are fitted in a least-squares sense to a straight line where the variable δ takes the value of +0.5 in the regions A and B and the value of -0.5 in region C. See also Section 10.4.1. This gives the groove depth d. This is equivalent to defining d as the distance of the mean of region C to the straight line between the means of region A and B. This definition aims to diminish the effects of geometrical imperfections as well

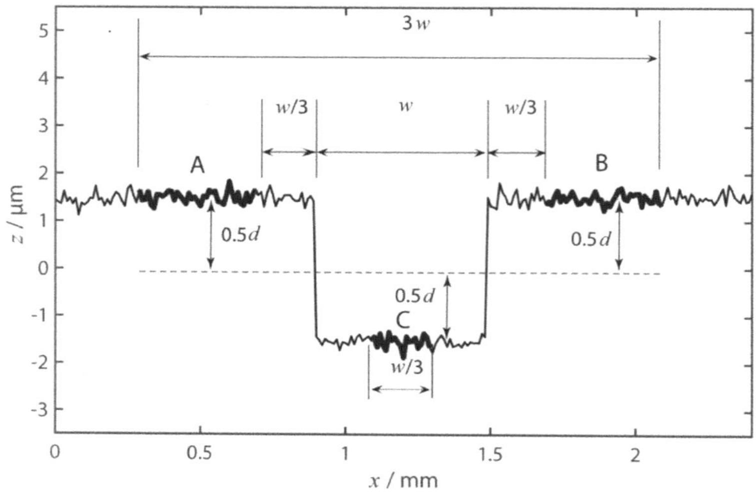

FIGURE 6.21 ISO-definition of groove depth; type A1 according to ISO 5436-1:2000; type PGR according to ISO 25178-70:2014.

as noise in the measuring instrument. The definition has been extended to areal surfaces in a similar way, see ISO 25178-700:2022.

6.4.2 Sinusoidal and Quasi-Random Standards

Sinusoidal standards, type C1 according to ISO 5436-1:2000; type PPS according to ISO 25178-70, give reference values for surface texture height parameters like *Ra* and *Rz*; as well as spatial parameters such as *RSm*. Also, they enable the determination of one point in the instrument transfer function of a surface topography measuring instrument (De Groot 2021). Sine-shaped specimens that are in the region where the profile can be transferred have both the advantage and disadvantage that the calculated parameters are only sensitive for the z-calibration for the height parameters. In order to have an overall calibration of the instrument characteristics, including probe- and filter effects, a specimen type, the so-called 'type D' standard was developed. For this specimen, an essentially random profile is repeated over the typical evaluation length l_e of a surface topography measuring instrument, for example 4 mm. The advantage of a random profile is that the overall performance of a measuring instrument is measured. The advantage of a profile that repeats over the evaluation length l_e is when a measuring instrument measures over this length, a profile with same features is measured independently of the starting point of the measurement; giving a high measurement reproducibility for averaging parameters such as *Ra* and *Rz* (see Section 7.9.2). This gives a smaller measurement uncertainty compared to a non-repeating random profile, while still giving an overall impression of the instrument's performance. Figure 6.22 gives

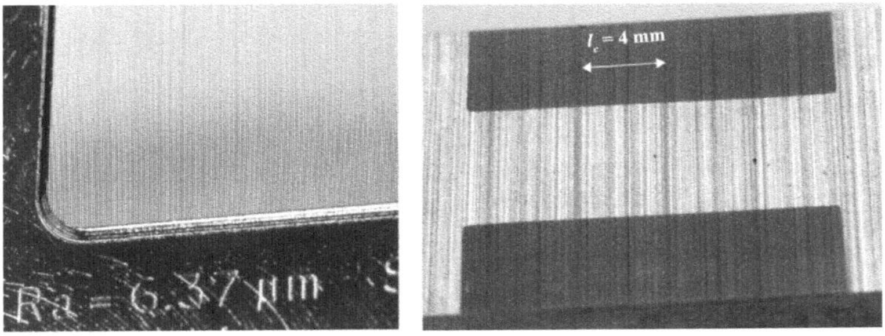

FIGURE 6.22 Images of surface texture reference specimen. Left: sinusoidal specimen with amplitude 10 µm and wavelength 135 µm. Right: quasi random specimen with $Ra = 1.5$ µm.

images of a type C1 standard (at the left-hand side) and a type D standard (at the right-hand side).

EXERCISES

1. Parallel gage blocks
   A measure of 50.004 mm must be assembled using gage blocks. Give the gage blocks that are needed from a forty-six, eighty-seven and 112-piece set and give the maximum deviation in the central length for class 1 gage blocks.
2. Angle gage blocks
   a. Sketch how an angle of 17°18’20″ can be assembled from a 13-piece gage block set as given in Section 6.2.1
   b. What is the maximum deviation in this angle if the maximum deviation in every gage block is 1″.
   c. What is the uncertainty in the angle if the maximum deviation in every gage block is 1″.
3. Sine bar
   Derive equation (6.6) from equation (6.5) and equation (3.28)
4. Sine bar
   A sine bar has the characteristic length $L = 200$ mm; the standard uncertainty in this length is 5 µm. An angle is set with a gage block with a nominal length of 2 mm. This gage block is class 0 according to ISO 3650:1998.
   a. Give the standard uncertainty $u(h)$ to be assigned to the length of the gage block.
   b. Calculate the generated angle α, expressed in degrees (g), minutes (m) and seconds (s) in the format g°m’s″.

FIGURE 6.23 Image of a flat surface in a Fizeau interferometer.

c. Calculate the standard uncertainty in the generated angle α in the small angle limit ($\sin \alpha \approx \alpha$).
d. Now the angle is set with a gage block with a length of 100 mm and a standard uncertainty of 2 μm. The $\sin \alpha \approx \alpha$ approximation no longer applies. Give the angle and the standard uncertainty in the angle.

5. Flatness
   a. Some specify flatness deviations, measured by optical interferometers, as a fraction of an interference fringe, others as a fraction of a wavelength. What is the difference?
   b. Figure 6.23 gives an image of a 40 mm diameter optical flat in a Fizeau interferometer, operating at $\lambda = 632.8$ nm. How would you assess the quality of this optical flat?

REFERENCES

ANSI/ASME B89.1.9, 2002, *Gage Blocks*. American National Standards Institute, Washington DC.

BS 4311. 2007 *Part 1. Gauge blocks manufactured to imperial specification – Specification and validation*. British Standards Institution. Milton Keynes, UK.

Carmignato S., De Chiffre L., Bosse H., Leach R.K., Balsamo A., Estler W.T. 2020. Dimensional artefacts to achieve metrological traceability in advanced manufacturing. *Ann. CIRP* **69** (**2**):693–716.

De Groot P.J. 2021. The instrument transfer function for optical measurements of surface topography. *J. Phys. Photonics* **3** (**2**):024004.

DIN 866. 2006. *Geometrical product specifications (GPS) – Graduated rules, rules for manufacturing purposes – Design, requirements*. Deutsches Institut für Normung, Berlin.

DIN 874. 2003. *Geometrical Product Specifications (GPS) – Straight edges – Part 1: Steel straight edges; Dimensions, technical delivery conditions. Part 2: Knife edge straight edges. Part 3: Granite straight edges.* Deutsches Institut für Normung, Berlin.

DIN 875. 2005 *Geometrical Product Specifications (GPS) – Squares 90° – Part 1: Steel squares 90°* and DIN 875-2. 2008 *Part 2: Granite squares 90°*. Deutsches Institut für Normung, Berlin.

DIN 876. 1984. *Surface plates; Part 1. Natural hard rock surface plates; requirements and testing. Part 2: Cast iron surface plates: requirements and testing.* Deutsches Institut für Normung, Berlin.

DIN 2273. 1979. *Sine bars.* Deutsches Institut für Normung, Berlin.

Eifler M., Hering J., Seewig J., Leach R.K., Von Freymann G., Hu X., Dai G. 2020. Comparison of material measures for areal surface topography measuring instrument calibration *Surface Topography: Metrology and Properties,* **8 (2):**025019.

ISO 3650. 1998 *Geometrical Product Specifications (GPS) – Length standards — Gauge blocks,* International Organization for Standardization, Geneva.

ISO 5436-1. 2000. *Geometrical Product Specifications (GPS) — Surface texture: Profile method; Measurement standards, Part 1: Material measures.* International Organization for Standardization, Geneva.

ISO 25178-70. 2014 *Geometrical product specification (GPS) — Surface texture: Areal Part 70: Material measures.* International Organization for Standardization, Geneva.

ISO 25178-700 2022 *Geometrical product specifications (GPS) — Surface texture: Areal Part 700: Calibration, adjustment and verification of areal topography measuring instruments.* International Organization for Standardization, Geneva.

Leach R., Giusca C., Rickens K., Riemer O. and Rubert P. 2014. Development of material measures for performance verifying surface topography measuring instruments *Surface Topography: Metrology and Properties*, **2 (2):**e025002.

7 Dimensional Measurement Equipment and Measurands

7.1 HAND TOOLS FOR DIMENSIONAL MEASUREMENTS

7.1.1 THE CALIPER

The caliper gage, commonly called a Vernier caliper (Figure 7.1) is the best-known measuring device in the workshop. There are different application options, which include:

- inner measurement of holes and slots;
- outside measurement of diameters and lengths;
- depth measurement;
- height measurement.

These options are schematically depicted in Figure 7.2. For specific tasks, there are calipers in a multitude of types. Measurement with a Vernier caliper takes place by clamping a product between the measuring jaws or by using one of the other elements (see Figure 7.2). The size is obtained by reading the measuring system. Common measuring systems are:

a. Scale with Vernier
On the support of the Vernier caliper, a graduation is provided with millimeter distribution, while a Vernier is arranged on the moving part. The reading of a Vernier is treated in Section 3.4.3.

b. ***Rack and pinion*** gear with clock indication
The support of the caliper is provided with a rack and the moving slide of a gear wheel with a dial. The translation of the gear is converted by this transmission into a rotation of the dial hand. This is also possible with a resolution of 0.05 mm.

c. ***Electronic displacement transducer with LCD-display***
The sensor principle that is used here is capacitive. Both on the carrier and slide of the caliper, small plates are arranged which, as a combination, have a certain

DOI: 10.1201/9780367821746-7

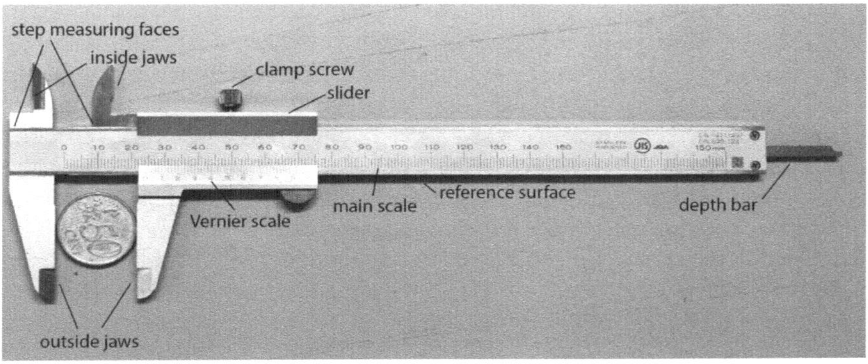

FIGURE 7.1 Photograph of a Vernier caliper with names of the parts.

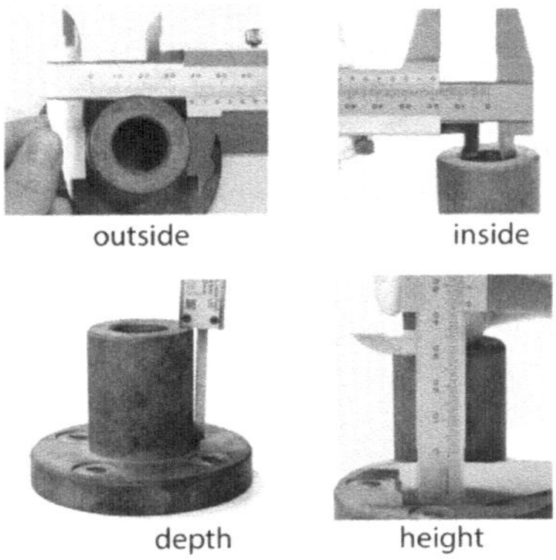

FIGURE 7.2 Measurement possibilities with a caliper: outside, inside, height, depth measurements.

capacity. By sliding the plates along each other, a position dependent periodic change of the capacitance is created, similar to the linear encoder in Section 5.5.3. This changing capacitance is a measure of the displacement, which is electronically displayed on the LCD. Commonly, a displacement is displayed with a resolution of 0.01 mm. An important advantage of this so-called digital caliper is the simple reading, which is both faster and leads to fewer errors. The measured value can also be set to zero at any position so that a comparative measurement is simplified. The absolute digital Vernier caliper is based on a similar principle where the absolute position is also encoded in the scale. In this case, the absolute position is also displayed after switching off and on, just as with the analogue caliper. The possible systematic deviations when measuring with a Vernier caliper result from the measuring system, the measuring force, tilting effects and the incorrect contact of the object to be measured on the measuring jaws. Temperature effects do not play a major role in this instrument. Due to the incremental measuring system (i.e., adjustable zero point), the zero position must be checked regularly. The most common check is whether the measuring faces close correctly when checked by looking to an enlightened surface. The default manufacturing dimensions and measurement tolerances of calipers (E_{MPE}, maximum permissible error) is specified in ISO 13385-1:2019. This ranges from 20 µm for sizes <50 mm measured with a resolution of 0.01 mm to 150 µm for sizes up to 1 m with a resolution of 0.05 mm, for outside measurements. Note that the Abbe principle is only satisfied with the depth measurement. The versatility combined with tight manufacturing tolerances makes up for this.

7.1.2 The Screw Gage/Micrometer

The screw gage, commonly called a screw micrometer or simply micrometer, is the most frequently used hand measuring tool for length measurements after the caliper. Measurement with a micrometer takes place by placing the workpiece between the jaws and screwing them against it, or in the case of a hole, screwing the specially shaped jaws up against the walls of the hole. The micrometer is also available in a large number of embodiments for internal and external measurements of thicknesses, lengths, diameters, grooves and chambers. In addition, there are also micrometers for special measurements, such as measurements on gears and screw threads. Figure 7.3 shows the names of different parts of a micrometer. The bracket micrometer from Figure 7.3 consists of a bracket to which the fixed measuring

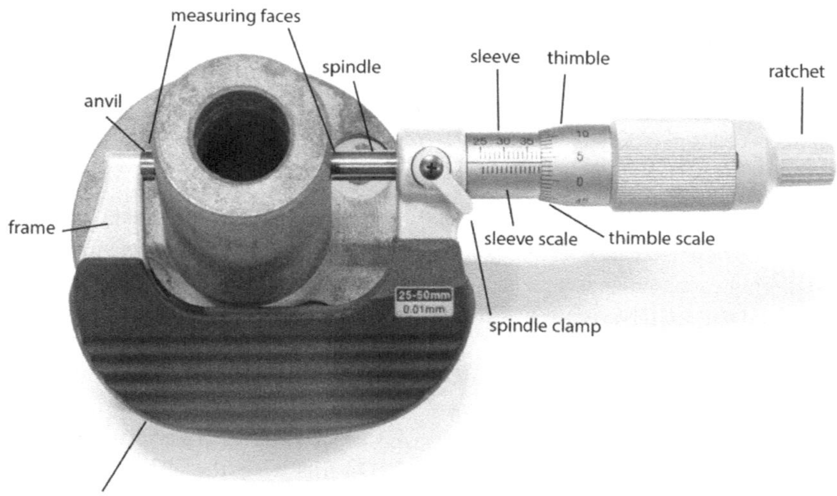

FIGURE 7.3 Photograph of a micrometer, measuring an outside size, with names of the parts.

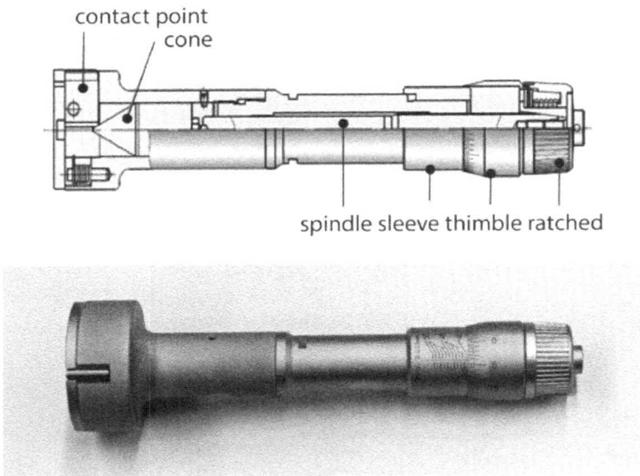

FIGURE 7.4 Schematic (top) and photograph (bottom) of a three-point inside micrometer.

surface is mounted, with the adjustable measuring surface opposite. This setting is done by means of a precision screw. It is important that the measuring surfaces are parallel and perpendicular to the measuring axis. The setting of the micrometer can be read either mechanically or digitally. The mechanical reading is a variation on the Vernier of a Vernier caliper; there is a scale on the measuring drum that interpolates between the fixed scale; see Figure 3.4 in Section 3.4.3. The accuracy of the fixed scale is of less importance here: the rotation mechanism and the thread within the screw size determine the accuracy of the reading. When clamping the product to be measured, the measuring force must not be too large. A measuring force limitation on the screw is provided for this purpose by means of a slip coupling. In connection with wear and tear, the measuring surfaces are generally made of hard metal. An important error source for length measurements with high accuracy is the temperature. In order to prevent expansion of the instrument due to heating, it is often provided with insulating material. In general, micrometers, contrary to calipers, meet the Abbe principle. The measurement range is commonly 25 mm, where the minimum size can be increased in steps of 25 mm. In those cases, the zero is set by a micrometer-setting gage. Default requirements and a classification system are given in ISO 3611:2023. Because the Abbe principle is followed, the measurement uncertainty can be smaller than with a caliper. The specification ranges from 1 μm maximum deviation for the highest accuracy class (class 1) with a 25 mm range, up to 23 μm for class 3 in the 975 mm to 1000 mm range. For inside measurements, such as cylindrical holes, slots and chambers, different embodiments are available. As an example, a three-point internal micrometer is given (Figure 7.4) for cylindrical holes. As the effective diameter determined by the contact point is not easy to define, a reference diameter is commonly set using a setting ring; from this setting, the diameter of a workpiece can be measured in a limited range. Figure 7.5 shows how a reference diameter is set using a setting ring.

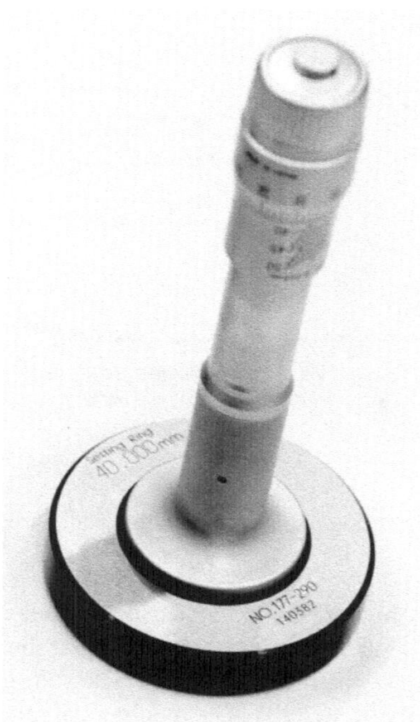

FIGURE 7.5 Photograph of a three-point internal micrometer, set by a setting ring.

7.1.3 Universal Angular Measuring Tool: Bevel Protractor

To measure the angle of an object over a wide range, a universal angle measuring device is often used, as shown in Figure 7.6. This angle measuring device has a so-called fixed blade and a rotatable blade, between which the object can be placed. The bevel protractor gage is available in various versions, as well as accessories such as stops that can be placed on the fixed or rotatable blade. The reading can take place via the following principles:

- mechanical with Vernier: the reading takes place on a mechanical scale using a Vernier. The typical resolution is 5'.
- mechanical with dial indicator: the reading of the mechanical scale goes via a dial indicator. The typical resolution is also 5'.
- optical: an optical scale is placed inside the instrument, which can be read with a magnifier. The typical resolution is 1'.

For angle measurements, the bevel protractor plays a similar role as the Vernier caliper for length measurements: it is versatile, however, not accurate enough to be used in highly accurate manufacturing processes. The bevel protractor is schematically shown in Figure 7.6.

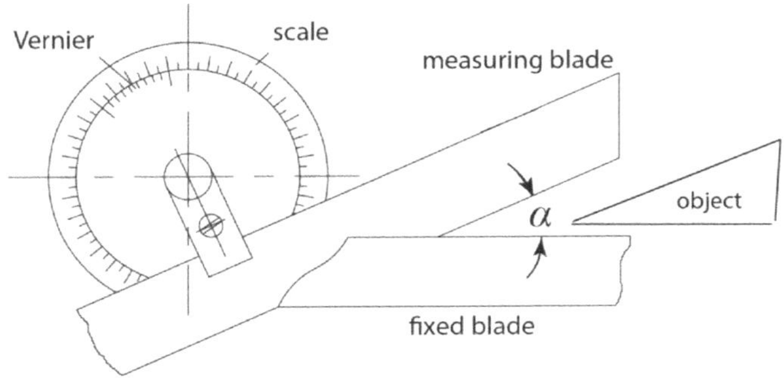

FIGURE 7.6 Schematic of a universal bevel protractor.

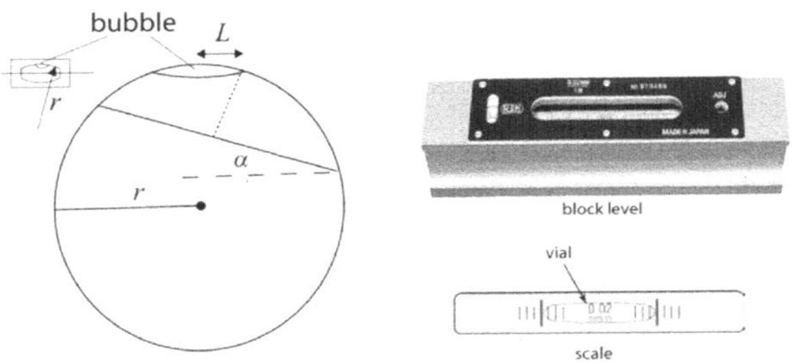

FIGURE 7.7 Schematic of a spirit level (left) and embodiment and scale of a block level (right).

7.1.4 Spirit Level

A spirit level is a well-known tool, present in most tool boxes, to establish level relative to the gravity vector. When the bubble is outside the center of the scale, a small angle is measured. A horizontal level is often used to set horizontal surfaces, such as machine guides. The tube, in which the air bubble is located, the vial, is generally barrel-shaped, so that the level can also be used 'upside down'. The lower side of the instrument is well finished and often vee-shaped, so that the level can also be placed on a cylindrical object, such as an axis. The inclination α is read on a scale along which the air bubble moves. The resolution of square levels is 0.01 mm/m to 0.25 mm/m that corresponds to milliradians. The sensitivity of the level is determined by the radius r of the barrel-shaped vial (Figure 7.7 left). For the angle α, as a function of the displacement L:

$$\alpha = \frac{L}{r} \text{ so } L = \alpha \cdot r. \tag{7.1}$$

This makes the sensitivity for an angular rotation α:

$$S = r. \tag{7.2}$$

The sensitivity thus increases with the radius that is applied in the glass. If the glass is perfectly straight, the radius is infinite, as is the sensitivity, and the measuring range is zero. Due to the limited precision of the glass this is never achieved, the realizable lower limit for the resolution is around 10 µrad. In addition to its use to check guides, and align general objects, the level is also used for flatness measurements. A block level is depicted in Figure 7.7 right. Concepts concerning the spirit levels and accuracy requirements are laid down in DIN 877:1986.

An obvious disadvantage is their manual read-out, and especially for accurate spirit levels it can take some time before the reading settles.

7.1.5 Clinometer

A clinometer is an extension of a spirit level for larger angles. It can either be a spirit level that can be rotated in a graduated shaft, or a rotary encoder that is oriented to a gravity reference. A simple concept that has been used in astronomy for centuries is shown in Figure 7.8-left. An embodiment is shown in Figure 7.8-right. The resolution of a clinometer is typically 1'. Applications include:

- tilting objects to a specified angle
- check the settings of machine tables
- measurement of angles of objects.

7.2 LARGER INSTRUMENTS FOR LENGTH MEASUREMENTS

7.2.1 Height Gage

The height gage is a measuring instrument for dimensional measurements beyond single measures of sizes, for example, the distance between two holes on a workpiece. Height gages are almost always used in combination with a surface plate. The principle consists of a vertical column with straight guidance on an air-bearing foot. This foot can move freely over the flat plate. A measuring trolley with a probe can move along the straight guide of the column. In addition to the straight guide, a measuring system is attached that determines the movement of the measuring trolley. Measurement objects can be placed on the surface plate. Different measuring points can then be taken with the probe. The height difference between measuring points and other simple calculations can be determined by the calculation unit of the height gage. It is also possible to program the height gage for certain measuring tasks. The resolution is typically 1 µm with a range of about 600 mm. The uncertainty depends on the design of the height meter and the surface plate. Because the height meter does not meet the Abbe principle, deviations can

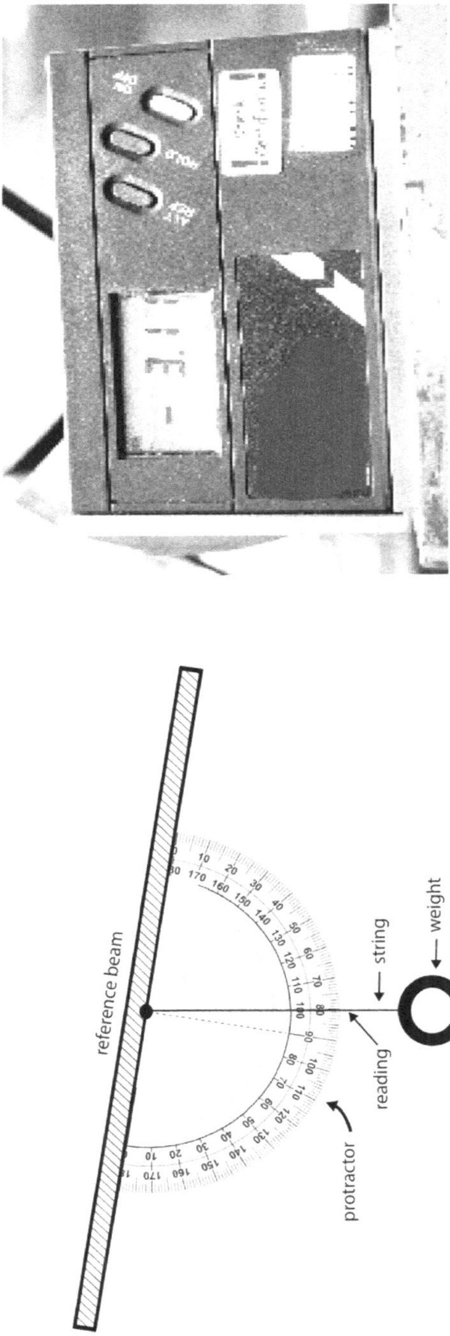

FIGURE 7.8 Left: simplified principle of a clinometer. Right: photograph of a clinometer with a resolution of 0.01°.

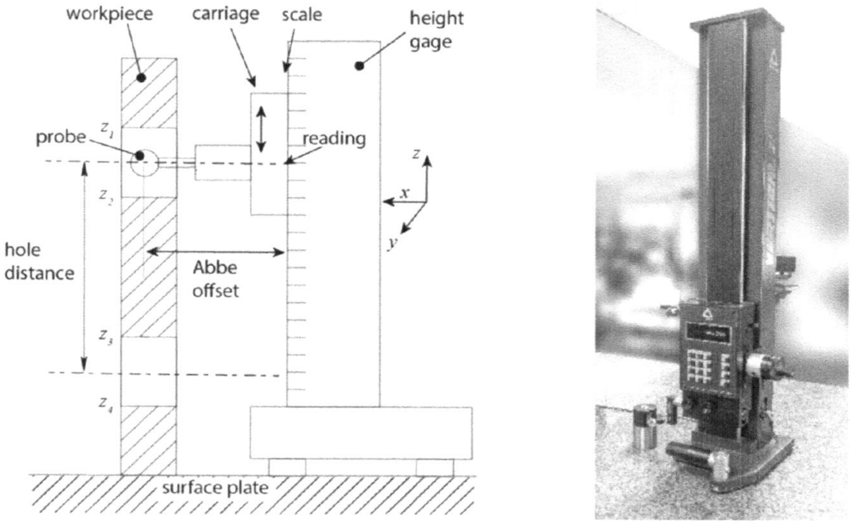

FIGURE 7.9 Schematic (left) and photograph (right) of a height gage on a surface plate.

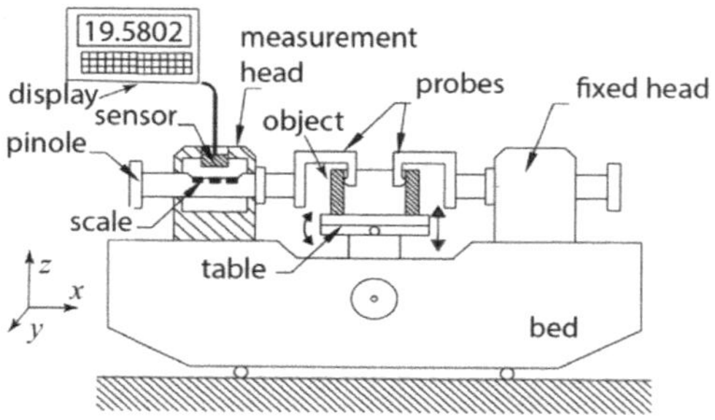

FIGURE 7.10 Schematic overview of a 1D horizontal measuring machine.

occur, especially when longer probes are used. The metrological characteristics are described in ISO 13225:2012. These include the length measurement error, the repeatability, the bi-directional error and the squareness. A schematic and a photograph of a typical height gage are given in Figure 7.9.

7.2.2 Horizontal 1D Measuring Machine

The horizontal measuring machine (schematic in Figure 7.10, photograph in Figure 7.11) consists of a bed, provided with two movable heads. A freely movable, height-adjustable workpiece table is arranged between the two heads.

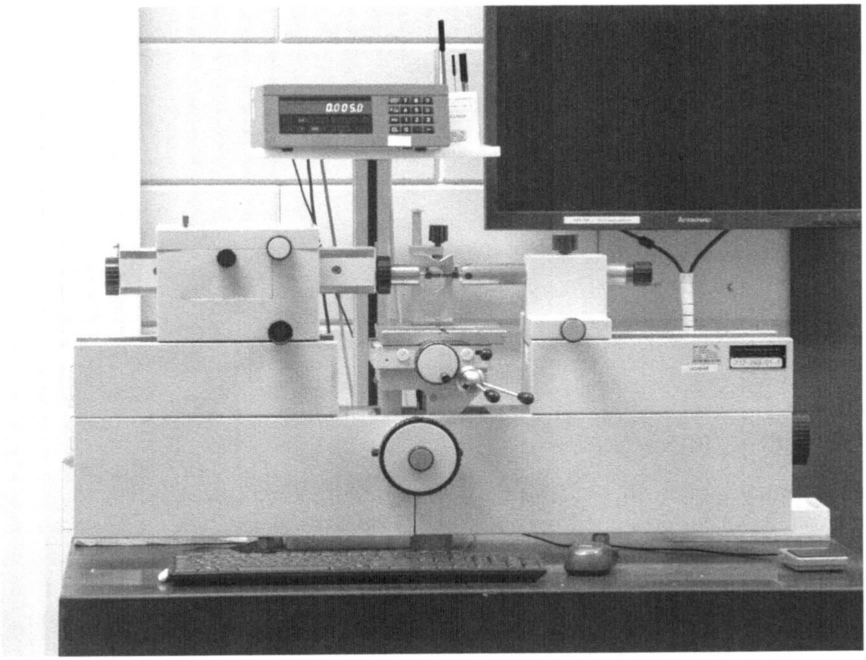

FIGURE 7.11 Photograph of a 1D measuring machine.

This allows the objects to be measured to be placed in the correct position between the heads. One of the heads contains a quill connected to a measurement system, commonly a linear encoder, so that the movement can be measured. In principle, such a measuring machine resembles an advanced screw gage, but the measuring machine is far more versatile. Different probes can be fitted in the heads: convex probes for measuring surfaces or cylinders, convex probes with a bracket for determining inside dimensions, cylindrical probes and flat probes for measuring balls or cylinders. These probes must always be aligned and adjusted for a zero or reference measurement. Also, the measurement force may be adjusted in direction (in/out) and in magnitude. As a result, the 1D measuring machine can be used to determine external dimensions and internal dimensions of objects with parallel, cylindrical or convex surfaces. In the operation, it may be necessary to move or tilt the measured object in order to obtain the correct minimum or maximum value. For example, a setting ring must be moved in the y-direction where the maximum must be sought, and it must be tilted around the y-direction where the minimum value must be sought. Some machines have automated these operations. As the measurement system follows the Abbe principle, this machine is sometimes called an Abbe-bench. In Figure 7.12, some measurement possibilities are illustrated. The outside thread measurement is treated in Chapter 3.

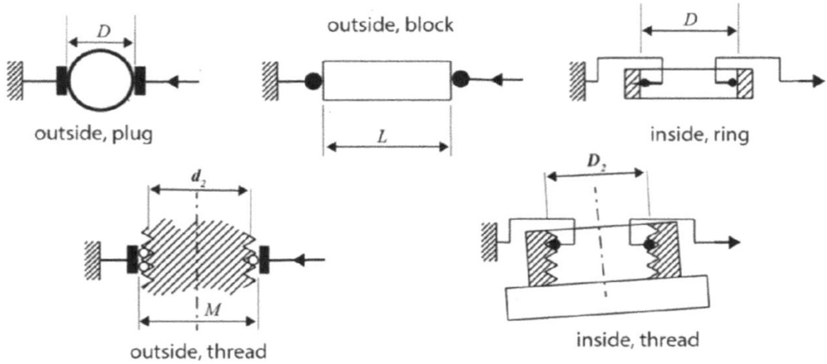

FIGURE 7.12 Some measurement possibilities of a 1D measuring machine.

7.2.3 Gage Block Comparator

As its name implies, this measuring device is designed to measure gage blocks, however, it can be considered as a dedicated 1D measuring machine for objects with parallel faces. The schematic design is given in Figure 7.13. In the measurement two LVDTs (see section 5.3) are used that can be lifted pneumatically so there is no side-force when moving the gage blocks. First, the reference gage block is probed at its center, the upper LVDT is moved so that a nominally zero displacement is measured and sum of the LVDT readings is recorded. This sum compensates small flatness or form variations between the gage blocks. Then the LVDTs are lifted, the gage block to be shifted is measured in the center and the four corners, as described in ISO 3650:1998, then the reference gage block is measured again in order to check whether any drift has occurred. As the gage block sizes are standardized, templates are available to position the gage blocks easily. For the reference gage block commonly a class K gage block is used that is calibration for its central length by interferometry, see Sections 4.6 and 6.1.1. The gage block to be calibrated has by default the same nominal length. A small uncertainty of about 30 nm in the length difference can be accomplished due to the following characteristics:

- Fast measurement, the gage blocks and the LVDTs have no time to 'drift'.
- Measurement values close to zero, enabling high amplification and accuracy of the LVDTs.
- Constant and same measuring force, giving the same indentation for reference the gage block and block to be measured.
- Same nominal temperature and nominal expansion coefficient of reference gage block and the gage block to be calibrated

For gage blocks of dissimilar materials, a correction must be made for the different indentations, and also for the difference expansion coefficient if the temperature deviates from 20 °C. For large gage block sets, this calibration can be labor-intensive, therefore, the following measures are sometimes taken:

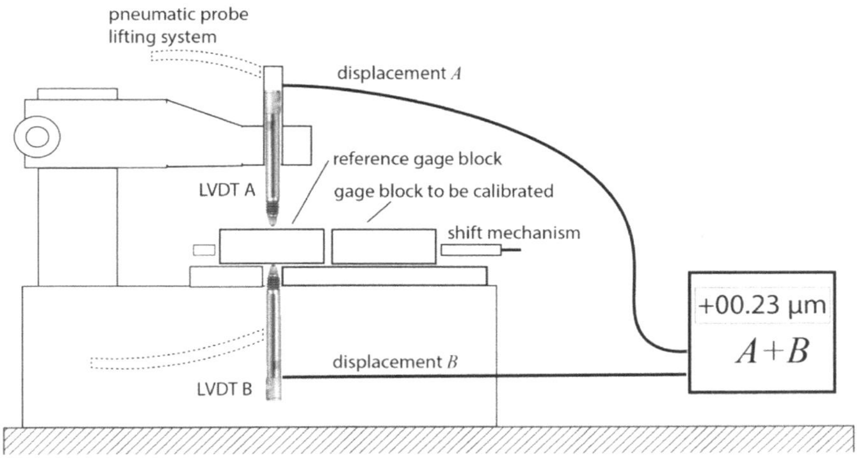

FIGURE 7.13 Schematic of a gage block comparator.

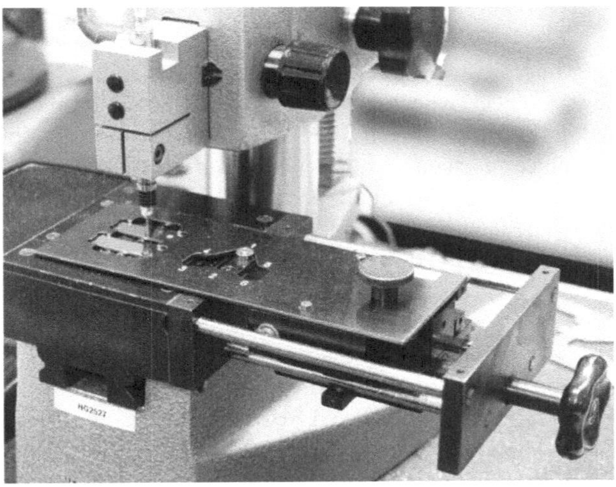

FIGURE 7.14 Photograph of a gage block comparator with two gage blocks and a template.

- Replace the LVDT by a linear encoder so that measurements can be taken in a wider range using a single reference gage block.
- Automate the measurement process by motorizing the gage block movements.

Figure 7.14 gives a picture of a gage block comparator with a template and a knob for manually manipulating the gage blocks. The measurement points are

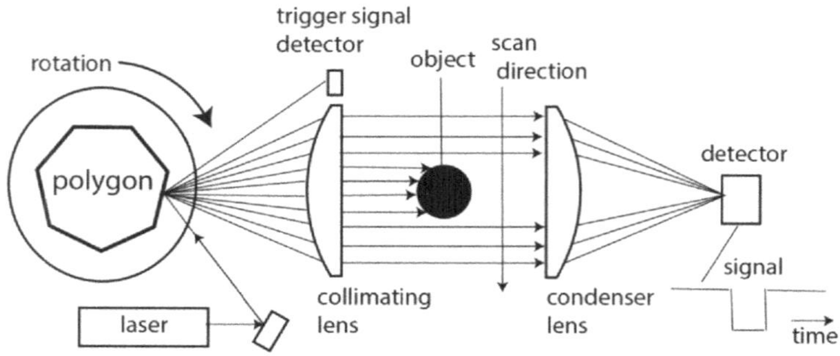

FIGURE 7.15 Schematic of the measurement principle of a laser scanning micrometer.

indicated in the template: position 1 measures the central length of the gage blocks, position 2 measures the central length of the gage block to be calibrated and points 3, 4, 5 and 6 measures the corners of the gage block to be calibrated, in order to determine its parallelism.

7.2.4 Laser Scanning Micrometer

A laser scanning micrometer is a length measuring device based on a laser beam that scans along an object, where the size is derived from the time the detector is in the shadow region of the object. The term micrometer in the device name is because it can replace mechanical micrometer measurements (see Section 7.1.2), especially the measurement of cylinder diameters.

A schematic of the instrument is given in Figure 7.15. A laser scans along the object and is focused on to the detector using a cylindrical lens. The time the beam is obstructed can be used to calculate the size of the object. Especially cylinder diameters or diametral variations, for example of a cable in a cable production line, can be almost continuously monitored. Because of diffraction at the edges, instabilities and non-linearities of the polygon rotation, etc., the accuracy can be limited, but these effects can be reduced by calibration using reference cylinders that are calibrated in a more classical way. Further uncertainty reduction is possible by comparison with a reference cylinder with a same nominal size as the object. Typical specifications are 3000 scans/s, 0.05 μm repeatability and 1 μm non-linearity in a range from 0.33 mm to 30 mm. Other applications include for example the measurement of tape width and the gap between rollers.

7.3 STRAIGHTNESS DEFINITIONS AND INSTRUMENTS FOR STRAIGHTNESS MEASUREMENTS

7.3.1 Definitions of Straightness

In general, the straightness (deviation) is defined as the deviation of a product, measured along a line, from a mathematically straight line. In practice, the straightness is derived from an extracted line, that is: the digital representation of

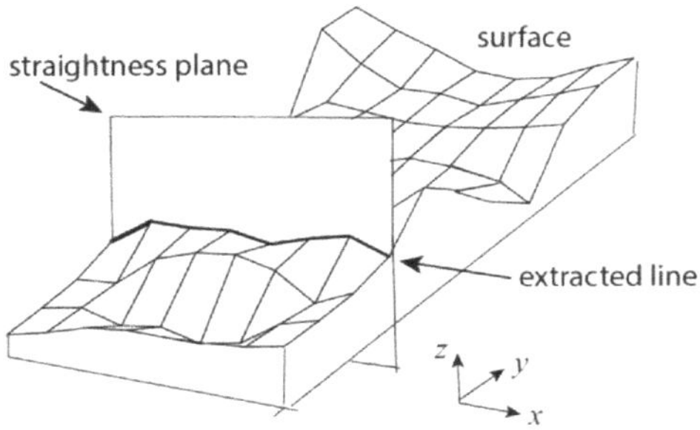

FIGURE 7.16 Definition of the extracted line $z(x)$ from which the straightness can be calculated.

the intersection of the real surface and the straightness plane, where the straightness plane is a plane nominally perpendicular to the actual surface. This is illustrated in Figure 7.16. The straightness deviation has the symbol(s) *STR*, with additional specification of the reference line and the parameter; for example *STR*t(MZLI) = 8 µm means that the total straightness (peak-valley) of the straightness profile referenced to the minimum-zone line is 8 µm. Where the straightness as just a function $z(x)$ has not yet a defined bias and direction, two kinds of straightness deviations are commonly defined and a third is given here for convenience.

1. Straightness relative to the least-squares line

This is the line defined such that the squared sum of differences of the straightness profile to the reference line is minimal. The straightness of the n transformed coordinates z'(x) can be defined as:

$$z'(x) = z(x) - a_0 - a_1 \cdot x, \tag{7.3}$$

$$\text{with: } a_1 = \frac{n\sum_{i=1}^{n} x_i \cdot z_i - \sum_{i=1}^{n} z_i \cdot \sum_{i=1}^{n} x_i}{n\sum_{i=1}^{n} x_i^2 - \left(\sum_{i=1}^{n} x_i\right)^2}, \text{ and } a_0 = \frac{\sum_{i=1}^{n} z_i}{n} - a_1 \frac{\sum_{i=1}^{n} x_i}{n}. \tag{7.4}$$

The derivation is given in Section A.4.1. In equation (7.4), only the vertical deviation from the line is considered; this is commonly justified if $z(x) << x$. The equations become more complex if the normal distance to the reference line is considered; in this case the calculation is called the total least-squares method.

2. Straightness relative to the minimum-zone line

This is the line defined such that the difference between the maximum and minimum z-coordinate of the profile is minimal. In equation (7.3), this means that the value of β is adjusted until the range of $z'(x)$ values is minimal. This is equivalent to the minimization of a function for which extensive mathematical techniques are available. A complication in this case is that it is not easy to make sure whether the minimum is unique, and the derivative of the function is discontinuous in the minimum. An alternative is searching a combination of points: search the two lowest points, connect these and check if all other points are higher than these points; or connect the two highest points, etc. and continue until the minimum range is found. This is called the combinatorial approach. Once the value of β is found, the value of α can be adjusted so that the minimum-zone line goes through the lowest points, the highest points, or in-between. This is related to the definition of straightness of a line: 'the minimum distance of the two parallel lines between which all points are contained'.

3. Straightness relative to the straight line between the first and last point of the profile

This way of straightness calculation is not in the relevant ISO standards (ISO 12780-1:2011 and ISO 12780-2:2011) but can be useful as a basis for calculating the least-squares or minimum-zone reference line, and for uncertainty evaluations. Figure 7.17 gives an example of a straightness profile with as reference the

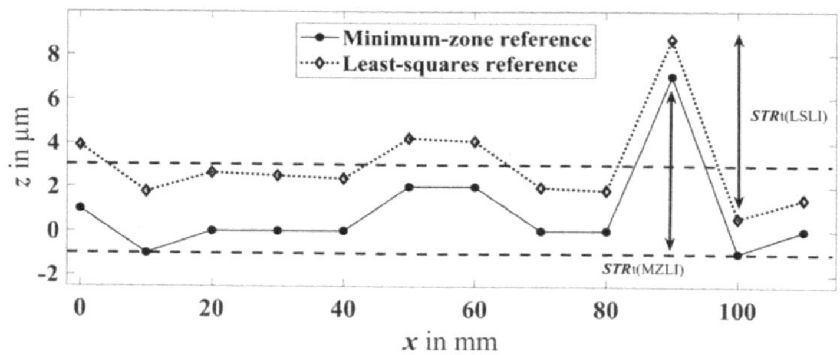

FIGURE 7.17 Straightness profile relative to the least-squares line (*STR*t(LSLI)=8.13 µm) and the same profile relative to the minimum-zone line (*STR*t(MZLI)=8 µm).

TABLE 7.1
Suggested S-Nesting Index, Point Spacing and Stylus Tip Radius When Filtering a Straightness Profile Using a Gaussian Filter

S-Nesting Index/mm	Maximum Sample Point Spacing/mm	Maximum Stylus Tip Radius/mm
8	1.1	5
2.5	0.36	1.5
0.8	0.11	0.5
0.25	0.036	0.15
0.08	0.011	0.05

least-squares plane and the minimum-zone line. When a straightness profile is probed in a quasi-continuous, scanning manner it can be useful to limit the bandwidth, that is the range of profile wavelengths that are considered. ISO 12780-2:2011 gives a table of suggested cut-off wavelengths (S-nesting indices); see Table 7.1. Such profile measurements can be taken by topography, form and contour measuring instruments, as well as by CMMs. Also, a straightness profile may be extracted from an interferometric surface measurement, see Section 4.5. Some more specific straightness measuring instruments are treated in the next sections.

7.3.2 Alignment Telescope

An alignment telescope is an optical instrument with which, over large distances, straightness deviations perpendicular to the optical axis can be measured in two directions. To achieve this, the viewer with a distance setting is focused on a target, within 0.5 m to infinity, see Figure 7.18.

A marking on the target is aligned with the cross plates and at this setting, both optical micrometers can be read. With the aid of two tilting glass plates, also known as optical micrometers, the incident light beam with a target image is aligned by the objective lens with the cross wires in the focal plane in front of the eyepiece. The angle φ by which the glass plates are rotated, is a measure for the perpendicular displacement d of the cross wire relative to the optical axis. For the shift d the following equation holds:

$$\Delta z = b \cdot \varphi \left(1 - \frac{1}{n} \right), \tag{7.5}$$

where Δz is the shift, b is the width of the glass plate, φ is the tilt angle of the glass plate ($\varphi < 0.1$ rad) and n is the refractive index of the glass, $n \cong 1.51$. The resolution is typically 0.02 mm. The measurement uncertainty can be less than 0.05 mm with a measuring range of ± 1.2 mm over a working distance of less than 10 m. In the

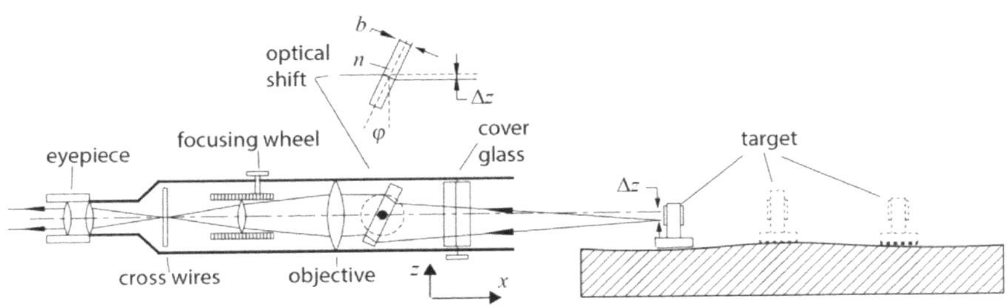

FIGURE 7.18 Schematic of an alignment telescope measuring a target.

design of the instrument, it is essential that the focus adjustment does not affect the location of the optical axis. In addition to measuring the straightness in two mutually perpendicular directions, an alignment telescope is also suitable for measuring the deviations of form, such as squareness and parallelism of, for example, two guide paths. Another important application is the alignment of bearings.

7.3.3 Laser Interferometer with Straightness Optics

A laser interferometer displacement measuring system can be used to measure straightness deviations using so-called straightness optics. This set-up consists of a laser interferometer system, a Wollaston prism and two reflecting mirrors at a fixed angle, see Figure 7.21. The Wollaston prism consists of double-refracting materials, which means that the refractive index in one direction is different from that in the other direction. This means that the refractive index is different for the two polarization directions of the laser interferometer. By creating a double prism of these materials, where the direction of the double-refraction is rotated 90°, the mutually polarized beams are split, reflected at two mirrors, are combined and return in the counting system. If this prism is moved in the x-direction, straight along the axis, the path length for both beams remains equal and the counting system detects no displacement. If the Wollaston prism is moved in the z-direction, the beam path lengths become different, and this amount is proportional to the straightness deviation. For this measurement, the sensitivity is far lower than for linear measurements: for 1 μm vertical displacement, the optical path difference is about 0.03 μm. This means that the optical path difference is amplified some 36× to obtain the straightness deviation value. Also, the effect of air turbulence and interpolation errors are amplified 36×. The same holds for the mirrors that act as the straightness references: a flatness deviation of 10 nm in one of the mirrors is measured as 0.36 μm measured straightness deviation from the horizontal line. This holds for the common 'short straightness' optics that has a range in the x-axis about 3 m and a few millimeters in the z-direction. This measurement set-up can be used to calibrate the straightness of guideways of machine tools and CMMs. When a probe is attached, it can also be used to measure the straightness of objects, see Figure 7.19.

7.4 SMALL ANGLE MEASURING INSTRUMENTS

7.4.1 Electronic Level

For the accurate measurement of tilting relative to the horizontal plane, much use is made of so-called electronic levels. The two most common principles according to which electronic levels work use inductance and capacitance. Both principles have already been discussed in Chapter 5.

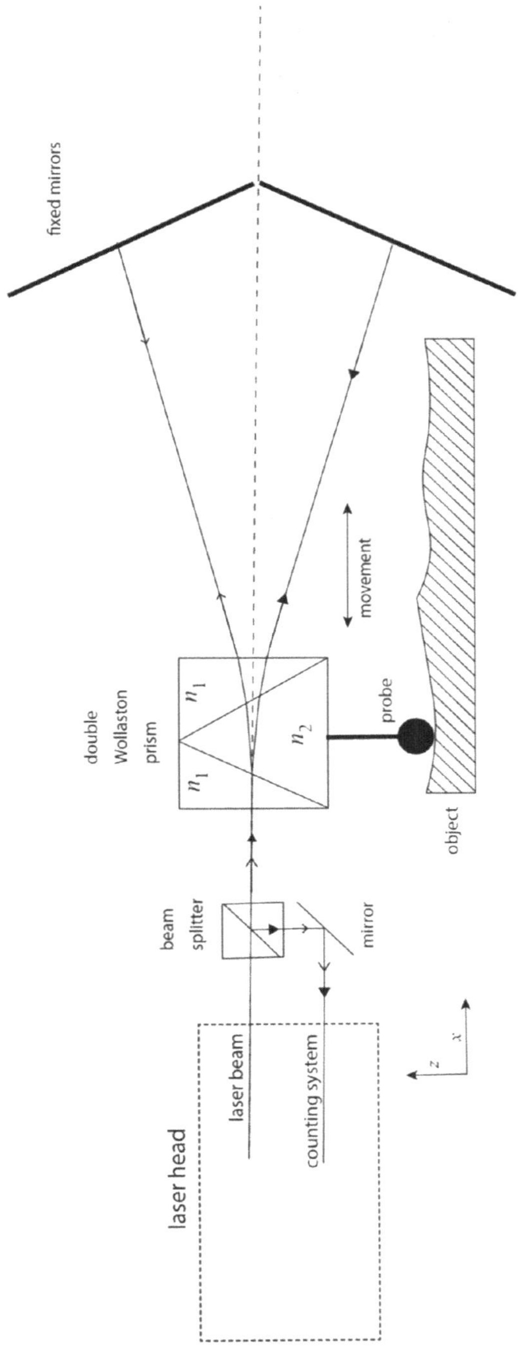

FIGURE 7.19 Schematic of straightness optics used with a laser interferometer to measure straightness.

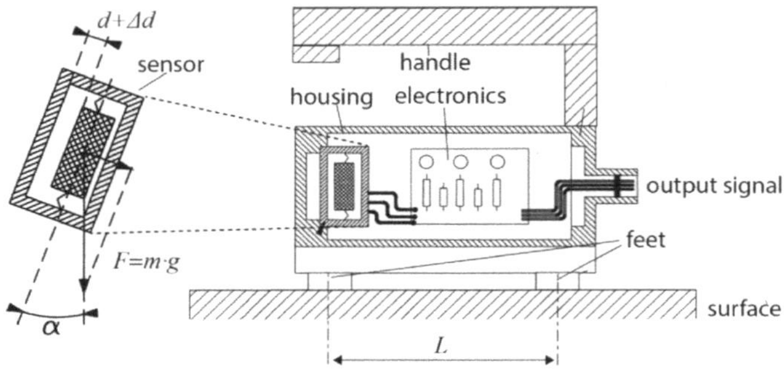

FIGURE 7.20 Schematic of an electronic level.

Applications include:

- flatness measurement
- straightness measurements
- rotation measurements on measuring and production machines (only around the horizontal axes)
- squareness measurements on guides.

Generally, two levels are used in these measurements; either for difference measurements, or for measurements where one level always serves as a reference to eliminate environmental influences (temperature, vibration, common bending or rotation). The capacitive measuring system (Figure 7.20) is based on a three-position capacitor in a housing. The center plate of the capacitor is suspended in a helical spring. This spring is stiff in the plane of the capacitor plate, but has a limited stiffness perpendicular to the plane. When the level is tilted, the center plate moves perpendicular to the spring. This changes the gaps between the capacitor plates and creates capacitance differences, see the electric scheme in Figure 5.8. In equation (5.12), the relationship between the relative displacement and the difference in tension has already been given, namely:

$$\frac{\Delta U}{U_0} \sim \frac{\Delta d}{d}. \tag{7.6}$$

The displacement Δd follows from the tilt α of the level, the mass m of the platen and the stiffness c_m of the springs:

$$m \cdot g \cdot \sin\alpha = c_m \cdot \Delta d. \tag{7.7}$$

Substitution of equation (7.7) in equation (7.6) gives for small angles ($\sin\alpha \approx \alpha$):

$$\frac{\Delta U}{U_0} \approx \frac{m \cdot g \cdot \alpha}{c_m \cdot d}. \tag{7.8}$$

This means that the output signal is proportional α. The resolution of such a measuring system can be as low as 0.2" (or 1 μm/m or 1 μrad). When the level rests on a surface with its feet having a distance *L,* and the angle changes with $\Delta\alpha$, the height difference Δh between the feet is changed by:

$$\Delta h = \Delta\alpha \cdot L. \tag{7.9}$$

This means that for a resolution of 1 μrad and for example a length between the feet of L = 100 mm, the resolution in the height difference is 0.1 μm. The measuring range can be approximately ±0.5° or larger but with less accuracy. The non-linearity is about 0.2%. A major advantage of the electronic systems with respect to the non-electronic systems for angle measurement is the possibility that they offer to measure the difference between two angles, so that jointly undergone rotations or vibration can be compensated. Moreover in electronic systems, interfacing and processing is possible.

7.4.2 Autocollimator

An autocollimator is an optical measuring instrument for measuring small angles (α <20'). The principle is as follows (see also Figure 7.21): with an objective lens, an illuminated crosswire is displayed from the focal plane and via a beam splitter imaged in a parallel beam. After reflection from a flat mirror, passing an objective lens and again the beam splitter, a crosswire image is created in the focal plane. With a mechanical micrometer, a template is moved in the focal plane so that the crosswire is centered in the template. The crosswire shift Δl thus measured is a measure of the angle α between the optical axis of the autocollimator and the

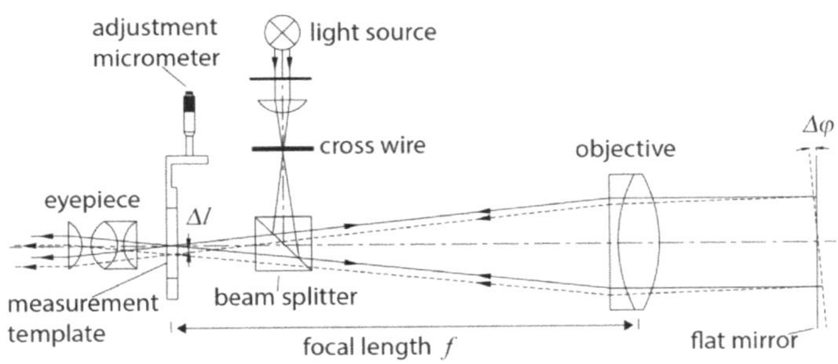

FIGURE 7.21 Optical scheme of an autocollimator.

normal on the mirror. The template may also be a scale enabling a direct reading, commonly in arc seconds. See also Figure 3.6. For small angles it can be written that

$$\Delta l = 2 \cdot f \cdot \Delta\varphi, \tag{7.10}$$

where Δl is the displacement of crosswire image, f is the focal distance of the objective and $\Delta\varphi$ is the rotation of the mirror over an angle $\Delta\varphi$. The angular resolution of an autocollimator is typically 0.1". The range is usually small, typically ± 15'. The useable working distance is limited by the air turbulence in the light path and can be up to 30 m. Depending on the implementation, the reading is done manually by a Vernier or another reading aid. Electronic autocollimators may detect the position of the reflected crosswire through a CCD or CMOS sensors (see Section 5.5.2) and can achieve a resolution as low as 0.005". If the mirror is mounted on a base with a pitch L, equation (7.9) holds. Applications include the use for comparative angle measurements with angular gage blocks as references, measuring tilt angles for measuring machines, measuring squareness, for the calibration of polygons and for the calibration of rotary tables using a polygon. When combined with a mirror mounted on a measurement base, straightness and flatness measurements of large objects such as surface plates can be carried out. The difference with the use of (electronic) levels is that the axis around which the angle is measured does not need to be nominally horizontal and that some autocollimators measure rotations about two axes simultaneously.

7.4.3 Laser Interferometer with Angular Optics

The laser interferometer, as discussed in Section 4.4, can be equipped with optics by which a small angular rotation can directly be measured, commonly called angular optics. Figure 7.23 gives a schematic of the principle of such a setup, consisting of an angular interferometer and angular corner cubes. The rotation angle of the corner cubes assembly while the interferometer system detects a displacement d is given by:

$$\sin\varphi = \frac{d}{a} \quad \text{so} \quad \varphi = \sin^{-1}\left(\frac{d}{a}\right), \tag{7.11}$$

where d is the detected displacement and a is the effective distance between the corner cubes. As the beams run parallel through nominally the same optical path through air, effects of air turbulence are commonly limited. The angle calculation in equation (7.11) gives little problems for small angles. When carrying out a measurement for larger angles, the angle between the beam splitter and the corner cube assembly, at the moment the system is zeroed, becomes relevant. Uncertainties in d and a propagate as with the sine bar (see Section 6.2.2). Based on a resolution of 10 nm in the displacement measurement and a typical for *a* of 30 mm, the resolution in the angle φ is 0.07", comparable to an autocollimator. The applications are also similar, as is the application to measure height differences using a measurement base, for which equation (7.9) applies.

7.5 STRAIGHTNESS MEASUREMENTS BY INTEGRATION OF SMALL ANGLE MEASUREMENTS

Using a measurement base with a defined pitch, a straightness measurement can be obtained by measuring the angular variation when moving this base over a surface with the pitch as the step size. This can be either an (electronic) level as illustrated in Figure 7.20, an autocollimator measuring a flat mirror mounted on a measurement base (see Figure 7.22) or the angular optics of a laser interferometer system, see Figure 7.23. This principle is illustrated in Figure 7.24. Summation of equation (7.9) gives the height of each measured point:

$$z_t(i \cdot L) = \sum_{j=1}^{i} \alpha_j \cdot L. \tag{7.12}$$

The straight reference line connecting the first and last point can be written for a total of *n* measurements as:

$$z_{ref}(i \cdot L) = \frac{i}{n} \sum_{j=1}^{n} \alpha_j \cdot L. \tag{7.13}$$

FIGURE 7.22 Photograph of a two-axis autocollimator with visual read-out, measuring a mirror on a measuring base.

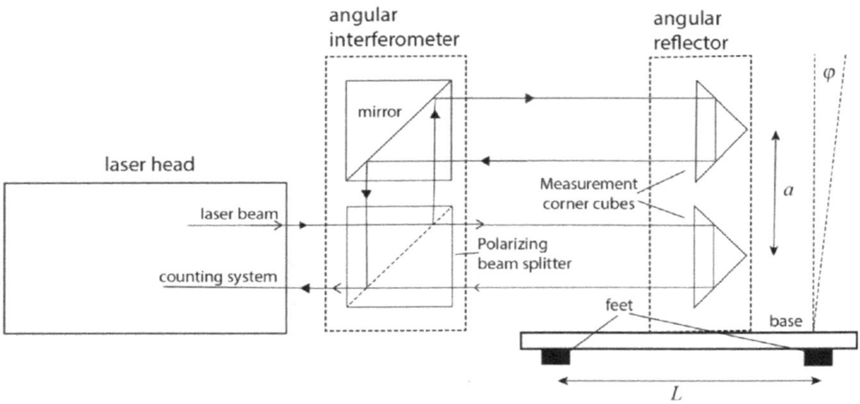

FIGURE 7.23 Schematic of measurement of small angles using a laser interferometer system and angular optics.

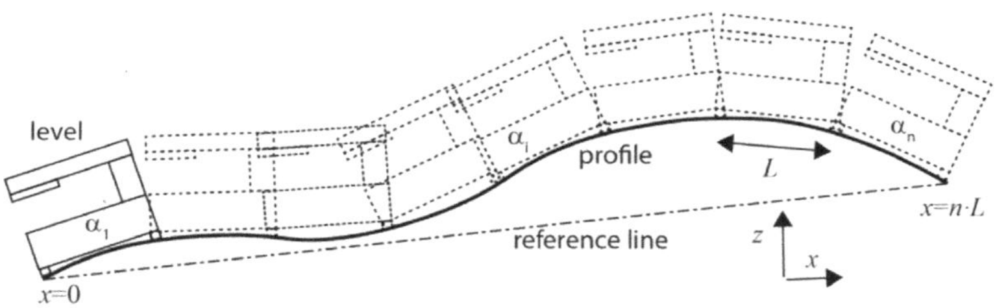

FIGURE 7.24 Straightness measurements by measuring small angular variations.

This gives for the straightness profile, referenced to the line between the first and the last point:

$$z(i \cdot L) = \sum_{j=1}^{i} \alpha_j \cdot L - \frac{i}{n} \sum_{j=1}^{n} \alpha_j \cdot L. \tag{7.14}$$

For surfaces that are optically specular, an arrangement consisting of an autocollimator and a pentaprism can be used, as shown in Figure 7.25. The use of the pentaprism guarantees a constant deflection angle of 90°, independent of the exact orientation of the pentaprism. See Section 6.2.5. In the application depicted in Figure 7.25, it is assumed that the surface is nominally flat over the length (area) *L*. All these methods implicitly assume that the dominant, or relevant, surface

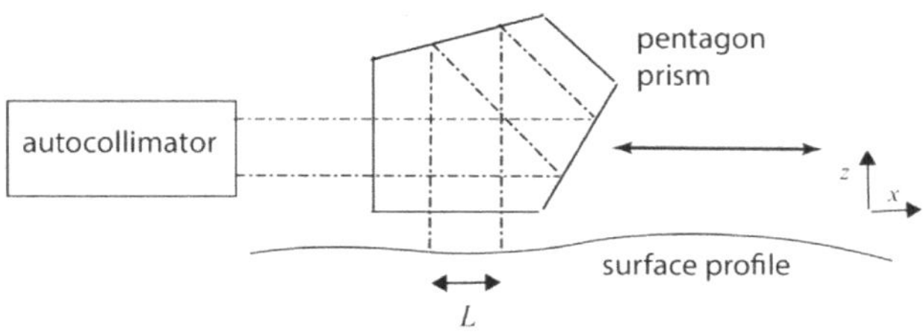

FIGURE 7.25 Schematic of measurement of a straightness profile using an autocollimator and a pentagon prism.

wavelengths are at least twice the pitch *L*. This is usually the case for bearings, granite workpieces and optical flats.

7.6 FLATNESS DEFINITIONS AND MEASUREMENTS

7.6.1 Definitions of Flatness

In general, the flatness (deviation) is defined as the deviation of an area on a material object, from a mathematically flat plane. In practice, the flatness is derived from an extracted plane, that is: the digital representation of the real surface as a number of coordinates that are probed on the surface, that can commonly expressed as $z(x,y)$, where the z-coordinate is the perpendicular distance from the surface to a reference plane, and the (x,y) coordinates are located in this reference plane. The flatness deviation has the symbol(s) *FLT*, with additional specification of the reference plane and the parameter; for example *FLTt*(MZPL) = 8 µm means that the total flatness (peak-valley) deviation of the flatness topography referenced to the minimum-zone plane is 8 µm. Where the flatness as just a function $z(x,y)$ has not yet a defined bias and direction, three kinds of flatness deviation are commonly defined.

a. Flatness relative to the least-squares plane

This is the plane defined such that the squared sum of differences of the flatness topography to the reference plane is minimal. The flatness deviation the n transformed coordinates $z'(x,y)$ can be defined as:

$$z'(x,y) = z(x,y) - \alpha - \beta \cdot x - \gamma \cdot y, \tag{7.15}$$

where the parameters α, β and γ are derived from the matrix equation:

$$\begin{pmatrix} \alpha \\ \beta \\ \gamma \end{pmatrix} = \begin{pmatrix} \Sigma 1 & \Sigma x_i & \Sigma y_i \\ \Sigma x_i & \Sigma x_i^2 & \Sigma x_i \cdot y_i \\ \Sigma y_i & \Sigma x_i \cdot y_i & \Sigma y_i^2 \end{pmatrix}^{-1} \cdot \begin{pmatrix} \Sigma z_i \\ \Sigma x_i \cdot z_i \\ \Sigma y_i \cdot z_i \end{pmatrix}, \tag{7.16}$$

where all summations are taken over the total of n points $z_i(x_i,y_i)$, with $i = 1, \ldots, n$. The derivation is given in Section A.5.1.2. In equation (7.16), only the vertical deviations from a plane are considered; this is commonly justified if $z(x,y) << x,y$. The equations become more complex if the normal distance to the reference plane is considered; in that case the calculation is called the total least-squares method. The total flatness referenced to the least-square plane is written as *FLTt*(LSPL).

b. Flatness relative to the minimum-zone planes

The minimum zone planes are the two parallel planes enclosing the extracted surface and having the least separation. In equation (7.15), this means that the value of β and γ *are* adjusted until the range of $z'(x,y)$ values is minimal. This is equivalent to the minimization of a function for two variables for which extensive mathematical techniques are available. A complication in this case is that it is not simple to prove that the calculated minimum is unique, and the derivative of the function is discontinuous in the minimum, see also Section A.5.1.4. An alternative is searching a combination of points: search three lowest points, level the topography to the plane containing the three points and check if all other points are higher than these points; or connect the three highest points, etc. Another possibility is that there are two highest and two lowest points that define the minimum zone planes. Finding these points is called the combinatorial approach. Once the values of β and γ are found, the value of α can be adjusted so that the minimum-zone plane goes through the lowest points, the highest points, or in-between. The total flatness referenced to the minimum-zone plane(s) is written as *FLTt*(MZPL).

c. Flatness relative to the plane that is oriented such that the end points of both diagonals have the same heights

This flatness representation is applicable to rectangular surfaces. It is not covered in the ISO standards (ISO 12781-1:2011 and ISO 12781-2:2011) but can be useful as a basis for calculating the least-squares or minimum-zone reference plane. Historically it is especially used for evaluations according to the Union-Jack method (see also Section 7.6.2) and some software is still based on this. These three representations of the same flatness topography are illustrated in Figure 7.26.

7.6.2 Measurement and Extraction Strategies

There are several strategies for obtaining measurement points. A logical choice is a square grid; this is automatically taken when taking interferometric measurements using a CCD camera (due to a grid of pixels). When using electronic levels, measuring the straightness along lines in two perpendicular directions will be sufficient to define a flatness, as the lines have a common reference to the gravity vector. Figure 7.27 gives two possible measurement strategies for combining straightness measurements taken by electronic levels, as illustrated in Figure 7.24 and equation (7.13). In the left-hand figure, the straightness measurements can be used right away; but the bias of each line has to be adjusted so the lines have the

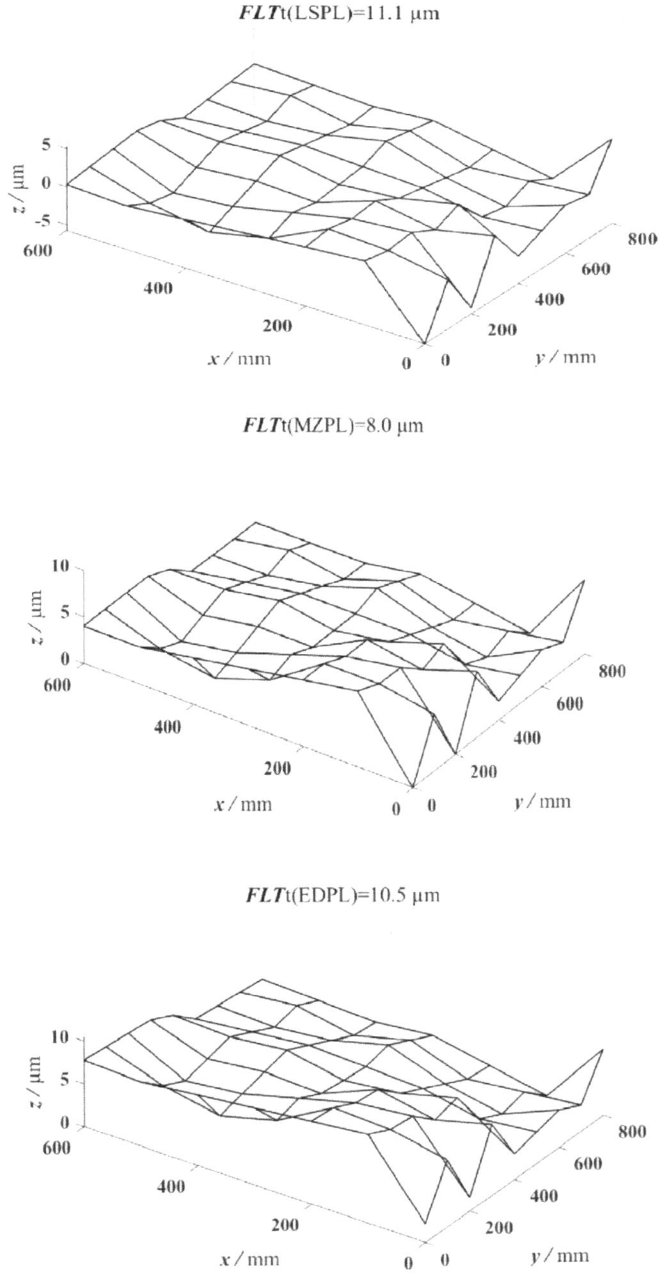

FIGURE 7.26 Three representations of the same surface topography; top: referenced to the least-squares plane (LS). Center: referenced to the minimum-zone plane (MZ). Bottom: diagonal end points are equal (ED).

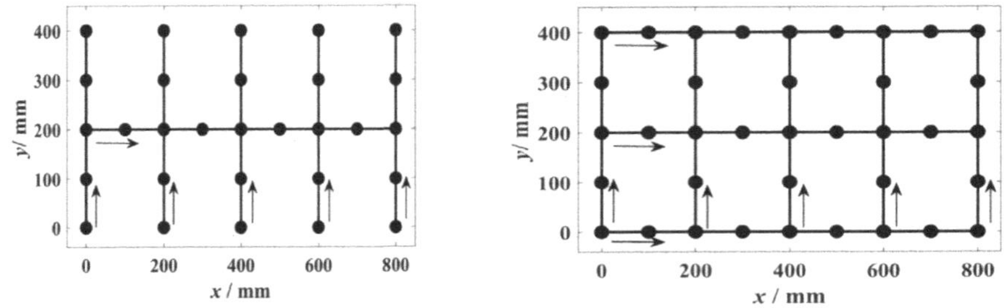

FIGURE 7.27 Examples of measurement strategies for flatness measurements using electronic levels.

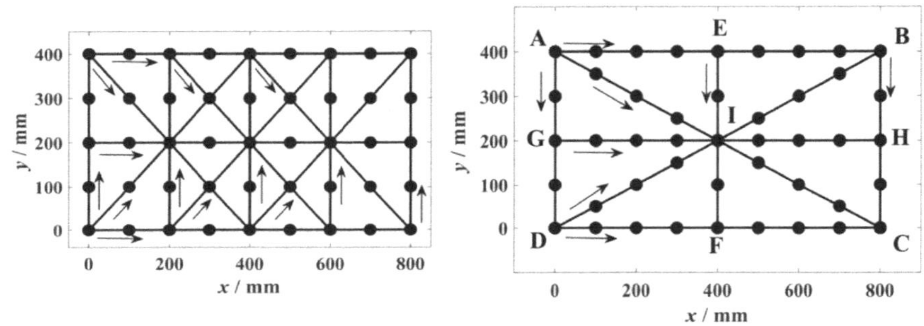

FIGURE 7.28 Flatness measurement strategy on a square grid including diagonals (left) and according to the Union Jack ('Moody') method (right).

same height where they intersect. After determining the topography, the height coordinates can be transformed to a reference plane as described in Section 7.6.1. In the right-hand figure, more measurements are taken in order to reduce the measurement uncertainty. Here the evaluation becomes more complex because of the redundancy in the measurements and generally a least-squares evaluation of the measured values is applied before the transformation to the reference plane is made, see Section A.5.2. If the measured lines have no common reference, as is the case with all straightness measurement methods discussed except for the electronic levels, diagonal measurement must be added. Without diagonal measurements a flat cannot be distinguished from a saddle-plane in which all lines in two orthogonal directions are still straight. Such a measurement strategy is illustrated in Figure 7.28. In the left-hand side, diagonal measurements are added. Clearly, the evaluation becomes more complex, as for every line measured, the reference line has to be adjusted in order to achieve a least-squares solution for the measured points (Meijer and Heuvelman, 1990). A configuration that allows a straightforward measurement and evaluation and requires a limited number of measurements is the measurement strategy proposed by Moody (Moody, 1955), also called the 'Union Jack' method. This is shown in the right-hand side of Figure 7.28, with the conventional denotation of the measurement points. The evaluation is as following:

- The lines A-C and D-B are evaluated to have equal endpoints; see equation (7.14).
- The lines A-C and D-B are shifted so that point I is the same (for example, value zero). This defines the height of points A, B, C and D.
- The lines A-B, D-C, A-D and B-C are oriented and shifted to the already calculated heights A, B, C and D.
- The lines G-H and E-F are shifted to have a zero value at point I and oriented to have a minimum deviation from the already calculated points G and H for the G-H line and E and F for the E-F line.
- The largest deviation of lines G-B and E-F from the already calculated/ defined point I is called the closure error and may act as a quality parameter; smaller is better.
- The calculated topography can be transformed to the least-squares or minimum-zone reference planes.
- The topography may be shifted so that the lowest point has zero height.

Despite its shortcomings (no least-squares calculation, measurement points not equally spread over the surface, measurement pitch may need adjustment for the diagonal lines) this method is widely used in surface plate flatness measurement. Other extraction strategies include a triangular grid, as is used in an STL ('standard triangular language') files that may be output from X-ray computed tomography measurements (XCT, see Section 8.5), or a random set of points as may be probed by a CMM.

7.7 INSTRUMENTS AND TECHNIQUES FOR GENERAL ANGLE MEASUREMENTS

7.7.1 Rotary Indexing Table

A rotary indexing table is a carefully supported platform for generating pure rotations over accurately known angles up to 360°. The reading can take place via a screw spindle with Vernier or via an optically readable glass disk in the table. The achievable accuracy is about 1". It is also possible to use a rotary encoder based on the same principle as the linear encoder, explained in Section 5.5.3. In such a case the resolution can be high (0.005"), although the absolute accuracy is rarely better than 0.2". Rotary indexing tables are used combined with other measurement means, such as an autocollimator and a flat mirror, angle gage blocks or a CMM, where the CMM probe takes the actual measurements. When only a limited number of angles can be rotated, an indexing table may be called an indexing plate.

7.7.2 Theodolite

A theodolite is an optical instrument that can rotate around a horizontal axis, measuring an angle β, and a vertical axis, measuring an angle α. The vertical axis

FIGURE 7.29 Photograph of a classical, manual-type theodolite.

is commonly aligned with the gravity vector using a level that is incorporated in the instrument. A theodolite is depicted in Figure 7.29. The scales for both angular rotations can be read optically. Although a theodolite is an angle measuring instrument, its application is mainly to determine distances between different points. The origin of the theodolite is, therefore, surveying. By measuring certain landscape characteristics or targets from different positions at known mutual distances, triangles can always be formed with measured angles and a known distance, so that other distances can be calculated from them. Accurate theodolites are also used for measuring machines or larger objects such as airplanes, ships and large ship propellers. The measuring range for both axes of such theodolites is 0° to 360° around both the horizontal and the vertical axes, and the smallest resolution is 0.5". The vertical angle is measured with respect to level, it is therefore essential that the theodolite in the x-y plane is well leveled. The theodolites for measuring geometries of objects in 3D are also applied in precision measurement technology. As the operation is based on triangulation, a few principles of triangulation are given here.

7.7.2.1 Simple Height Triangulation

A classic example of simple height triangulation is given in Figure 7.30. The height of an object is derived from the angle(s) from the horizontal plane. In fact for this

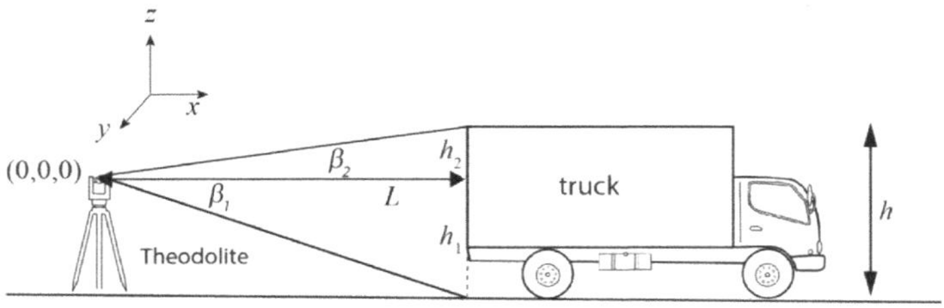

FIGURE 7.30 Determination of the height of a truck.

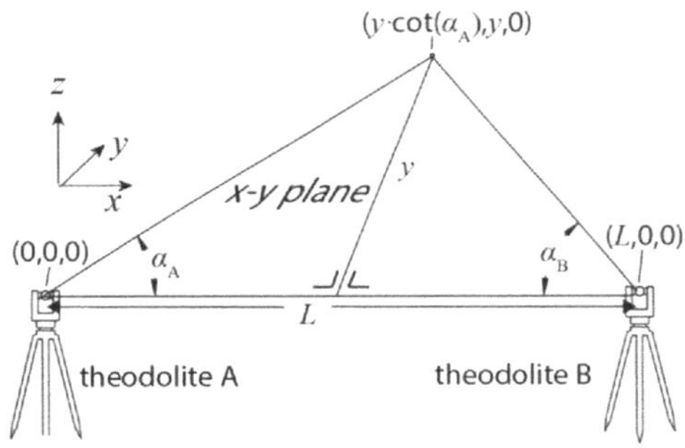

FIGURE 7.31 Illustration of the triangulation principle in a plane.

case, the theodolite is reduced to a telescope equipped with a clinometer. For the height h of the truck in Figure 7.30 it can be written:

$$h = h_1 + h_2 = L(\tan\beta_1 + \tan\beta_2). \tag{7.17}$$

The length of the baseline L can be obtained by measuring a calibrated line scale – called a leveling staff – with a known height h, and applying the inverse of equation (7.17). This establishes the traceability to the unit of length.

7.7.2.2 Triangulation in a Plane

The example of triangulation that has been used for thousands of years is the triangulation in a plane. This is illustrated in Figure 7.31. Two theodolites measure the same point. From the two angles and the distance between the theodolites, the triangle is fully defined, the distance of the point to the line connecting the theodolites is given by:

$$y = L\frac{\sin\alpha_A \sin\alpha_B}{\sin(\alpha_A + \alpha_B)}. \tag{7.18}$$

7.7.2.3 Triangulation in Three Dimensions

In practice, the points to be measured will have different heights, and both theodolites may not be at the same height. This gives the situation as in Figure 7.32 For the coordinates x,y and z we find:

$$y = L\frac{\sin\alpha_A \sin\alpha_B}{\sin(\alpha_A + \alpha_B)}, \tag{7.19}$$

$$x = \frac{y}{\tan\alpha_A} = L\frac{\cos\alpha_A \sin\alpha_B}{\sin(\alpha_A + \alpha_B)}, \tag{7.20}$$

$$z_1 = \frac{x}{\cos\alpha_A}\tan\beta_A = L\frac{\sin\alpha_B \tan\beta_A}{\sin(\alpha_A + \alpha_B)}, \tag{7.21}$$

$$z_2 = h + \frac{L - x}{\cos\alpha_B}\tan\beta_B = h + \frac{L\left(1 - \frac{\cos\alpha_A \sin\alpha_B}{\sin(\alpha_A + \alpha_B)}\right)}{\cos\alpha_B}\tan\beta_B = h + L\frac{\sin\alpha_A \tan\beta_B}{\sin(\alpha_A + \alpha_B)}. \tag{7.22}$$

The two measurements β_A and β_B give a redundance in the determination of z, leading to equations (7.21) and (7.22) where the mean can be taken as final value. Once this triangle is fully defined, one theodolite can be placed on the third point and a triangle can be formed with a fourth point, and so on. In this way, an area can be covered with measured points. Also, this 3D measuring method is an alternative for 3D CMMs, especially for large objects that are difficult to move around, such as ships, satellite antennas, industrial robots and even aircraft and parts thereof.

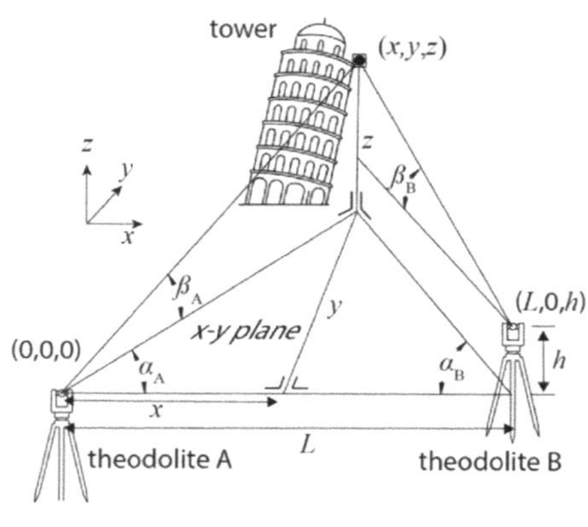

FIGURE 7.32 Illustration of the triangulation principle in three dimensions.

The measurement uncertainty can be about 0.1 mm, depending on the environment and the distances measured. To perform the calculations, theodolites are often equipped with electronic readout and data processing. Sometimes, a direct opto-electronic distance measurement (EDM) is also integrated. Such a system is then called a 'total station'. Combining the distance and angle information makes the measurements easier to evaluate, and makes the distance calibration by a leveling staff obsolete. A further advancement can be made by a mirror that rotates around a horizontal axis, so that measurements are taken along a vertical line, recording the measured distance d as a function of the angle β. Once these distances at one angle α have been taken, the instrument is rotated in steps around the vertical axis and other vertical lines can be measured. In this way, a total image of the surroundings can be taken. When doing this at various locations, data can be merged and areas that cannot be observed from a single position can be included. Inclusion of special marks or balls that are probed from various directions improves the accuracy. These instruments are called 3D terrestrial (laser) scanners.

7.8 MEASUREMENT OF ROUNDNESS AND CYLINDRICITY

7.8.1 Definition of Roundness

The roundness (deviation) of an object is the deviation of a measured roundness profile, taken in a roundness plane, from a perfect circle. The roundness plane is, for example, a plane perpendicular to the axis of a cylinder or the surface of revolution of an object; see ISO 12181-1.

A typical representation of a roundness measurement is in a roundness diagram, that is a polar plot of the roundness deviations, magnified to display the deviations visibly in a circular area, displayed relative to a circle with a standard diameter. $r(\theta)$ is the measured signal of the workpiece, that is the radius r as a function of the angle θ. In order to visualize the out of roundness in a polar diagram, the deviation must be amplified. The signal which is displayed in a roundness diagram is:

$$\rho(\theta) = M\left(r(\theta) - L\right) + R \tag{7.23}$$

where M is the magnification, L is the mean workpiece radius and R is the standard radius for plotting the deviations. Now, $\rho(\theta)$ has a different graphical form to $r(\theta)$, which is caused by the effect of radius suppression and magnification. This is shown in Figure 7.33 for the case of a coin.

This coin has a constant diameter that illustrates that roundness is not the variation in diameter, but the variation in radius. When the circumference is displayed as a roundness diagram, the deviations from the mean circle are amplified, which has the effect that convex parts of the coin look concave in the roundness diagram. Parameters derived from the roundness diagram are denoted by *RON*. Similar to the straightness and flatness, there are several roundness definitions depending on

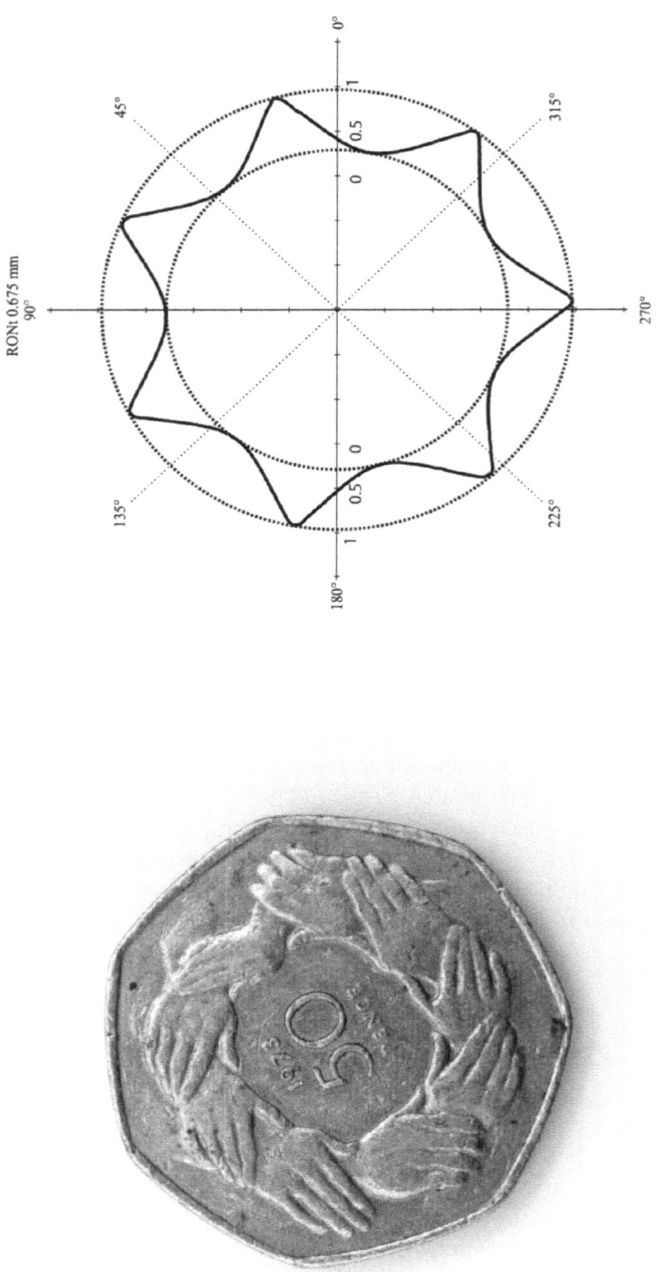

FIGURE 7.33 Photograph of a coin (left) and the circumference displayed as a roundness diagram (right).

how the center and, related to that, the reference circle is defined. The possible choices for the reference circle(s), as specified in ISO 12181-1:2011 are as follows.

1. Least-squares reference circle (LSCI)

The least-squares reference circle is the circle such that the sum of the squares of the local roundness deviations is minimal. The center (x_c, y_c) coordinates of the profile $r(\theta)$ consist of the amplitudes of the cosine and sine Fourier components of the first harmonic respectively:

$$x_c = 2\frac{\sum x_i}{N} = 2\frac{\sum r_i(\theta)\cos(\theta_i)}{N}, \tag{7.24}$$

$$y_c = 2\frac{\sum y_i}{N} = 2\frac{\sum r_i(\theta)\sin(\theta_i)}{N}. \tag{7.25}$$

This can be calculated from a series of $r(\theta_i)$ values at equally spaced angles, and a correction can be applied so that $(x_c, y_c) = 0$. For small deviations, $r(\theta) - L \ll L$, $\rho(\theta)$ can be used instead of $r(\theta)$ in equations (7.24) and (7.25). The least-squares circle radius is the mean value of the corrected diagram $\rho(\theta)$. The related roundness parameter *RON*t(LSCI) is calculated from the roundness diagram as the range of measured points $\rho(\theta)$.

2. Minimum zone reference circles (MZCI)

The minimum zone reference circles are two concentric circles enclosing the roundness profile and having the least radial separation. The circles are determined by either three maximum points and one minimum point, or three minimum points and one maximum, or two minimum and two maximum points. These points can be found by considering the range of local roundness variations as a function and minimizing this function as a function of the center coordinates x_0 and y_0. A direct search for the points mentioned is another possibility. Also, here the requirement: $r(\theta) - L \ll L$ applies and additionally, $\rho(\theta) - R \ll R$, for using $\rho(\theta)$ instead of $r(\theta)$. The related roundness parameter *RON*t(MZCI) is calculated from the roundness diagram as the range of measured points $\rho(\theta)$.

3. Minimum circumscribed reference circle (MCCI)

The minimum circumscribed reference circle is the smallest possible circle that can be fitted around the roundness profile. It can be found by varying the center coordinates (x_0, y_0) and searching for the smallest circle that still encloses the measured roundness diagram. Also, the requirement $r(\theta) - L \ll L$, applies and additionally, $\rho(\theta) - R \ll R$, for using $\rho(\theta)$ instead of $r(\theta)$. The related roundness parameter *RON*t(MCCI) is calculated from the roundness diagram as the range of measured points $\rho(\theta)$.

4. Maximum inscribed reference circle (MICI)

The maximum inscribed reference circle is the largest possible circle that can be fitted inside the roundness profile. It can be found by varying the center coordinates (x_0, y_0) and looking for the largest circle that is enclosed by the measured roundness diagram; and maximize the radius of this largest circle. Also here the requirement: $r(\theta) - L \ll L$, applies, and additionally $\rho(\theta) - R \ll R$, for using $\rho(\theta)$ instead of $r(\theta)$. The related roundness parameter *RON*t(MICI) is calculated from the roundness diagram as the range of measured points $\rho(\theta)$. Figure 7.34 gives the same roundness measurement referenced to the four defined reference circles. The roundness deviation *RON*t(MZCI) is, by definition, smallest. The difference between the roundness types is typically 5% to 10%.

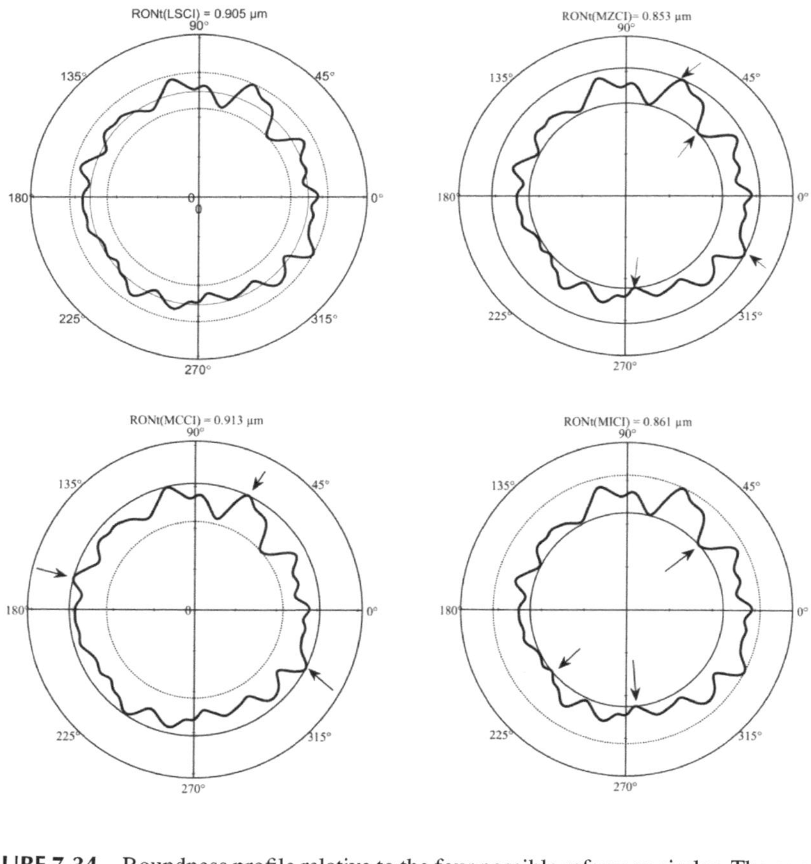

FIGURE 7.34 Roundness profile relative to the four possible reference circles. The arrows indicate the points that determine the reference circle(s).

7.8.2 Harmonic Components in Roundness Measurements

The roundness measurement, displayed in a roundness diagram, is an ideal signal for harmonic (Fourier) analysis, as it is perfectly periodic: $\rho(\theta)=\rho(\theta+n\cdot 2\pi)$, where n is an integer number. An amplitude spectrum of the signal may reveal harmonic components that are not directly visible in the signal. The interpretation of these harmonic components is summarized in Table 7.2. As an illustration, Figure 7.35 shows the harmonic components of the roundness measurement of the coin displayed in Figure 7.33. The harmonic components represent the amplitude of angular frequencies that are commonly denoted as undulations per revolution, abbreviated as 'upr'. The figure shows the dominant seventh harmonic component that is obvious from the roundness diagram of Figure 7.33, but also the higher harmonics that appear as multiples of seven, and indicate the deviation of the form from a sine wave.

TABLE 7.2
Interpretation of the Harmonic Components of a Roundness Profile r(θ)

Angular Frequency/upr	Interpretation of Corresponding Amplitude
0	Constant term, L in equation (7.23), converted to R in the diagram. Also called the 'piston term'. Commonly omitted in the amplitude spectrum.
1	Decentering: deviation of the center from the least squares circle (LSCI) center.
2	Ellipticity
3	3-point out-of-roundness, for example, due to clamping a workpiece with a three-claw
n	n-point out of roundness

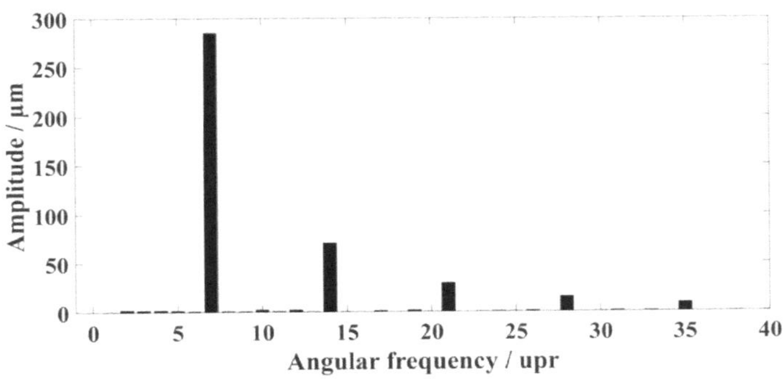

FIGURE 7.35 Amplitude of the harmonic components of the roundness diagram of the coin as displayed in Figure 7.33.

7.8.3 Default Filtering and Probing in Roundness Measurements

As a roundness measurement is a typical form measurement, it is common to filter the higher harmonic components out of the profile; this means that a long-wave-pass filter is commonly used. This is specified in ISO 12181-2:2011. The default filter is a Gaussian filter according to ISO 16610-21:2011 (see also Section 9.3), where the cut-off frequency is taken from the series 15 upr, 50 upr, 150 upr, 500 upr and 1500 upr. Figure 7.36 gives the transmission characteristics of the default filters. Requirements for the maximum probe tip diameter and the minimum amount of points to be taken depending on the filtering, are given in ISO 12181-2:2011 and reproduced in Table 7.3. In general, a filter setting of 150 upr is taken, however, 50 upr is taken for diameters < 20 mm and 15 upr is taken for objects diameters < 5 mm. Additionally, a short-wave pass or a bandpass filter can be used, but this is rare in practice. As the LSCI-centering is equivalent to defining the

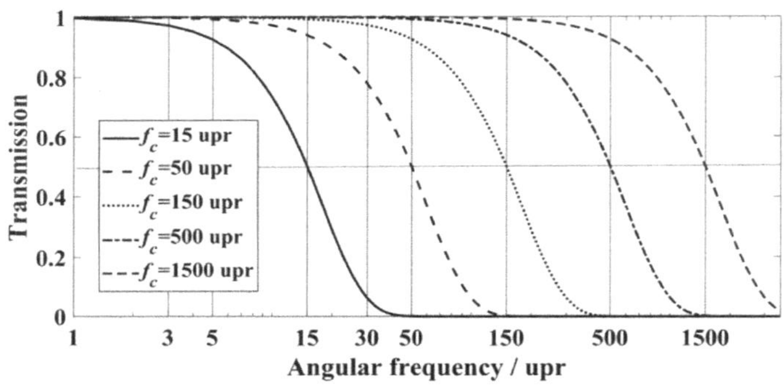

FIGURE 7.36 Transmission characteristics of the default filters for roundness measurements.

TABLE 7.3
Recommended Filter Settings, Sample Points and Probe Radius for Roundness Measurements

Angular Cut-Off Frequency/upr	Minimum Number of Sample Points	Minimum Ratio *d*/*r* Between the Object Radius *d* and the Probe Radius *r*
15	105	5
50	350	15
150	1050	50
500	3500	150
1500	10 500	500

first harmonic component equal to zero, the filtering, including centering, may be referred to as '1-15', where the 1 stands for the centering according to the least-squares center and the '15' is the cut-off frequency in upr where a sinusoidal signal is reduced by 50 %.

7.8.4 Roundness Reversal

A method to distinguish the spindle deviation from the roundness deviation of a specimen is called the roundness reversal method, or the Donaldson reversal method (Donaldson, 1972). It consists of measuring a roundness specimen in the usual way, see Section 7.8.7, and then rotating the specimen 180° and probing the specimen from the opposite side, which is called the reverse position. This is depicted in Figure 7.37. The probe measures a positive displacement when moved from the rotation center. Denoting the spindle deviation as $S(\theta)$ and the deviation of the roundness standard as $R(\theta)$, the measurements, both centered to the least-squares circle, see Section 7.8.1, can be written as:

$$r_A(\theta) = R(\theta) + S(\theta) \text{ and } r_B(\theta) = R(\theta) - S(\theta). \tag{7.26}$$

This gives:

$$R(\theta) = \frac{r_A(\theta) + r_B(\theta)}{2} \text{ and } S(\theta) = \frac{r_A(\theta) - r_B(\theta)}{2}, \tag{7.27}$$

where both $R(\theta)$ and $S(\theta)$ are referenced to the least-squares center. This is an example of a reversal method where an absolute form deviation is obtained without the need for an external reference. It is also called 'error separation' as the errors of both the spindle and the roundness standard are separated. Another term is 'absolute testing': the roundness standard is measured in an absolute way,

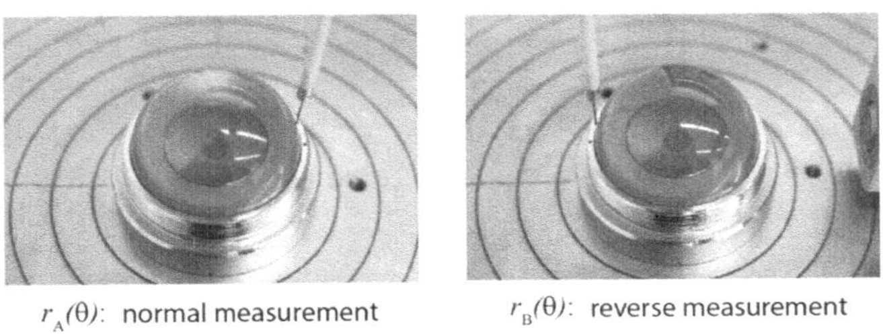

$r_A(\theta)$: normal measurement $r_B(\theta)$: reverse measurement

FIGURE 7.37 The positions of a roundness standard on a roundness tester for the Donaldson reversal measurement. At the left-hand side the normal position, at the right-hand side the roundness standard rotated 180° and probed from the opposite side.

without any systematic deviation caused by an external reference standard. See also Section 1.5.6.

7.8.5 Definition of Cylindricity

The cylindricity (deviation) of an object is the deviation of a measured cylindrical surface from a right circular cylinder; this means that the nominal cylinder has a right angle between the cylinder axis and every circular cross-section of a perfect cylinder. In practice, the cylindricity is derived from extracted coordinates measured along the surface, that is the digital representation of the real surface as a number of coordinates that are probed on the surface, that can commonly be expressed as $r(\theta,z)$, where the z-coordinate is the distance from the ground-plane and the angle is the angle around the circumference relative to a defined direction. Points on the surface can also be represented as 3D (x,y,z) coordinates. The cylindricity deviation has the symbol(s) ΔC *or* CYL, with additional specification of the reference cylinder and the parameter; for example, $CYLt$(MZCY) = 8 µm, means that the total cylindricity (peak-valley) deviation of the cylinder topography referenced to the minimum-zone cylinders is 8 µm. Similar to the roundness deviations, four kinds of reference cylinders are commonly defined: the least-squares cylinder (LSCY), the minimum-zone cylinder (MZCY), the minimum circumscribed cylinder (MCCY) and the maximum inscribed cylinder (MICY); see ISO 12180-1:2011 and ISO 12180-2:2011.

Any radial cross-section can be treated as a roundness measurement and any measurement of the surface parallel to the cylinder axis can be treated as a straightness measurement.

7.8.6 Cylinder Measurement and Extraction Strategies

In practice, it is often difficult to achieve a complete covering of the cylindrical feature given by a minimum density of points. In these situations, more limited extraction strategies are employed that give specific rather than general information concerning the assessment of cylindrical form.

Points at the circumference of a cylinder can be taken as a random pattern (for example, using a CMM) or patterned on the surface, using a roundness tester. Possibilities are a 'bird cage' strategy, using both roundness and straightness measurements, a roundness profile extraction strategy and a straightness profile extraction strategy, see Figure 7.38. More complete and evenly distributions can, for example, be obtained by XCT measurements. Elementary cylindrical forms can be considered as a combination of harmonic components of the angular coordinate θ, similar to roundness measurements, and polynomial profiles along the z-direction. Figure 7.39 gives some elementary cylindrical form deviations. Reference cylinders can be used as form standards as well as squareness standards, in which case, the squareness relative to the ground plane will be relevant.

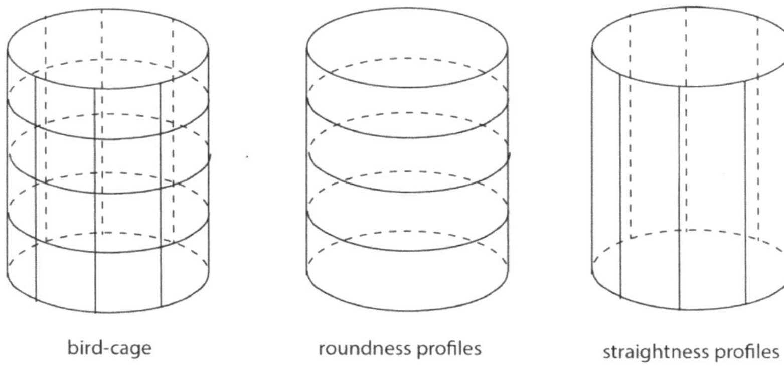

FIGURE 7.38 Extraction strategies for measuring cylindricity at patterned point series.

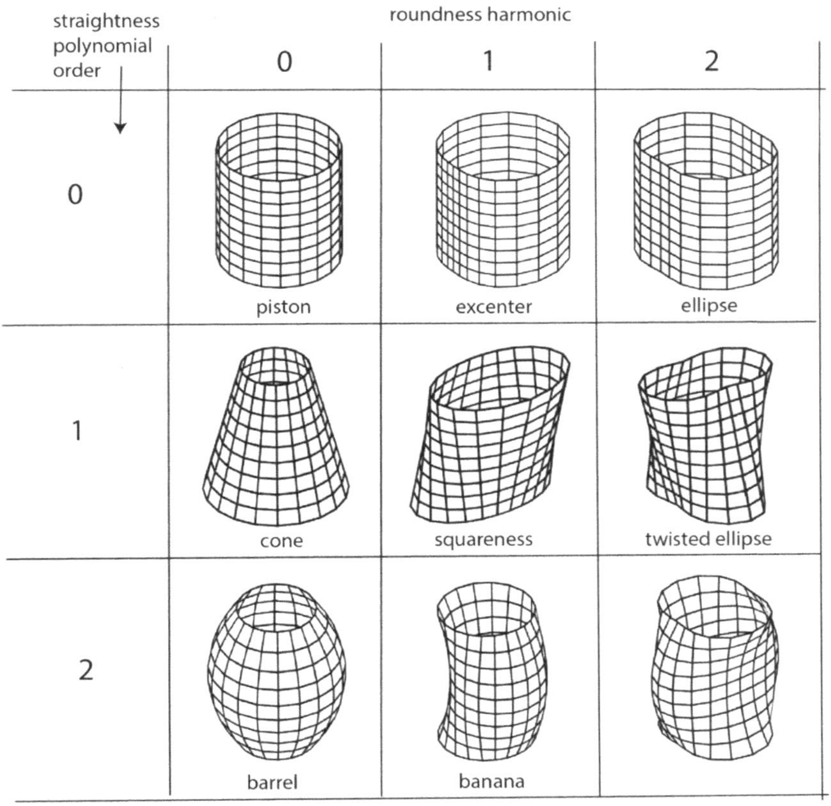

FIGURE 7.39 Elementary cylindrical form deviations.

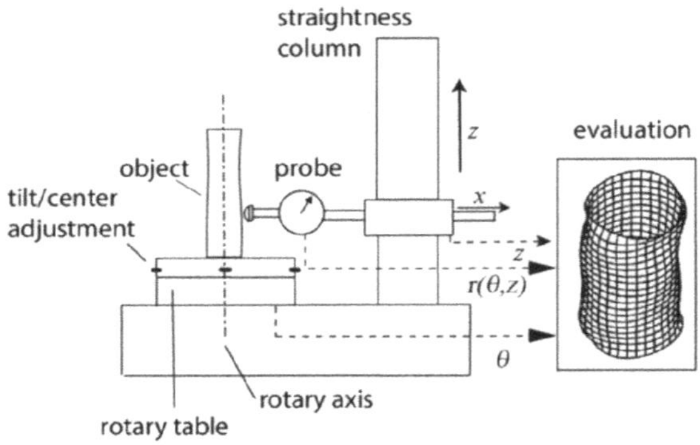

FIGURE 7.40 Schematic of a roundness tester based on a rotating workpiece.

7.8.7 Form Measuring Instruments

Measuring instruments designed for measuring form deviations are called form testers or roundness testers. Roundness measuring instruments are classified by measurement with a stationary probe and a rotating workpiece or measurement with a rotating probe and a stationary object. The instrument with rotating workpiece is most common. It is schematically shown in Figure 7.40 and is depicted in Figure 7.41. The heart of the instrument is the rotary table that rotates with a radial and axial accuracy < 0.1 µm. The table can rotate with a speed of a few rotations per minute and the angle is read by an angular encoder. The form deviations are measured using an inductive probe. In order to enable the probe to have a maximal magnification, the object needs to be accurately centered, and in case of straightness and cylinder measurements, it must also be levelled. Some instruments can do this centering and levelling automatically. If a straightness column is present, straightness and cylindricity measurements can be taken in the vertical direction. A linear encoder then measures the vertical position. The reference straightness deviation is commonly smaller than 1 µm. Some instruments can operate in CNC mode; this means that a measurement sequence can be programmed and can run automatically once a workpiece is positioned and aligned.

7.9 MEASUREMENT OF SURFACE TEXTURE

7.9.1 Surface Topography, Texture, Waviness and Roughness

As the concepts of surface topography, surface texture, waviness and roughness often lead to confusion, these concepts will be briefly treated here. In practice, the topography is measured as the digital representation of the real surface as a number of coordinates that are probed on the surface, that can commonly expressed as (x,y,z) coordinates in the reference plane of the measuring instrument. The surface topography does not depend on the orientation, for instance, on how the workpiece

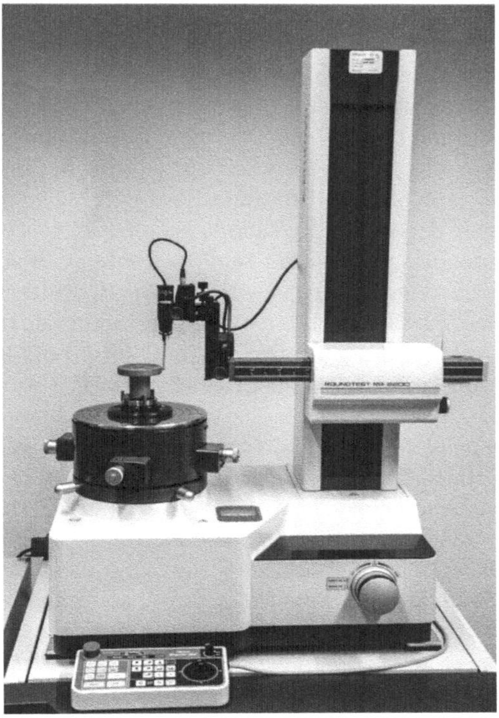

FIGURE 7.41 Photograph of a roundness and form tester with rotating workpiece and a straightness column.

is oriented on the measuring machine. For example, the flatness relative to the least-squares plane or relative to the minimum-zone plane represents the same surface topography. The surface texture consists of the geometrical irregularities contained in a scale limited surface (see ISO 21920-2:2021). This means that the surface texture does not include geometrical irregularities that contribute to the form or the shape of the profile or topography.

The primary profile or primary surface is the surface texture where the highest spatial frequency components that are at the limit of the instruments transfer function and are difficult to discriminate are filtered. Such a filter is commonly denoted as the S-filter. The spatial wavelength below which spatial wavelength components are reduced by more than 50% is called the S-nesting index N_{is} or, more commonly, the cut-off wavelength λs. The other filter commonly employed reduces the longer surface wavelengths and is denoted as the L-filter with a corresponding L-nesting index N_{ic} or long-wavelength cut-off wavelength λc. The form removal operation is the F-operation, and the F-operator can also have an index, for example, denoting the polynomial degree of the form removed: $N_{i\mathrm{F}} = 2$ may denote a second-order polynomial (a parabolic form). The P-profile (P stands for primary) is the profile where the form is removed and the small-wavelength components are filtered using an S-filter. The R (roughness) profile is the profile where the wavelength components are limited by an L-filter for the longer surface wavelength and an S-filter for the shorter wavelength components. The W (waviness) profile is the profile where the shorter wavelength components are limited by an S-filter having an S-nesting index equal to the L-filter in the corresponding roughness profile, and is indicated as N_{ic}, and an F-operator that removes the form or even longer wavelengths. These aspects are illustrated in Figure 7.42. The parameters derived from these profiles are denoted by *P*, *R* and *W* parameters respectively. For surface areas instead of profiles, roughly the same conventions hold, but parameters are denoted as *S*-parameters independent of the applied filtering, which always needs to be specified. Surface texture measurement and evaluation is extensively standardized. For profiles, the indication of limits for profile texture parameters on a drawing are standardized in ISO 21920-1:2021. The parameter definitions are given in ISO 21920-2:2021 and the measurement conditions are given in ISO 21920-3:2021. The standards series ISO 16610 gives an extensive number of defined filters, where ISO 16610-21:2011 describes the Gaussian filter that is the default for most parameters. A more detailed background of filtering theory and methods is given in Chapter 9. The measurement conditions in ISO 21920-3:2021 give the filter settings dependent on the specifications given in the technical drawing, but next to that also requirements for the probe radius and sampling distance with a mechanical measuring instrument must be considered. The instrument settings are summarized in five setting classes denoted as SC1, 2, .., 5 that can be applied to

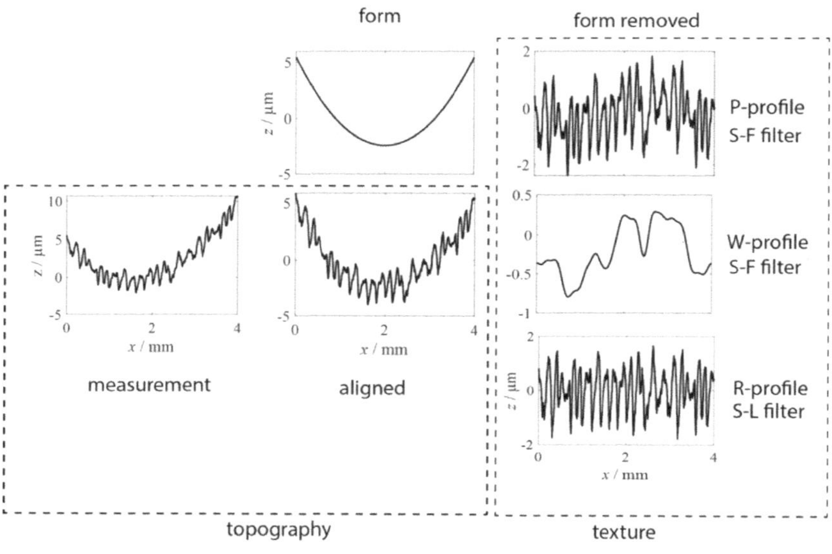

FIGURE 7.42 The sequence of measurement to scale limited texture and corresponding example profiles.

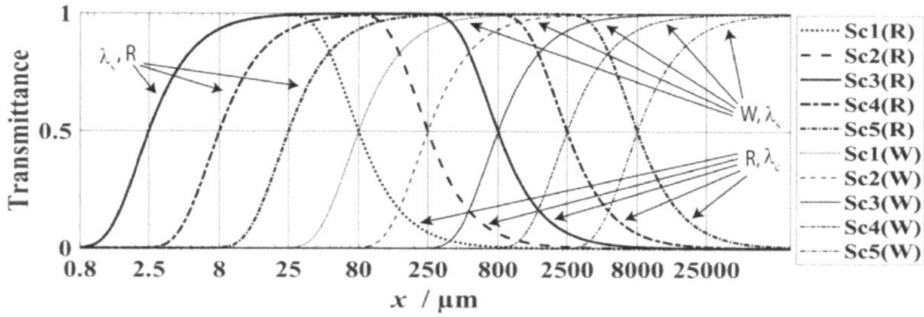

FIGURE 7.43 Surface wavelength transmission characteristics for the filter settings defined as Sc1 to Sc5 in ISO 21920-3:2021.

TABLE 7.4
Default Filter Settings for Some Surface Texture Parameter Based on the Upper Tolerance Limit

Setting Class	Sc1	Sc2	Sc3	Sc4	Sc5
S filter, λ_s/µm	2.5	2.5	2.5	8	25
L filter, λ_c/mm	0.08	0.25	0.8	2.5	8
Maximal nominal tip radius r_{tip}/µm	2	2	2	5	10
Ra/µm	≤0.02	0.02<*Ra*≤0.10	0.10<*Ra*≤2	2<*Ra*≤10	*Ra*>10
Rq/µm	≤0.032	0.032<*Rq*≤0.16	0.16<*Rq*≤3.2	3.2<*Rq*≤16	*Rq*>16
Rt/µm	≤0.26	0.26<*Rt*≤1.3	1.3<*Rt*≤26	26<*Rt*≤130	*Rt*>130
Rz/µm	≤0.16	0.16<*Rz*≤0.8	0.8<*Rz*≤16	16<*Rz*≤80	*Rz*>80

both roughness (*R*) and waviness (*W*) parameters. The filter characteristics, that is, the transmittance of surface wavelength components for the five setting parameters is given in Figure 7.43. Further requirements are a maximum probe diameter of 2 µm for Sc1 to Sc3, of 5 µm for Sc4 and 10 µm for Sc5. The maximum sampling distance is one fourth of the probe diameter and the default evaluation length l_e is $5{\cdot}\lambda_c$; see also Table 7.4. The most common filter setting is λ_s = 2.5 µm, λ_c = 800 µm, where the evaluation length l_e is set as $l_e = 5{\cdot}\lambda_c$ = 4 mm. In general, there is an areal equivalent for profile filters; for example, the areal Gaussian filter can be applied by applying the Gaussian profile filter to all profiles in the x- and *y*-direction. Chapter 9 gives some more background to filtering techniques.

7.9.2 Surface Texture Parameters

A plethora of surface texture parameters is defined in ISO 21920-2:2021 for profile parameters and in ISO 25178-2:2021 for areal parameters. In general, the parameter types can be further classified as amplitude parameters (depending on the profile height), spatial parameters (depending on the length of profile elements), hybrid parameters (depending on both the (y,x) and z coordinates) and parameters that are dependent on the direction for areal parameters. We just give a few of the most commonly used parameters below.

- *Ra* (*Wa, Pa*) for profile, *Sa* for area; arithmetic mean height

$$R_a = \frac{1}{l_e}\int_0^{l_e} |z(x)|\,\mathrm{d}x; \; S_a = \frac{1}{A}\iint_A |z(x,y)|\,\mathrm{d}x\mathrm{d}y. \tag{7.28}$$

The arithmetic mean height is the arithmetic mean of the absolute of the ordinate values within an evaluation length l_e (profile) or area (A). Numerically, this means that for all z-values, the absolute value is taken, of which the mean is taken; $Ra = \Sigma|z_i|/n$. This parameter is by far the most used in industry and is often referred to as just 'the roughness'. Note that the added 'a' in *Ra* is written on the line in running text, but as a subscript in an equation, this holds for all surface texture parameters.

- *Rq* (*Wq, Pq*) for profile, *Sq* for area; root mean square height

$$R_q = \sqrt{\frac{1}{l_e}\int_0^{l_e} z^2(x)\,\mathrm{d}x}\,; \; S_q = \sqrt{\frac{1}{A}\iint_A z^2(\mathrm{x,y,})\mathrm{d}x\mathrm{d}y}. \tag{7.29}$$

The root mean square height is the root mean square value of the ordinate values within an evaluation length l_e (profile) or area (A). This parameter is equivalent to the standard deviation. The parameter value is mostly somewhat higher (typically 15%) than the arithmetic mean height value; by definition, it cannot be lower.

- *Rt* (*Wt,Pt*) for profile; *Sz* for area: total height

The total height parameter is the sum of the largest height (peak, top) and the largest depth (pit, bottom). It is calculated as:

$$R_t = \max(z(x)) - \min(z(x)); \; S_z = \max(z(x,y)) - \min(z(x,y)). \tag{7.30}$$

- *Rz* for profile: maximum height

The maximum height profile parameter is the mean value of the *Rt* parameter evaluation over each section length, which is commonly the cut-off length λ_c. This parameter is defined as such for historical reasons: in an older standard ISO 4287:1997, all parameter values were defined as the mean parameter value calculated over each cut-off length separately. For continuity, the definition of this parameter has not changed. This is also the case for the profile parameters *Rp* (the peak, maximum value of the ordinates), and for *Rv* (the valley, also called pit, the absolute value of the minimum of the ordinates). In ISO 21920-3:2022, default filter settings for these parameters are defined, based on their specifications. For a few parameters specified as a maximum value, these are summarized in Table 7.4. Default means: unless otherwise specified. When checking for tolerances as indicated on technical drawings it is logic to follow these setting. It the case of investigating process influences or functional surface properties, the settings must be considered and chosen with care, taking into account that the values given in Table 7.4 are by no means 'the law'. The settings must be reported in the non-default cases as parameter values depend on them.

Optically smooth surfaces are commonly within the Sc1 tolerance, well finished surfaces may be within Sc2, normal workpieces from the workshop will be within class Sc3 or Sc4. Sc5 is for very rough surfaces, for example, those from additive manufacturing.

7.9.3 Instruments for Measuring Surface Texture

Instruments for measuring areal surface texture are standardized in a series of ISO-standards that describe the common metrological characteristics (ISO 25178-600:2019) as well as specific characteristics depending on the measuring principle (ISO 25178-601 to 609). These standards in the series 'Geometrical Product Specifications (GPS) – surface texture –areal' cover:

ISO 25178-600:2019 Metrological characteristics for areal topography measuring methods
ISO 25178-601:2025 Contact (stylus) instruments.
ISO 25178-602:2025 Confocal chromatic probe instruments.
ISO 25178-603:2025 Phase shifting interferometry (PSI) instruments.
ISO 25178-604:2025 Coherence scanning interferometry (CSI) instruments.
ISO 25178-605:2025 Point autofocus probe instruments.
ISO 25178-606:2015 Focus variation instruments.
ISO 25178-607:2019 Confocal microscopy instruments.

Extensive textbooks have been published about these instrument types (Leach, 2011); in this section, the main technologies and characteristics will be given. Other techniques such as atomic force microscopy (AFM) and scanning electron microscope (SEM) are fields on their own and will not be treated here.

7.9.3.1 Mechanical Contact Stylus Instrument

The mechanical contact stylus instrument is by far the most used instrument for measuring surface texture along a profile. The basic design and characteristics are given in ISO 25178-601:2025. The probe is commonly of the LVDT type (see Section 5.3) and has the following characteristics (see also Section 7.9.1):

- Nominal tip radius r_{tip}= 2 µm, 5 µm or 10 µm.
- Cone angle: $\gamma = 60°$ or $90°$.
- Nominal static measuring force 0.75 mN.

The combination of tip radius and measuring force is taken such that in steel, a negligible scratch is made, however, in aluminum and other softer materials, scratches may appear. Commonly, the drive unit contains a linear scale that records the x-position and provides the reference for a straight movement in the x-direction. The x-range may vary from some 20 mm for hand-held instruments to 200 mm for laboratory-type instruments. The measurement speed is limited by the probe dynamics: at a too high speed 'stylus flight' may appear when the probe loses contact with the workpiece. Therefore, a larger force may be applied for the 5 µm and 10 µm radius styli. The z-range may be limited to the sub-millimeter region to allow for a sufficient amplification of the LVDT probe. This may require alignment of the workpiece surface with the x-movement of the drive unit. For this reason, the drive unit can be rotated around the y-axis, or the fixture has some tilt-arrangement. As an alternative for the LVDT, probes with an internal linear scale or even an internal laser interferometer system may be applied, which can enlarge the measurement range. Such instruments may combine surface texture measurements with form measurements. Because of the relatively slow measurement speed, about 1 mm/s, these instruments are rarely used for areal measurements. If a y-axis table is present, areal measurements can be taken by moving the y-table in small increments after each profile measurement. In that case, the areal measurement consists of a number of profile measurements that are connected by the common measurement frame. A y-axis table can also be useful for assessing different areas of a workpiece without the need of touching, displacing and re-aligning the workpiece. A schematic of a laboratory-type instrument for mechanical contact surface texture measurements is depicted in Figure 7.44. Widespread in industry are the handheld roughness testers. These can be placed directly on the workpiece to be measured and may directly output the parameter on a display. These may be equipped with a so-called skid, that is a guidance with a large effective radius that guides the probe on the surface. This makes the instrument robust against external vibrations, at the expense of some additional mechanical filtering effect of the skid itself.

7.9.3.2 Lateral Scanning Instruments with an Optical Probe

The mechanical lateral scanning instrument as discussed in Section 7.9.3.1 has disadvantages in that the surface must be physically touched and that the measurement speed is limited by the probe-surface dynamics. Using optical probes instead of the mechanical stylus has the advantage that the object does not need to be touched and that the signal can be obtained opto-electronically at a far higher speed. Optical probes that can be used for this purpose are the confocal chromatic probe, as described in Section 5.7 and the confocal point sensor described in Section 5.6. Their use in areal or profile measurements is described in ISO 25178-602:2025 and ISO 25178-605:2025 respectively. The x-y movement can also be achieved by using a stacked x-and y- positioning unit and a probe head

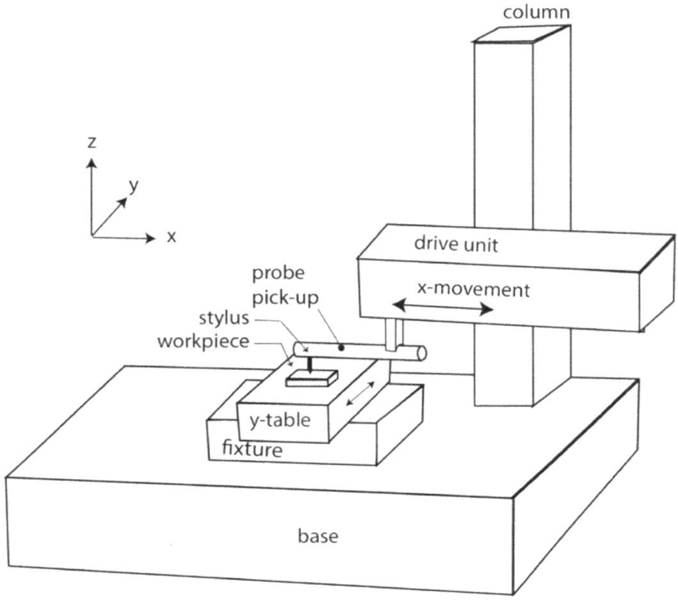

FIGURE 7.44 Common design of a contact stylus instrument for profile and areal measurement.

that enables z-scanning for vertical scanning instruments or a z-movement if a confocal point sensor is used in feedback mode. This is illustrated in Figure 7.45.

7.9.3.3 Vertical Scanning Instruments with an Imaging Optical System

A popular type of areal topography measuring instrument is the instrument shown in Figure 7.45. The optical probe is replaced by a microscope system and a vertical scan is made where the topography of the region of the workpiece that is within the areal field of view of the objective is calculated from a stack of images collected during the vertical scan of the microscope system. The performance of these instruments highly depends on the used microscope objective, which can be either an interference objective, as described in Section 4.7, or a microscope objective without additional interference features. A schematic overview of the features that describe a microscope objective, omitting the internal assembly of multiple optical elements, is given in Figure 7.46. Objectives are commonly denoted by their magnification, for example 10×; this relates to their focal distance. For metrology applications, the following characteristics are most relevant. The lateral resolution is mostly used for imaging, but for topography measurements, a related but more relevant concept is the lateral period limit D_{LIM}, which is the lateral wavelengths below which less than 50% of the amplitude of a measured sine wave is transmitted. In general, this depends on the amplitude and the form of the measured object, the optical system and the coherence of the light, but for surface amplitudes << λ it can be approximated by (de Groot, 2015)

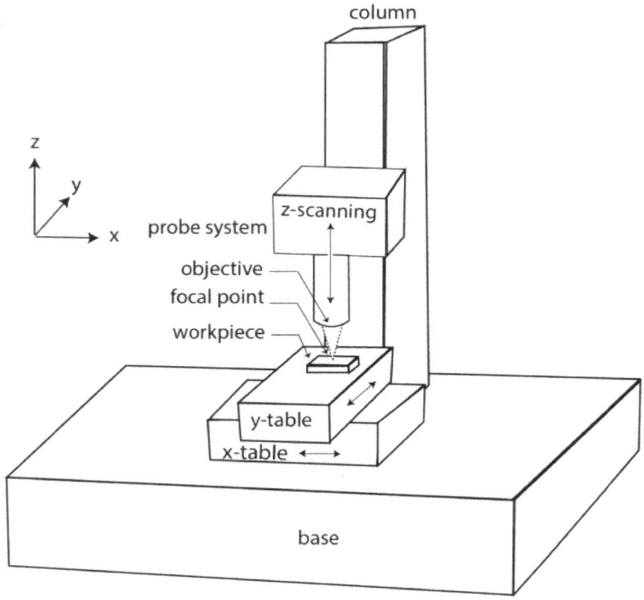

FIGURE 7.45 Common design for optical instruments for measurement of surface texture.

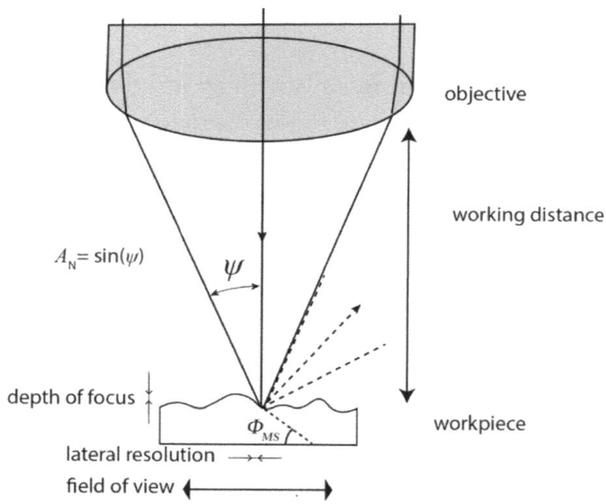

FIGURE 7.46 Schematic of a microscope objective with the most relevant characteristics for metrology.

$$D_{LIM} = \frac{1.2 \cdot \lambda_0}{A_N}, \tag{7.31}$$

where λ_0 is the measurement optical wavelength, that is the effective (central) wavelength of the light source and A_N is the numerical aperture (*NA*) of the objective. Related terms are the Rayleigh criterion, Abbe resolution limit, Sparrow criterion, topographic spatial resolution and the width limit for full height transmission W_1, which is the width of the narrowest rectangular groove whose step height can be properly measured, see ISO 25178-600:2019. If axial discrimination is important, for example, in confocal microscopy or in focus variation microscopy, the depth of focus d_f is relevant for which the classical expression is:

$$d_f = \frac{\lambda_0}{A_N^2}. \tag{7.32}$$

However for confocal systems, the axial distance from the focal point for which the signal intensity drops by 50% is given by (ISO 25178-607:2019):

$$d_{FWHM} = \frac{0.88 \cdot \lambda_0}{1 - \sqrt{1 - A_N^2}}. \tag{7.33}$$

The maximum measurable local slope Φ_{MS} depends on the measurement principle, the specimen and the illumination. For a perfectly reflecting flat surface and illumination through the objective, it is given by:

$$\Phi_{MS} = \sin^{-1} A_N. \tag{7.34}$$

Table 7.5 gives typical specification of objectives used for surface topography measuring instruments. An important subject in optical surface topography measurements is the appearance of outliers and missing points in measurements. Where easily some million surface data points are output, some pixel signals have insufficient intensity to give a sensible output, and the software marks it as a 'missing

TABLE 7.5
Typical Specifications of Objective Lenses Used in Surface Topography Measuring Systems

Magnification	2.5×	5×	10×	20×	50×
NA	0.08	0.15	0.20	0.40	0.70
Field of view/mm²	6 × 4	3 × 2	1.5 × 1	0.75 × 0.5	0.3 × 0.2
Working distance/mm	8	25	8	4	2

point'. On the other hand, there are diffraction effects or neighboring points with a high intensity that affect the signal and give as a height value far outside the actual surface (and appear as high peaks or deep valleys). The mechanical contact stylus method suffers less from these effects as the stylus cannot physically deviate from the actual surface so easily, although outliers may appear and need to be considered.

The stacked x-y positioning system as depicted in Figure 7.44 is used to position the workpiece. In some cases, the x-y range of the topography is increased by measuring several areas on the object and merging these into a larger range topography; this process is referred to as 'stitching'. In this merging of different areas, the relevant parameters are the overlapping regions on the object and the number of degrees of freedom that are allowed to the topographies to be stitched together: just the z-coordinates and rely on the instruments coordinate frame, or take the measured workpiece as its own reference and stitch the topographies allowing three translations and rotations for an optimal result. The error propagation for stitched measurement areas is complicated and can easily be underestimated; for example a systematic spherical flatness deviation propagates quadratically with the length of a stitched area.

7.9.3.4 Phase Shifting Interferometry (PSI) and Coherence Scanning Interferometry (CSI) Instruments

The principles of phase shifting interferometry and coherence scanning interferometry on a microscopic scale were treated in Section 4.5.5, more specific details are given in ISO 25178-603:2025 and ISO 25178-604:2025. Often the principles are combined in one instrument. In comparison with other microscopic topography measuring methods these instruments have the characteristic that the vertical resolution can be in the sub-nanometer range, independent of the NA of the objective. During vertical scanning, the interference must be detected; this means that an image must be taken at every $\lambda/8 \approx 0.08$ μm, which limits the measurement speed. The reference mirror must be well focused, aligned, flat and free of defects. The maximum measurable slope is limited by equation (7.34), but also by the pixel size of the camera used: if more than one fringe is imaged on a single pixel there is no contrast.

7.9.3.5 Confocal Microscopy Instruments

In confocal microscopy instruments, the intensity of every focal point on the surface is recorded as a function of the scanning z-coordinate, similar to the principle of the confocal point sensor, see Section 5.7. Laterally scanning the probe in x- and y-directions over the surface, without changing the z-coordinate, gives a confocal image of the surface with a high transmitted intensity for the points that are in focus and a lower intensity for the points that are out of focus. Doing this for multiple z-positions gives an image stack where, for every pixel that corresponds to a point on the surface, the maximum intensity can be determined that defines the z-height. In a confocal microscope system, this principle is used, however, measures are taken such that either the lateral scanning is extremely fast, or many points are illuminated in parallel and processed in a way that they do not affect each other. Some techniques are:

- Laser scanning configuration

A laser beam is focused on the surface, as in Section 5.7, however, the x/y scanning takes place using small mirrors that steers the beam over the surface in the x- and y-directions. Combined with a high laser intensity, in a fraction of a second a confocal image can be stored, and the z-scanner can take a new position, and so on.

- Disk scanning configuration

A disk containing a multi-pinhole configuration is scanned (rotated) in a focal plane of the light source and imaged on the object, such that every pinhole acts both as a light source for the illumination and a diaphragm for the light reflected from the surface. The pinhole configuration is such that every point on the surface is illuminated during one revolution of the disk, while the CCD sensor is exposed. This gives a confocal image. The system is scanned through a number of z-positions where the confocal image is recorded.

- Programmable array configuration

In a programmable array configuration, a digital mirror device (DMD) is used to control the illumination of the object and/or the illumination of the detector. The case of a programmable illumination of the object is depicted in Figure 7.47. The DMD can be programmed to display a specific pattern on the workpiece, for example, a pattern of dots, lines or a grid. In the case of illuminating the object with a sinusoidal line pattern, an evaluation method similar to the evaluation of interference fringes can be used. For example, at one z-position, the lines are moved in four steps of one quarter of the line pitch P. For every pixel of the CCD camera, the four intensities $I(x,y)$ can be denoted as $I_{A..D}(x,y)$. The effective modulation intensity I_{MOD} can be calculated for every z-height of the scanning system as in equation (4.28):

$$I_{MOD} = \sqrt{\left(I_D(x,y) - I_B(x,y)\right)^2 + \left(I_A(x,y) - I_C(x,y)\right)^2}. \tag{7.35}$$

For every pixel this modulation intensity will be maximal if the corresponding point on the surface is in focus. If the movement of the grid is carried out while the z-scan is made, i.e. the z-scanning does not stop at a specific number of z-height, but moves continuously while the CCD-sensor records the images, the pixel intensity as a function of the z-scanning height will resemble that of a white-light interference pattern, as shown in Figure 4.11b. If the z-scanning is moving

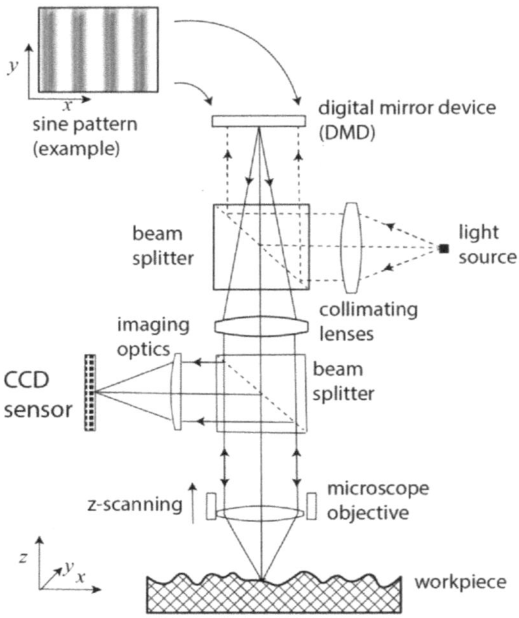

FIGURE 7.47 Scheme of a confocal microscopy system based on a digital mirror device, as it can be attached to the system depicted in Figure 7.45.

with steps Δ while the pattern is moving in steps of $P/4$, equation (7.35) can be modified as

$$I_{MOD}(x,y,z) = \sqrt{\left(I(x,y,z+\Delta)-I(x,y,z-\Delta)\right)^2+\left(I(x,y,z)-\left(\frac{I(x,y,z+2\Delta)+I(x,y,z-2\Delta)}{2}\right)\right)^2}. \tag{7.36}$$

Also, this function will have a maximum when the corresponding point on the workpiece is in focus. In this way, from the stack of images $I(x,y,z)$, a stack of 'confocal' images $I_{MOD}(x,y,z)$ can be obtained, where for every (x,y) pixel, the maximum in the z-direction corresponds to the z-height of the corresponding illuminated point on the workpiece. A stable way to find the z-position of maximum intensity (and with that the z-coordinate on the surface) is to determine the centroid in the region from z_{min} to z_{max} where there is a significant signal I_{MOD}:

$$z_O(x,y) = \frac{\sum_{z_{min}}^{z_{max}} z \cdot I_{MOD}^2(x,y,z)}{\sum_{z_{min}}^{z_{max}} I_{MOD}^2(x,y,z)}. \tag{7.37}$$

Note that this is just an example of pattern projection and evaluation. In practice, manufacturers will use proprietary methods and algorithms. In confocal systems, it is possible to have a trade-off between vertical scanning speed and accuracy. The vertical resolution will largely depend on the objective (see equation (7.33)) and may reach nanometer-levels with high-NA objectives. The flatness reference is the flatness of the focal plane that may be less well defined than the physical reference in an interference objective.

7.9.3.6 Focus Variation Instruments

In focus variation instruments, the surface height is obtained from the local contrast in an image as a function of the z-height of the object relative to the measuring system. An advantage is that a normal microscope system can be used without additional adaptation of objectives or the illumination system. Another advantage is that the illumination does not need to be restricted to 'through-the-lens' illumination of the other systems, so the limitation of the maximum measurable local slope Φ_{MS} does not hold if the workpiece can be illuminated by an external source. The local contrast of an image is supposed to be maximal if an area is in focus: then the corresponding features are sharp. For a measure of the contrast/sharpness, there are several possibilities, here we mention one example: for the pixel whose height is to be determined, a measure proportional to the variance of the intensities in the region around that pixel is determined. For pixel (x_i,y_j) recorded at height z, the local contrast can be calculated as:

$$C(x_i,y_j,z) = \sum_{i-k}^{i+k}\sum_{j-k}^{j+k}\left(I(x_i,y_j,z) - \frac{\sum_{i-k}^{i+k}\sum_{j-k}^{j+k} I(x_i,y_j,z)}{(k+1)^2}\right)^2. \tag{7.38}$$

From a stack of images, for every (x,y) pixel, the maximum of $C(x,y,z)$ in the z-direction corresponding to the z-height of the corresponding illuminated point on the workpiece can be calculated, for example, similarly to equation (7.37):

$$z_O(x,y) = \frac{\sum_{z_{min}}^{z_{max}} z \cdot C(x,y,z)}{\sum_{z_{min}}^{z_{max}} C(x,y,z)}. \tag{7.39}$$

Depending on the features on the surface, the region can be enlarged for low-contrast regions on the surface, at the expense of the lateral period limit. As with confocal systems, it is possible to have a trade-off between vertical scanning speed and accuracy. The vertical resolution will largely depend on the objective (see equation (7.33)) and the contrast features on the workpiece. The flatness reference is the flatness of the focal plane that may be less well defined than the physical reference in an interference objective. For perfectly flat and homogeneous surfaces, it is difficult to measure any contrast at all, so for such surfaces this method fails. In some instruments, this is countered by projecting a pattern on the surface. This makes the system resemble a confocal system, where the only difference is in the axial or lateral evaluation of the intensities. However, a user usually is not aware of what technique is used in the manufacturers' software. Even a combination can be used, for example, a confocal system uses the confocal evaluation as expressed in equations (7.36) and (7.37), but for the 'missing points' equations (7.39) and (7.38) are applied.

7.9.4 Common Metrological Characteristics for Surface Topography Measuring Methods

The instruments treated in Section 7.9.3 have metrological characteristics in common that are defined in ISO 25178-600:2019 and are summarized here in Table 7.6. ISO 25178-700:2022 gives methods to asses these characteristics. The metrological characteristics may be used to specify or compare instruments of similar or different measurement principles. Also they can be used as a basis for an uncertainty evaluation (Leach et al, 2020). They are shortly summarized as following:

- Amplification coefficient $\alpha_x, \alpha_y, \alpha_z$
  These are the calibration coefficients in the axial (z) and two lateral (x and y) directions. Commonly the α_z is most relevant, by default this can be determined by a calibrated step height, see Section 6.4.1. For the calibration in the lateral directions, calibrated line scales, see Section 6.1.4, can be used.

TABLE 7.6
Metrological Characteristics for Surface Texture Measurement Methods, according to ISO 25178-600

Metrological Characteristic	Symbol	Main Potential Error Along
Amplification coefficient	$\alpha_x, \alpha_y, \alpha_z$	x,y,z
Linearity deviation	lx, ly, lz	x,y,z
Flatness deviation	z_{FLT}	z
Measurement noise	N_M	z
Topographic spatial resolution	W_R	z
x–y mapping deviations	$\Delta_x(x,y), \Delta_y(x,y)$	x,y
Topographic fidelity	T_{FI}	z

- Linearity deviation l_x, l_y, l_z
  This concerns the linearity in the three directions, see Section 1.6.3. A default calibration method is the measurement of several step height standards for the z- direction and measure a calibrated line standard at several points for the x- and y directions. Measuring a tilted optical flat may be an alternative to determine l_z.
- Flatness deviation z_{FLT}
  The flatness deviation is the flatness deviation of the measured topography of an ideally flat object. It is expressed as the *Sz* parameter (see Section 7.9.2), that is measured by the instrument when a perfectly flat and levelled object is measured. The default calibration method is the measurement of a levelled optical flat (see Section 6.3.2), where the measurement is levelled to the least-squares plane (see Section 7.6.1). The influence of the optical flat can be reduced by measuring this flat at several locations and average the *z*-values for every *x,y* coordinate in the image.
- Measurement noise N_M
  Measurement noise is the noise added to the output signal occurring during the normal use of the instrument. It is expressed as the root mean square (rms) standard deviation in all $z(x,y)$ coordinates. The default way to determine this quantity is to take two measurements $z_1(x,y)$ and $z_2(x,y)$ of the same surface levelled to the least-squares plane (see Section 7.6.1), and take the *Sq* parameter (see Section 7.9.2) of the difference topography divided by $\sqrt{2}$:

$$N_M = \frac{1}{\sqrt{2}} S_q\left(z_1(x,y) - z_2(x,y)\right) \tag{7.40}$$

- Topographic spatial resolution W_R
  The topographic spatial resolution describes the ability of a surface topography measuring instrument to distinguish closely spaced surface features. This may be the stylus tip radius for stylus-based instruments or one of the optical resolution measures as described in Section 7.9.3.3 for microscope-objective based instruments. Material measures for calibration of the topographic spatial resolution include structures with variable width, grating constants or spatial wavelengths, including parallel lines, chirped material measures or star-shaped groove types.
- x-y mapping deviations $\Delta_x(x,y), \Delta_y(x,y)$
  The x-y mapping deviations are given as a gridded image of *x*- and *y*-deviations of actual coordinate positions on a surface from their nominal positions. The mapping deviations can be used to calculate the *x*- and *y*- linearity deviations, the *x-y* axis perpendicularity and field deformations. The default method for the determination of the amplification factors and mapping deviations of the *x*- and *y*-axes is with material measures with calibrated distances in the *x-y* plane.

- Topographic fidelity T_{FI}
  The topographic fidelity is the closeness of agreement between a measured surface profile or measured topography and one whose uncertainties are insignificant by comparison. The topography fidelity accounts for all remaining effects where the other metrological characteristics have been evaluated and accounted for. A structured material measure with known and/or calibrated shapes having known uncertainty may be used for the determination of the topography fidelity.

EXERCISES

1. Polygon
   A classical calibration method for a polygon is using two autocollimators and determining the difference in reading of the autocollimators for every two adjacent polygon faces. This is depicted in Figure 7.48 for a six-sided polygon. Two autocollimators are at a fixed position and directed to two adjacent faces, with an angle θ. The polygon can be rotated around its center using a rotary table. Every polygon face is rotated such that Autocollimator 1 has a zero reading, then the reading is taken from Autocollimator 2. These six measurements are denoted by $M_{1..6}$. For example, $\alpha_1 = M_1 + \theta$.

   a. Give an expression for θ in terms of $M_{1..6}$, using the fact that $\sum_{i=1}^{6} \alpha_i = 360°$
   b. Calculate the angles α_i for M_i = 0.2"; 0.4"; 0.0"; 0.2"; 0.4"; 0.0.
   c. Give a general expression for the angles α_i for an n-sided polygon and n measurements $M_{1..n}$.
   d. Give the uncertainty in the angle α_1 based on an uncorrelated standard uncertainty of 0.2" in every autocollimator reading. Use equation (3.28)

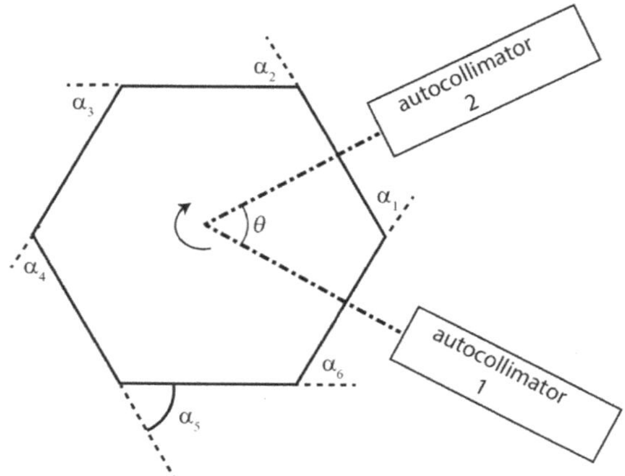

FIGURE 7.48 Calibration set-up for a six-sided polygon.

and check it with equation (3.29). Note that the expression derived in c. should be (re) written in separate terms for every measurement M_i

e. Calculate $\alpha_1+\alpha_2$ and the uncertainty in $\alpha_1+\alpha_2$

f. Give the general expression for the sum of m angles for an n-sided polygon and n measurements

g. Calculate the uncertainty in the sum of m angles for an n-sided polygon and n measurements $M_{1..n}$. Check the result for $m=n$

2. Straightness

A machine guidance is measured over a length of 1200 mm using an electronic level with a pitch of 100 mm.

The measured angles are given in Table 7.7.

a. Sketch the measured form in a graph

b. Sketch the measured form in graphs relative to three references: zero start and end points, least-squares line and minimum zone lines

c. Calculate the total straightness deviation, i.e. the peak-valley value in micrometers for the three cases

d. Take the straight line through the first and last point as reference and consider the point with the highest deviation from this line. Calculate the uncertainty in the height of this point, assuming an uncorrelated standard uncertainty of 1 µrad in every measurement. Note some similarity with exercise 1.

e. Suppose this measurement is taken with the same instrument, but the pitch is reduced to 50 mm. How would this affect the uncertainty in the profile?

TABLE 7.7
Angle Measurements with a 100 mm Base to Determine Straightness

Position Center of Level/mm	Measured Angle/µrad
50	1
150	2
250	1
350	1
450	2
550	2
650	0
750	0
850	0
950	1
1050	1
1150	2

3. Objective specification
   Add to Table 7.4 the lateral period limit, the focal depth for a confocal system, and the maximum measurable local slope Φ_{MS} for the mentioned objectives. Look on the internet for objective specifications (choose from Bruker, Mitutoyo, Nikon, Olympus, Sensofar, Zygo or any other) and check with the equations of Section 7.9.3.3 what criteria/equations are used.
4. Roughness of a turned surface
   In Figure 7.49, a roughness profile is sketched as obtained by fine turning. Determine the texture parameter *Pt* as a function of the pitch *P* and the tool radius *R* in general, and specifically for $R = 0.5$ mm and $P = 10$ μm. What filter settings should be used to obtain useful values of *Ra* and *Rt*?
5. Roughness parameters of a sinusoidal specimen
   Figure 7.50 depicts a roughness profile given by:

$$z(x) = A\sin\frac{2\pi x}{\lambda}.$$

   Determine *Pa*, *Pq* and *Pz* of this profile, assuming that the evaluation length is a multiple of the wavelength. What are the requirements for λ_s and λ_c to measure this profile properly as an R-profile?
6. Flatness measurement
   A 300 mm square surface plate is measured at nine points over the central 200 mm by 200 mm using electronic levels with a pitch of 100 mm. The readings are in μrad, also expressed as μm/m. Figure 7.51 depicts the measurement as seen from above with the measurement values indicated along the lines. For example, the (positive) angle between points (-100,-100) and (-100,0) is 4 μrad.

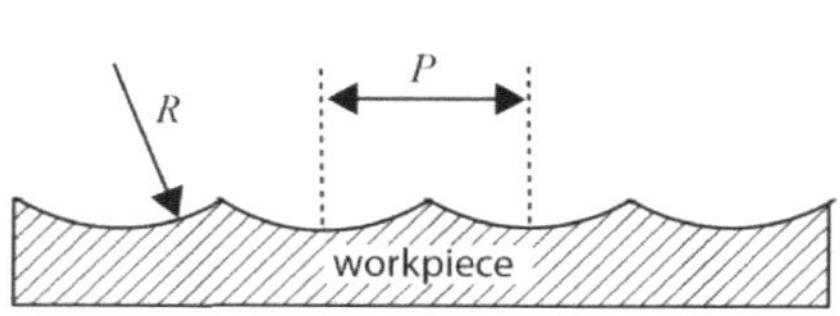

FIGURE 7.49 Workpiece surface after fine turning with tool radius *R* and pitch *P*.

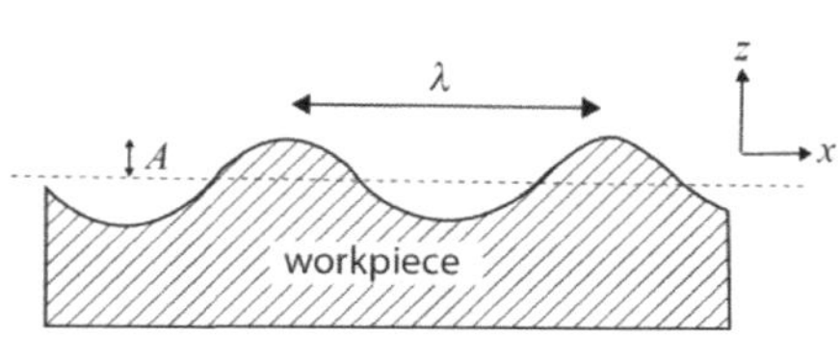

FIGURE 7.50 Roughness profile with a sinusoidal shape.

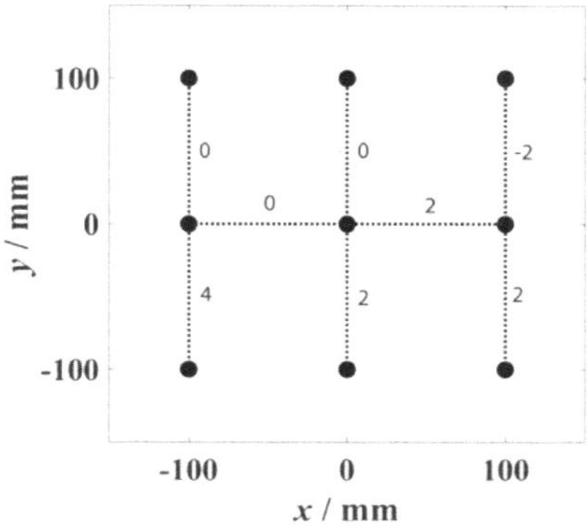

FIGURE 7.51 Lay-out of inclination measurements in μrad and measured points on a 300 mm by 300 mm surface plate.

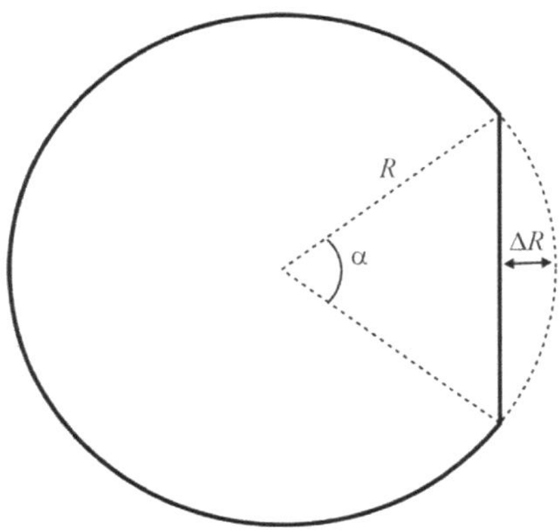

FIGURE 7.52 Design of a cylindrical standard for the calibration of roundness measuring instruments, the so-called flick-standard.

a. Calculate the z-height of the points, defining the point (0,0) as z=0
b. Transfer the points to the least-squares plane
c. Consider any difference with the minimum-zone plane or the plane defined with equal diagonal end points
d. Give the total flatness deviation and check whether this is within class 00 of DIN 876.

7. Roundness measurement

A flick standard is a roundness standard that can be used to check or calibrate the magnification of a roundness profile. It is produced by grinding a flat part onto a cylinder to create the cross-section as displayed in Figure 7.52. A perfectly round cylinder of 10 mm diameter is ground so that the roundness deviation from the diameter of the original cylinder is 100 μm, this means that in Figure 7.52 R = 5 mm and $\varDelta R$ = 100 μm. The roundness

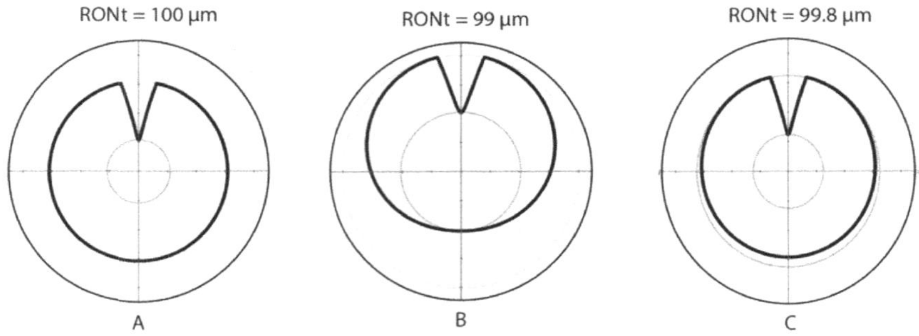

FIGURE 7.53 Roundness diagrams of a flick-standard for different reference circles.

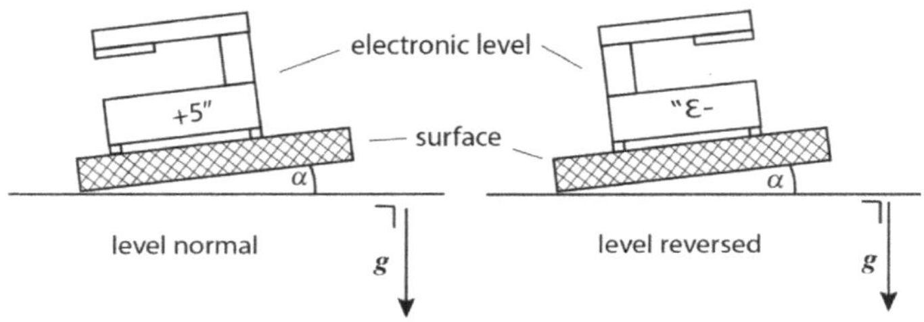

FIGURE 7.54 Electronic level measuring a surface in normal and reversed orientation – meaning it has been rotated over 180° around the vertical axis. The surface has a tilt α. The level measures + 5" in the normal position and –3" in the reversed position.

diagram with respect to common reference centers can take three forms as in diagrams A, B and C in Figure 7.53.

a. Calculate the angle α in Figure 7.52 for this case
b. Determine for diagram A in Figure 7.53 the values of M, L and R in equation (7.23)
c. For the four common reference circles: MCCI, MICI, MZCI and LSCI, indicate to which diagram (A, B or C) these belong; there is one diagram that is found with two reference circle definitions.

8. Level reversal

Commonly the zero setting of an electronic level is not strictly defined, therefore the absolute angle relative to the gravity vector can be adjusted. To determine if a surface is properly levelled and to adjust such a level, the level reversal method is used, that is considered the simplest reversal method. The principle is depicted in Figure 7.54. The surface is nominally levelled, but still has a (small) angle deviation α from true horizontal that has an angle of 90° with the gravity vector **g**. This surface is measured by an electronic level that has a bias of β, this means that it will indicate an angle β on a truly horizontal surface.

a. Give the measurement values of the level, M_1 and M_2, as a combination of angles α and β in the normal and reversed position as indicated in Figure 7.54.
b. Give α and β for $M_1 = 5''$ and $M_2 = -3''$, and the standard uncertainty in α and β if the standard uncertainty in M_1 and M_2 is 1".

REFERENCES

DIN 877, 1986, Inclination Measuring Instruments (Levels). Berlin, German Institute for Standardization.

Donaldson, R.R., 1972, A Simple Method for Separating Spindle Error from Test Ball Roundness Error. *Ann. CIRP* **21(1):**125–126.

Groot, P. de, 2015, Principles of Interference Microscopy for the Measurement of Surface Topography. *Adv. Opt. Phot.* **7(1):**1–65.

ISO 3611, 2023, *Geometrical Product Specifications (GPS) – Dimensional Measuring Equipment: Design and Metrological Characteristics of Micrometers for External Measurements.* Geneva, International Organization for Standardization.

ISO 3650, 1998, *Geometrical Product Specifications (GPS) – Length Standards – Gage Blocks Parameters.* Geneva, International Organization for Standardization.

ISO 4287, 1996, *Geometrical Product Specifications (GPS) – Surface Texture: Profile – Part 2: Terms, Definitions and Surface Texture Parameters.* Geneva, International Organization for Standardization. Replaced by ISO 21920-2:2021.

ISO 12180-1, 2011, *Geometrical Product Specifications (GPS) – Cylindricity – Part 1: Vocabulary and Parameters of Cylindrical Form.* Geneva, International Organization for Standardization.

ISO 12180-2, 2011, *Geometrical Product Specifications (GPS) – Cylindricity – Part 2: Specification Operators.* Geneva, International Organization for Standardization.

ISO 12181-1, 2011, *Geometrical Product Specifications (GPS) – Roundness – Part 1: Vocabulary and Parameters of Roundness.* Geneva, International Organization for Standardization.

ISO 12181-2, 2011, *Geometrical Product Specifications (GPS) – Roundness – Part 2: Specification Operators.* Geneva, International Organization for Standardization.

ISO 12578-600, 2019, *Geometrical Product Specifications (GPS) – Surface Texture: Areal – Part 600: Metrological Characteristics for Areal Topography Measuring Methods.* Geneva, International Organization for Standardization.

ISO 12578-601, 2025, *Geometrical Product Specifications (GPS) – Surface Texture: Areal – Part 601: Design and Characteristics of Contact (Stylus) Instruments.* Geneva, International Organization for Standardization.

ISO 12578-602, 2025, Geometrical Product Specifications (GPS) – Surface Texture: Areal – Part 602: Design and Characteristics of Non-Contact (Confocal Chromatic Probe) Instruments. Geneva, International Organization for Standardization.

ISO 12578-603, 2025, *Geometrical Product Specifications (GPS) – Surface Texture: Areal – Part 603: Design and Characteristics of Non-Contact (Phase Shifting Interferometry) Instruments.* Geneva, International Organization for Standardization.

ISO 12578-604, 2025, *Geometrical Product Specifications (GPS) – Surface Texture: Areal – Part 604: Design and Characteristics of Non-Contact (Coherence Scanning Interferometry) Instruments.* Geneva, International Organization for Standardization.

ISO 12578-605, 2025, *Geometrical Product Specifications (GPS) – Surface Texture: Areal – Part 605: Design and Characteristics of Non-Contact (Point Autofocus Probe) Instruments.* Geneva, International Organization for Standardization.

ISO 12578-606, 2015, *Geometrical Product Specifications (GPS) – Surface Texture: Areal – Part 606: Design and Characteristics of Non-Contact (Focus Variation) Instruments.* Geneva, International Organization for Standardization.

ISO 12578-607, 2019, *Geometrical Product Specifications (GPS) – Surface Texture: Areal – Part 607: Design and Characteristics of Non-Contact (Confocal Microscopy) Instruments.* Geneva, International Organization for Standardization.

ISO 12578-700, 2022, *Geometrical Product Specifications (GPS) – Surface Texture: Areal – Part 600: Calibration, Adjustment and Verification of Areal Topography Measuring Instruments.* Geneva, International Organization for Standardization.

ISO 12780-1, 2011, *Geometrical Product Specifications (GPS) – Straightness – Part 1: Vocabulary and Parameters of Straightness.* Geneva, International Organization for Standardization.

ISO 12780-2, 2011, *Geometrical Product Specifications (GPS) – Straightness – Part 2: Specification Operators.* Geneva, International Organization for Standardization.

ISO 12781-1, 2011, *Geometrical Product Specifications (GPS) – Flatness – Part 1: Vocabulary and Parameters of Flatness.* Geneva, International Organization for Standardization.

ISO 12781-2, 2011, *Geometrical Product Specifications (GPS) – Flatness – Part 2: Specification Operators.* Geneva, International Organization for Standardization.

ISO 13225, 2012, *Geometrical Product Specifications (GPS) – Dimensional Measuring Equipment, Height Gauges – Design and Metrological Characteristics of Callipers.* Geneva, International Organization for Standardization.

ISO 13385-1, 2019, *Geometrical Product Specifications (GPS) – Dimensional Measuring Equipment Part 1: Design and Metrological Characteristics of Callipers.* Geneva, International Organization for Standardization.

ISO 16610-21, 2011, *Geometrical Product Specifications (GPS)-Filtration-Part 21: Linear Profile Filters: Gaussian Filters* I. Geneva, International Organization for Standardization.

ISO 21920-1, 2021, *Geometrical Product Specifications (GPS) – Surface Texture: Profile – Part 1: Indication of Surface Texture.* Geneva, International Organization for Standardization.

ISO 21920-2, 2021, *Geometrical Product Specifications (GPS) – Surface Texture: Profile – Part 2: Terms, Definitions and Surface Texture Parameters.* Geneva, International Organization for Standardization.

ISO 21920-3, 2021, *Geometrical Product Specifications (GPS) – Surface Texture: Profile – Part 3: Specification Operators.* Geneva, International Organization for Standardization.

ISO 25178-2, 2021, *Geometrical Product Specifications (GPS) – Surface Texture: Areal – Part 2: Terms, Definitions and Surface Texture Parameters.* Geneva, International Organization for Standardization.

Leach, R., Haitjema, H., Su, R. and Thompson, A., 2020, Metrological Characteristics for the Calibration of Surface Topography Measuring Instruments: A Review. *Measurem. Sci. Technol.* **32(3):**032001.

Leach, R. (ed)., 2011, *Optical Measurement of Surface Topography.* Berlin, Springer.

Meijer, J. and Heuvelman, C.J., 1990, Accuracy of Surface Plate Measurements - General Purpose Software for Flatness Measurement. *Ann. CIRP.* **39(1):**545–548.

Moody, J.C., 1955, How to Calibrate Surface Plates in the Plant, The Tool Engineer. October 1955.

8 Coordinate Measurements

8.1 PRINCIPLE AND AIM OF COORDINATE MEASUREMENTS

The aim and the principle of coordinate measurements are schematically shown in Figure 8.1. On the one hand, there is the design of the workpiece, the division of the workpiece into elements to which geometrical tolerances are assigned. On the other hand, there is the produced workpiece, where from each element, a number of coordinates is measured, from which statements can be made about the actual location, shape and size of the measured elements. In the end, these features can be compared to the nominal values and tolerances on a technical drawing. Often it is assumed that the contribution of the errors of the co-ordinate measuring machine (CMM) are negligible compared to the tolerances, but this is not necessarily the case. Also, the deviations of the ideal form of the geometrical elements contribute to the measurement uncertainty which may further complicate a thorough uncertainty evaluation.

8.2 2D MEASURING MACHINES

As the name indicates, this group of measuring machines have basically two measuring axes which are nominally orthogonal. The two measurement axes determine the measurement plane within which objects can be measured. For the observation of the object boundaries, use is usually made of optical or opto-electronic means. In principle, there are two systems, namely the measuring microscopes and the measuring projectors. The first uses a microscope to determine the object boundaries, while the second projects an image of the object onto a screen, which is then measured with special tools, or directly compared to a technical drawing transparency.

DOI: 10.1201/9780367821746-8

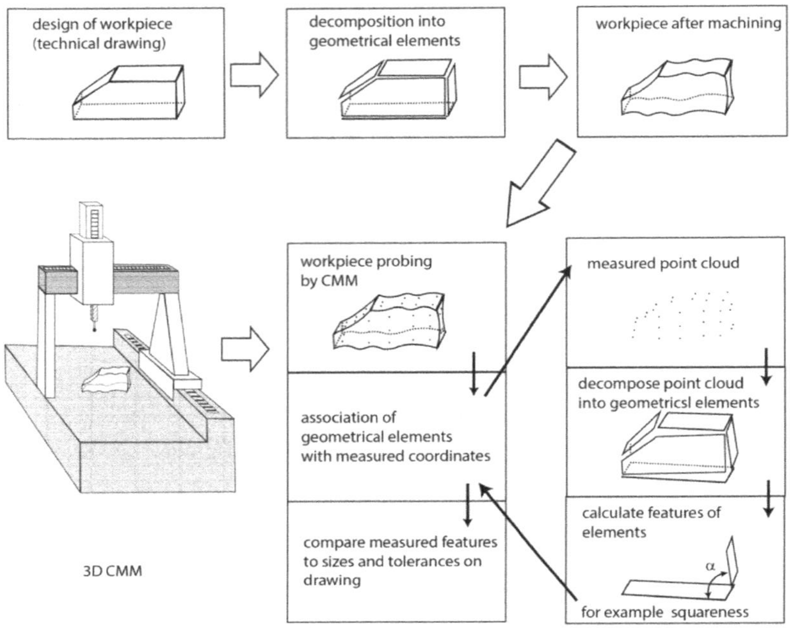

FIGURE 8.1 Schematic display of the role of coordinate measurements when measuring a workpieces.

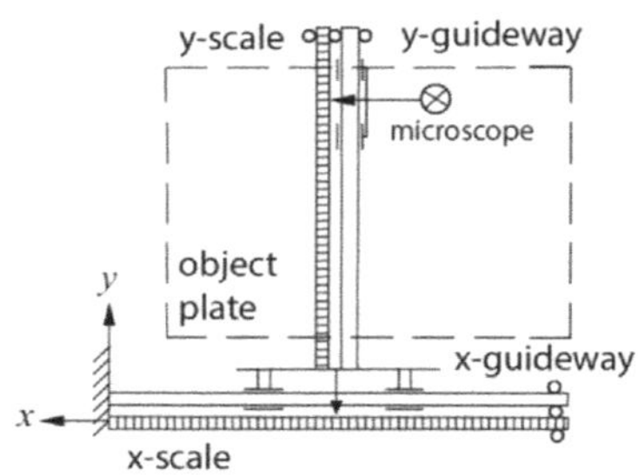

FIGURE 8.2 Schematic design of a measuring microscope.

8.2.1 The Measurement Microscope and Vision Systems

A basic design of a measuring microscope is shown schematically in Figure 8.2. The basis of the measuring microscope is formed by a table on which the workpiece to be measured can be placed. The two axes each consist of a straight guideway along which a linear measuring system (see Section 5.5.3) is mounted, with which the movement of the table with the workpiece along the relevant axis is measured.

When operated in a basic way, the microscope acts as an optical probe; the object is moved in such a way that the reading of a feature, for example, the edge of a workpiece is taken at the central point in the microscopes field of view. The resolution of the measuring systems is 0.1 μm to 1 μm. Typical measurement uncertainties are in the order of 1 μm to 10 μm. The range that can be measured with a measuring microscope is typically 200 mm by 300 mm. An embodiment of a measuring microscope is shown in Figure 8.3. The microscope is attached to one of the machine's axes above the measuring table. The two axes determine the measuring plane in which the workpiece can be moved. The microscope can be moved in the vertical z-direction for focusing, and can be tilted over a small angle, which is important for screw thread measurements. The measuring microscope has a large number of measurement options in the field of 2D measurements, some of which are sketched in Figure 8.4 and listed below.

- length;
- angle;
- coordinates;
- threads;
- profile comparisons, for example for gears.

The measuring takes place by recording the points of the workpiece in coordinate pairs (x, y) with a microscope or camera. For observing the workpiece points it is necessary that the workpiece is illuminated, which can be done in several ways – by means of illumination from the top (through the lens), from the bottom

FIGURE 8.3 Photograph of a measuring microscope.

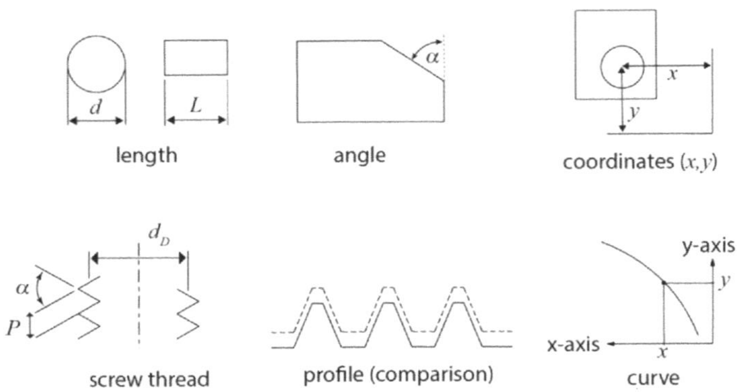

FIGURE 8.4 Some measurement possibilities of a measuring microscope.

or from the side. It depends on the lighting whether a light or a shadow image of the object is made. When using a microscope, an observer places a measuring point in the field of view of the microscope and with the aid of the measuring device, the measuring eyepiece, is accurately positioned with respect to a reference, for example cross wires, which are also visible in the eyepiece. The coordinate values (x, y) are determined by reading the linear measuring systems of both axes. The measuring eyepiece can contain various optical aids (references) to accurately determine different types of object boundaries. This way, radii, thread profiles and gear profiles can also be compared directly with standard shapes. The measuring procedure runs as follows with measuring microscopes.

- placing objects on the measuring table;
- choice of correct optical aids;
- setting the measuring microscope to the desired object boundaries and reading the measured values (x_i, y_i);
- processing the measured values (x_i, y_i) to measures or sizes.

When measuring in the manner described above, one encounters the problem that in general the zero point of the workpiece does not match that of the measuring machine and that the axes of the workpiece are not parallel to the coordinate axes. This means that the desired dimensions cannot be determined immediately. Using coordinate transformations, the measuring points defined in machine coordinates must be transformed into workpiece coordinates. However, this problem can be reduced by mechanically aligning the workpiece along the coordinate axes so that the coordinates coincide. Figure 8.5 shows an example of a measurement on a mechanically aligned workpiece. The dimensions can be calculated from the measuring points. For example, the coordinates of the center point $C(x,y)$, in relation to the zero point of the workpiece (x_1, y_1) are calculated from:

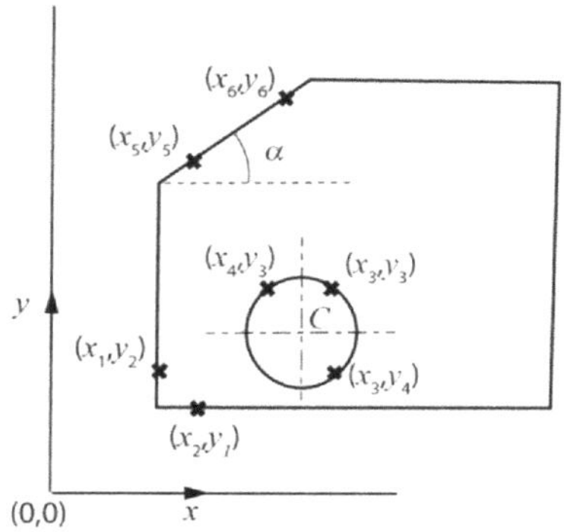

FIGURE 8.5 Example of a 2D measurement of a workpiece.

$$x_C = \frac{x_3 + x_4}{2} - x_1 \tag{8.1}$$

and

$$y_C = \frac{y_3 + y_4}{2} - y_1. \tag{8.2}$$

For the diameter D of the hole we find:

$$D = \sqrt{(x_3 - x_4)^2 + (y_3 - y_4)^2}, \tag{8.3}$$

and the angle α is given by:

$$\alpha = \tan^{-1} \frac{y_6 - y_5}{x_6 - x_5}. \tag{8.4}$$

The reading of the measured values is usually automated via a computer coupled to the measuring systems. This also makes it possible to carry out the necessary calculations and transformations by means of software. Standard features of such software are generally:

- computational alignment of the workpiece (transformation);
- calculate line or other function by measuring points;
- determine the angle between lines;
- determine circle through measuring points;
- determining the distance from a measuring point to a line;
- calculate various intersections of circles and lines.

The modern measuring microscopes use CCD camera systems instead of observation by eye to determine the measuring points. Within the CCD field of view, the pixels define an (x,y) coordinate system of its own and this must be calibrated (for example, in pixels/mm) with the machine axis. Rather than measuring microscopes, such systems are denoted as vision systems. Commonly these systems have provisions for detecting circles, partial arcs and lines within the field of view. In addition to the advantages of automatically determining measuring points, higher resolution and accuracy can often be achieved with such camera systems. If the measuring axes are CNC-controlled, that is to say the positioning of the x-y table is performed by a computer, similar workpieces can be measured fully automatically after the desired measuring points have been programmed globally for one workpiece. The camera measuring system can then automatically determine the desired measuring points of the workpieces after alignment.

For the measurement uncertainty, the threshold definition when measuring an edge is important, see section 6.1. The evaluation of the point on the workpiece corresponding to the CCD pixel measured may be affected by out-of-focus neighboring points closer by or further away from the point being measured. If this happens at the edges of the field of view of the objective, additional deviations may arise due to the fact that these regions are imaged at some angle. This can be countered using telecentric objective lenses.

8.2.2 The Profile Projector

The construction of the profile projector is similar to that of the measuring microscope, with regard to the x-y table, the straight guides and the measuring systems. The essential difference with a measuring microscope is that the shadow image, or a projection image of the workpiece, is magnified on a projection screen. A picture of a benchtop type projector is shown at the left-hand side of Figure 8.6; the right-hand side of Figure 8.6 shows a vertical or floor type projector. The source produces light with a high intensity, so that a clear image can be observed even at large magnifications. After passing the condenser, a parallel light beam of uniform intensity is produced, with which the workpiece is exposed through the mirrors M1 and M2 (semi-transparent). The workpiece is on the table and can move in the plane perpendicular to the optical axis. The reflected light passes through the beamsplitting mirror M2, the lens, the aperture and the mirror M3 on the projection screen. With this optical system, magnifications up to 200× can be achieved. Measuring with the aid of the measuring projector can be done in various ways, given below.

- Comparison of the projected workpiece on the screen with a transparent contour drawing of the same scale. The drawing can be fixed using special clamps. This enables fast visual inspection of the workpiece and is the original use of this instrument.

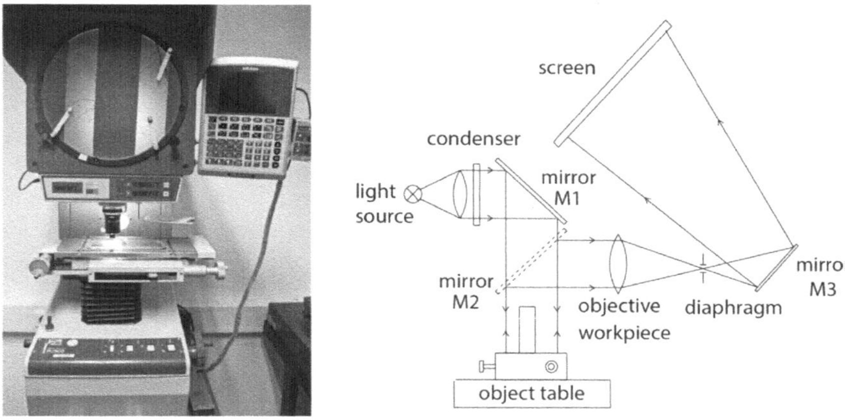

FIGURE 8.6 A photograph of a benchtop type profile projector (left-hand side) and schematic of vertical- or floor-type projector (right-hand side).

- Comparison of the projection with measurement plates on the screen. The measuring plates can, among other things, be provided with millimeter lines and different profiles.
- Measure the projected image with a glass ruler with, for example, a 0.1 mm pitch. Because of the magnification, much higher resolutions can be achieved provided the ruler and the magnification are calibrated with sufficient accuracy.
- Measuring with linear measuring systems along the guides, as described in the measuring microscope with cross wires.
- Automatic detection of edges of the projection with a sensor (photodiode or CCD camera), linked to an automatic reading of the measuring systems by a computer.

The possibilities for processing the measurement data with the aid of software are the same as with measuring microscopes. Measurement accuracy depends on the measurement method used and depends on factors such as image accuracy (0.2 % – 1.5 % within an image), influence of object edges (1 μm to 3 μm) and the resolutions of measuring systems (0.1 μm to 1 μm).

8.3 3D COORDINATE MEASURING MACHINE

The 3D coordinate measuring machines (CMMs) are the machines that can be most universally used for measuring industrial workpieces. With the advent of CMMs, it became possible to determine sizes of complex products, which was not or hardly possible due to the complexity of such products in another way. By linking CMMs to a computer with measuring software, objects can be measured relatively easily, quickly and accurately. An important advantage of using, for example, a computer and processing measurements, is the reduced influence of the operator. The calculation is carried out via a software package, the retrieval of the measured values is done by the machine control, whereby a high degree of objectivity is realized. The measuring accuracy is, however, highly dependent on the construction of the CMM and the different error sources that influence the measuring process. Since the insight into the reliability of the measured results and the possible error sources is an important aspect of this book, a separate section will be devoted to this. First of all, however, the principle of 3D measurement technology, the construction of a CMM and the collection and processing of the measurement data will be discussed.

8.3.1 Principle of the 3D CMM

The principle of 3D measurement technology is the determination of boundaries and dimensions of an object in a Cartesian, orthogonal reference coordinate system that has been realized in a 3D measuring machine. This description may seem formal, but the fact that measuring points of a certain object can all be fixed with respect to the same 3D coordinate system, brings with it major advantages and is the essence of the 3D measurement technique. With the aid of software, all desired calculations can now be carried out with the collected measuring points. This means that, in principle, the most complex measurement problems can be addressed. The type of products that can be measured is diverse and can vary from a simple product, such as a plate with a hole, to for example, a complete car body or even parts of an aircraft. Figure 8.7 shows an example of a simple 3D measurement set up with a workpiece on a machine. With the aid of a 3D measuring machine, the position of a number of points of the workpiece can be determined with respect to the Cartesian coordinate system of the measuring machine (x_m, y_m, z_m). Measurands can be calculated from this, such as flatness of the planes, the mutual distance between planes or the location hole centers. In general, one is not interested in the position of elements with respect to the measuring machine but with respect to a reference point on the workpiece itself. To this end, a coordinate system can be defined on the workpiece with base vectors (x_w, y_w, z_w). The transformation from machine coordinates to workpiece coordinates takes place by first translating the machine coordinates to the workpiece coordinates and then subsequently rotating the coordinates around the z-, y-, and x-axis respectively. For example, at first a rotation around the z-axis by the angle α between the machine x-axis and the workpiece x-axis in the x-z plane, then a rotation around the y-axis by the angle β between the x-machine axis and the x-workpiece axis in the x-y plane and then a rotation around the x-axis by the angle γ between the machine z-axis and the z-workpiece axis in the y-z plane. These operations are summarized here:

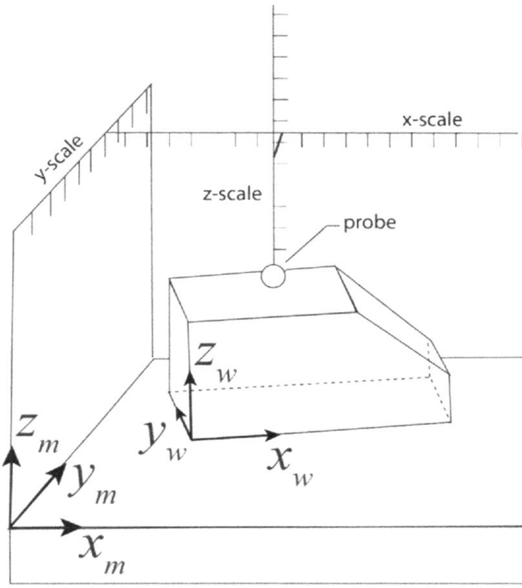

FIGURE 8.7 Schematic of a 3D coordinate machine with axis and coordinate systems indicated.

$$\begin{pmatrix} x_w \\ y_w \\ z_w \end{pmatrix} = \begin{pmatrix} 1 & 0 & 0 \\ 0 & \cos\gamma & -\sin\gamma \\ 0 & \sin\gamma & \cos\gamma \end{pmatrix} \begin{pmatrix} \cos\beta & 0 & -\sin\beta \\ 0 & 1 & 0 \\ \sin\beta & 0 & \cos\beta \end{pmatrix} \begin{pmatrix} \cos\alpha & -\sin\alpha & 0 \\ \sin\alpha & \cos\alpha & 0 \\ 0 & 0 & 1 \end{pmatrix} \begin{pmatrix} x_M - x_{0,w} \\ y_M - y_{0,w} \\ z_M - z_{0,w} \end{pmatrix}, \tag{8.5}$$

where $x_{0,w}$,$y_{0,w}$ and $z_{0,w}$ are the origin of the workpiece coordinates, expressed in machine coordinates. The six constants in equation (8.5) must be determined by probing the workpiece at strategically chosen points. If the technical drawing indicates reference lines of planes these can be used. For example, the normal of a plane can be defined as one axis by probing this plane at three positions. By probing two points in another plane, the rotation of the other two (orthogonal) axes can be recorded first. This determines the orientation of the three workpiece axes. A last probing point (or more) can determine the workpiece zero point. All measuring points can now be transformed into workpiece coordinates. The CMM will give a warning if the workpiece coordinate system cannot be determined from the measurements taken.

The realization of a 3D CMM consists of:

- a mechanical structure of three coupled, orthogonal guides with slides, each equipped with a linear measuring system. To one of the slides, a probing system is coupled for the recording of measuring points (x_p, y_p, z_p). Sometimes, rotation axes are also added, for instance in the form of turntables which can rotate the workpiece over well-known angles;
- a linked computer for the control and processing of the measurement data.

Figure 8.7 shows schematically a possibility for mechanical construction, where for the sake of simplicity, the guides coincide with the scales. The touch probe is connected to the z-guide, with which the workpiece can be probed. As far as mechanical construction of CMMs is concerned, there are various embodiments. The most important and most frequently used forms of construction are: column, moving portal, bridge and C-type (horizontal arm). The most appropriate CMM design for a specific measurement task or group of tasks depends on the mass and size of the product, the desired measurement speed, and in connection with the accuracy, the accessibility of the product. Mechanical probing is almost always done by means of a ball. The advantage of a ball is that measurements of a geometric element can almost always be corrected for the diameter of the probe ball. The other aspect enabling the general use of a ball is the fact that these can be produced with tight tolerances (form errors < 0.5 µm) for reasonable costs. The principle of the probe correction was already illustrated for the case of a circle in Figure 6.8 and applies as well to ball measurements. A spherical surface measured with a spherical probe will give a sphere as a result, with the same center point and as a radius the sum of the sphere and probe radius. Similar principles hold for – for example – the measurement of cylinders, one- and two-sided linear sizes (see Figure 6.2.) or a flat.

8.3.2 Construction of 3D CMMs

In the construction of 3D CMMs, a number of important elements can be distinguished, namely:

- guides and slides;
- measuring systems;
- probing systems.

These construction elements will be briefly explained in the following.

- **Guideway systems**

The guideway systems that can be used with CMMs can be divided into three main groups:

a. Guideway systems based on sliding friction as used in, for example, conventional prismatic guides for turning and milling machines;

b. Guideway systems in which the carriage is moved via rolling bearings along the straightness guide;
c. Aerostatic guideway systems in which a rigid air cushion forms between the carriage and the straight guide, such systems are called air bearings.

For the straightness guides itself, use is often made of granite or ceramic materials. These materials, which can be processed well with the currently available tools, have a number of favorable properties compared to steel, such as being corrosion-free, having a low specific mass and a lower thermal expansion coefficient. The thermal conductivity coefficient, however, is considerably lower, so that gradients are easily formed which cause form changes, unless suitable settling times are followed and/or the laboratory is temperature controlled; see also Section 8.3.4.2.

- **Measurement systems**

The most common measuring systems for installation in 3D CMMs are linear encoders; see section 5.5.3. These can deliver an absolute measurement accuracy with deviations being limited to a few micrometers per meter with a resolution of 0.1 μm to 0.5 μm. Because of the sensitivity to contaminants, these systems must be protected during installation and operation in an industrial environment. The position of the measuring systems is important because their distance to the object to be measured is effectively the Abbe deviation. For the most accurate CMMs in a stable and clean laboratory environment, displacement laser interferometers are also used, see section 4.4. The construction of a typical portal-type CMM is sketched in Figure 8.8.

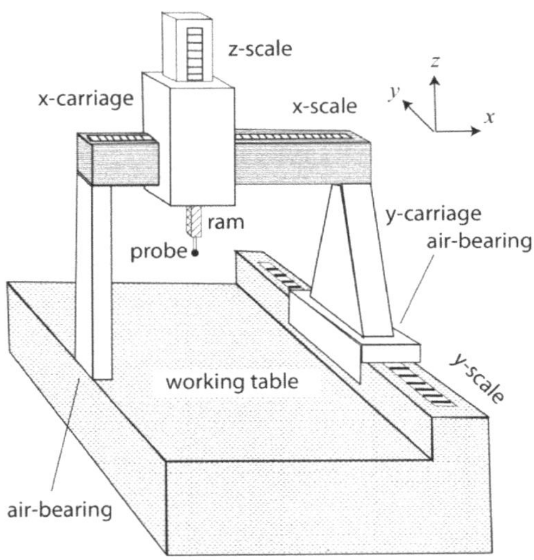

FIGURE 8.8 Schematic of a common CMM: a portal type based on linear scales and air-bearings. The 'ram' is the component of a CMM that carries the probing system.

- **Probing system**

Mechanical touch probes are most often used because of their robustness and reliability. The function of the touch probe is to determine the edges of the measurement object, to read out the measuring systems at the right time and, in case of a motor drive, to stop the slides of the measuring machine in time. The probing can in principle be carried out with tactile systems that make mechanical contact with the object and the systems that determine the positions of the boundaries without contact. As optical probes, triangulation probes can be used, or the laser line scanner as described in section 5.9. A comprehensive review of probe systems was given by Weckenmann (Weckenmann et al., 2004). Probe systems based on the mechanical contact between the object and the probe are still most common. They can be divided into four main groups:

a. *Rigid mechanical "probes"*: these probes, which can only be used with hand-operated machines, are pressed against the object with a certain measuring force, after which the measuring systems are read out.
b. *Touch trigger probes*: a circuit is opened or closed, at which time the measuring system is automatically read. If the object is in the electrical circuit, it is a resistance circuit; in case of probing, the total resistance in the circuit is low. The most commonly used is the type of touch probe where a contact is opened when the probe touches, after which the measuring system is read out, usually electronically, while the drive systems are also switched off. This touch probe is referred to as the dynamic touch probe, since the machine is still in motion when reading the measuring system. Figure 8.9 gives a schematic of the principle. Because the measuring force when the electronic circuit is opened is somewhat dependent on the probing direction, a 3-point lobe may occur when a perfectly round object is measured. This is called the 'lobing effect' (Cauchik-Migual and King, 1998).
c. *Measuring touch probes*: these have, in three directions, an inductive measuring system that measures the position of the probe and takes over the control of the machine from a certain position until the zero point of the probe measurement system is approximately reached, after which a signal is issued on which the measuring systems are read from both the probe and the CMM. The CMM is at rest at that moment, hence the name: static touch probe. As the stiffness of the measurement system will be known, the measurement force can be controlled. However, measuring touch systems can also be used dynamically. For example, to "scan", where the probe is constantly in contact with the object and the measuring systems are read continuously. This allows a complete contour to be defined. Effectively three coordinates are measured inside the measuring touch probe. These must be combined with the readings of the linear scales of the CMM. Figure 8.9 gives a possible mechanical embodiment of such a probe, where flexure hinges enable small movements in x, y and z.

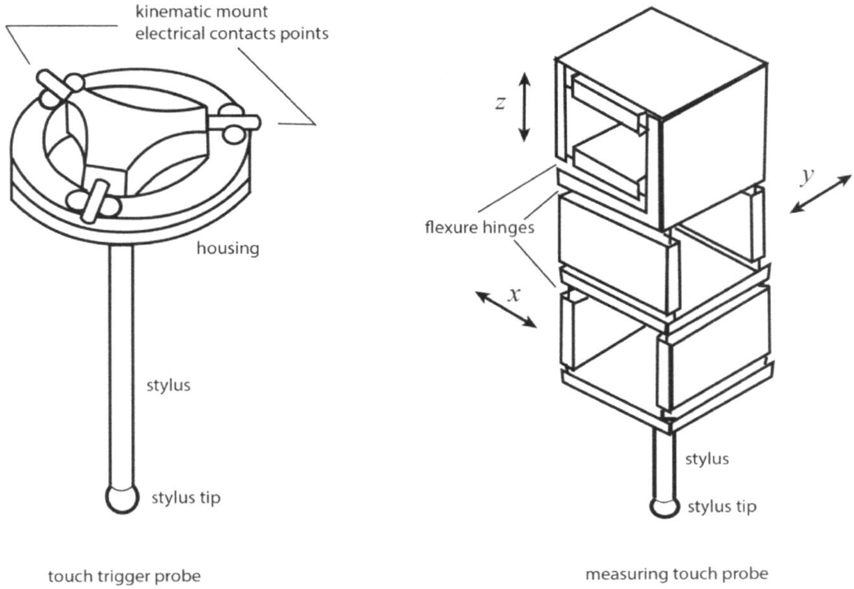

FIGURE 8.9 Schematic of a touch trigger probe system (left) and a measuring touch probe (right). The springs and the three measuring systems inside the measuring touch probe are not drawn.

d. *Motorized head systems*: these have a rotary moving system around two axes so the probe can be moved in any direction. This may save the use of a star-form with multiple balls. One step further is that this probe can scan small feature without moving the CMM, enabling high measuring speeds. For such systems, the control and measuring evaluation are complex.

The number of touch points (balls) does not have to be limited to one, usually the touch probe can be expanded to a maximum of five balls. With some measuring machines it is even possible to expand up to five groups of five balls. Figure 8.10 gives an example of a multi-stylus probing system. Such a configuration requires additional calculations in the software: the position of each ball must be known in three coordinates and the diameter of the probe ball, with the position of one of the stylus tips acting as a reference point. These positions must, before the measurement starts, be calibrated with a reference ball. Before any probe can be used, its effective diameter and other properties, such as the bending of the shaft at various measurement forces, must be determined. Commonly, this is carried out after any probe change, even when a known probe is used again. For this measurement, a reference sphere is used; most common this is a ceramic ball of 25.4 mm nominal diameter. This can commonly be recognized to be present on all CMMs. Basically, this procedure is proprietary for any manufacturer, however, in

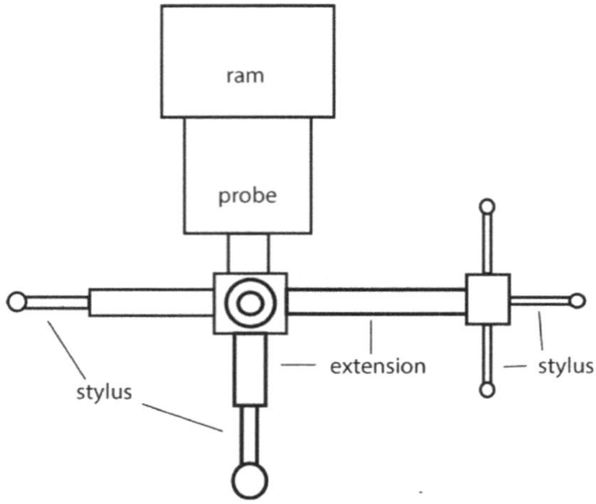

FIGURE 8.10 Example of a multi-stylus probing system.

the ISO standard series 10360, common measurement points are defined that must be used for qualification of the probe system, see Figure 8.15.

8.3.3 Control and Data Processing

As far as the control of 3D CMMs is concerned, in principle three levels (from the perspective of automation) can be distinguished.

a. Fully manual

The operator controls the drives and positions the touch probe against the measurement object. The reading of the measured values is carried out by the operator from the displays of the measuring systems. A computer is usually not used here.

b. Manual control

The machine controls are controlled by the measuring technician, where the machine automatically stops when a probing is complete. The measured values are read out via the computer. Extensive use is made here of software procedures.

c. CNC control

With this form of measuring, the machine can be pre-programmed for the measuring process to be followed, after which the machine can repeat the measuring cycle fully once or several times. Also, the desired software procedures can be "inserted" which are mainly used for calculations and output. The manual control mentioned under a. is rarely used in practice and can be regarded as outdated. Here attention will be paid to the other two principles, in which the CMM is motor-driven. With

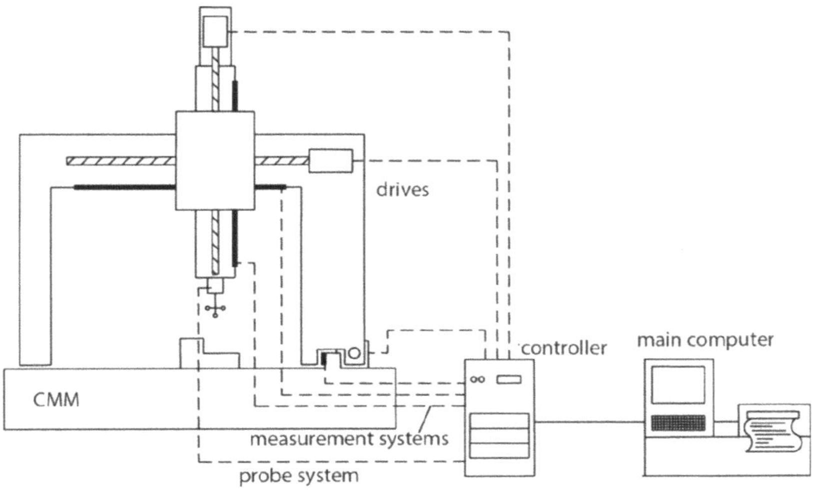

FIGURE 8.11 Schematic of a CMM with control and data processing.

both these principles there is a computer link, which is schematically shown in Figure 8.11.

Generally, at least two computers are used: one for control (the controller) and one main computer for data processing and communication with the user. Broadly speaking, the course of the measuring process is as follows: via the main computer with user software or via a joystick control, the control computer receives the commands to move the measuring machine via the drives. If the probe tip touches a workpiece, the touch probe emits a signal and the machine stops. At the same time, the measuring systems of the CMM, and possibly of the probe (if it concerns a measuring probe), are read out. The CMM is automatically moved back a small amount to account for stylus bending and geometry, after which the next point can be probed. The measurement data collected by the control computer, i.e. coordinate values (x_m, y_m, z_m), is read by the main computer for further processing. In this way, the coordinates can be determined from any point in the operating range of the CMM. When processing the measurement data by the main computer, i.e. determining the desired dimensions, use is made of the calculation part of the user program. If this part of the software is well developed, use is made of a significant advantage of the 3D CMM, namely that it gives access to a highly functional measuring machine on which various types of measurements can be performed, for which a number of different machines would be required if carried out without a CMM. As a result, a relatively expensive measuring machine can still be economically attractive. The calculation program has three main tasks:

- correct measurement points for the geometric deviations of the axes if software correction is applied;

- performing transformations for the software alignment of the workpiece;
- correcting measurement points for the probe diameter, depending on the probing direction and the measurement task;
- determining the desired dimensions of the workpiece on the basis of the measuring points;
- displaying the determined measurement information via a measurement protocol.

8.3.4 Errors and Error Corrections in Coordinate Measuring Machines

The systematic errors in CMMs; 1D, 2D and 3D measuring machines, come from the same deviation sources. The 3D CMMs, possibly provided with a turntable for workpiece rotation, have the most error sources due to the largest number of movement axes. The main features of the deviation structure of 3D CMMs will, therefore, be discussed here. The deviation structure of 1D and 2D measuring machines, also for shape and surface texture measurement, can be derived from this.

First, the error structure of a 3D CMM on the basis of possible sources of deviations is considered. As indicated in Figure 8.12 the following main error types are distinguished

- deviations caused by the mechanical system;
- deviations caused by the electronic system;
- deviations caused by the software in the computer system;
- influence of the environment.

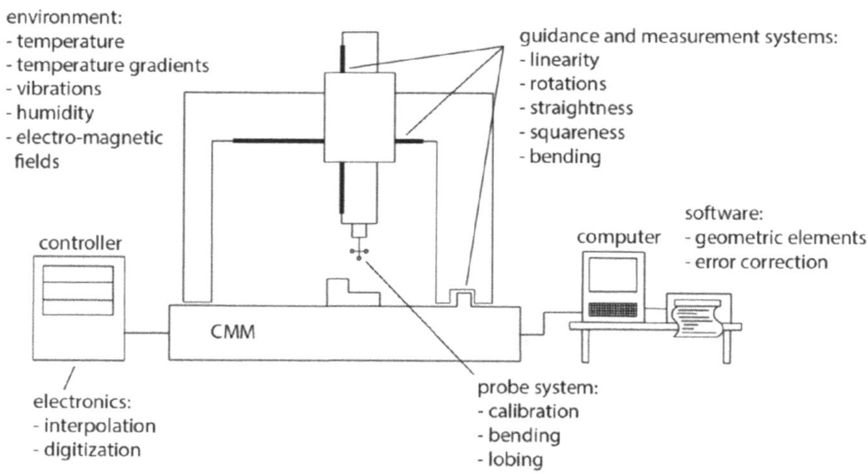

FIGURE 8.12 Error sources of a 3D CMM.

8.3.4.1 Deviations Originating from the Mechanical System

Here we distinguish the guides, measuring systems and the touch probe.

8.3.4.1.1 Deviations Originating from Guides and Measurement Systems

The guides and measuring systems usually form a Cartesian system in the case of more axes, i.e. an orthogonal coordinate system along which carriages can be defined. The touch probe is used to define the boundaries of the object in coordinate values (x, y, z). These coordinate values are read out by the machine measurement systems that are mounted along the guides. In general, the actual position of the touch probe will not be equal to the measured values (x, y, z). This is caused by, in particular:

- tilting effects, because the machine for several axes does not comply with the Abbe principle;
- straightness deviations in the guides;
- squareness deviations between the guides;
- deviations in the measuring systems;
- deviations caused by elastic deformation of machine parts, the so-called finite stiffness.

The influence of a non-perfect guide on the position of the touch probe can be deduced from the movement of a carriage along a straight guide, as illustrated in Figure 8.13. As a result of imperfections in the guideway, the position of the touch probe will not show exactly the same displacement as the displacement measured along the y scale. A deviation arises between the actual position and the measured position caused by small tilting $_jR_i$ (rotations) and small movements $_jT_i$ (translations) of the carriage. Each carriage exhibits this behavior and basically has the following sources of deviations:

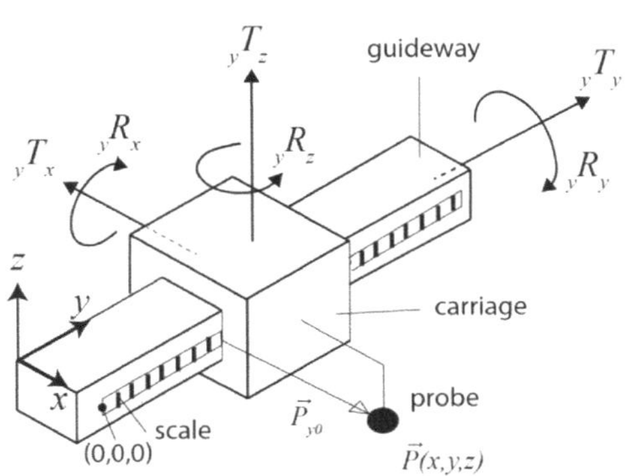

FIGURE 8.13 Guideway with carriage and error sources.

${}_jT_i$: a translation in the i-direction when moving the carriage along the j-direction

${}_jR_i$: a rotation around the i-direction when moving the carriage in the j-direction with $i,j = x,y,z$

For the y-axis this becomes:

$$\vec{T}_y = \begin{pmatrix} {}_yT_x \\ {}_yT_y \\ {}_yT_z \end{pmatrix} \quad \text{and} \quad \vec{R}_y = \begin{pmatrix} {}_yR_x \\ {}_yR_y \\ {}_yR_z \end{pmatrix} \tag{8.6}$$

where ${}_yT_y$ is the positioning deviation in the measurement system. The translations and rotations occur (except ${}_yT_y$) by unwanted movements of the carriage, which act on the spatial position $P(x,y,z)$. A common designation of the rotational errors is:

- ${}_yR_x$ is called **pitch**
- ${}_yR_z$ is called **yaw**
- ${}_yR_y$ is called **roll**

The influence of the translations and rotations on the probe position is different here:

- The translations ${}_yT_x$ and ${}_yT_z$ act directly on the position as well as the measurement system deviation ${}_yT_y$. This means that these can easily be added to the probe position (with the correct sign).
- However, the effect of rotations are proportional to the arm length between the probe and the rotation point at the scale. For example, the rotation ${}_yR_z$ propagates, via the x-coordinate, as $x \cdot {}_yR_z$ on the y-direction and the rotation ${}_yR_x$, via the z-coordinate, as $z \cdot {}_yR_x$, also in the y-direction.

The error vector due to rotations $d\vec{P}_y(R)$ can be found from the vector product (the external product) of the rotation vector $\vec{R}_y({}_yR_x, {}_yR_y, {}_yR_y)^T$ and position vector of the probe with respect to the pivot point on the y-scale $\vec{P}_{y0} = (x_0, y_0, z_0)^T$. This means that the error vector $d\vec{P}_y(R)$ can be calculated as following

$$d\vec{P}_y(\vec{R}) = \vec{R}_y \times \vec{P}_{y0} = \begin{vmatrix} \boldsymbol{i} & \boldsymbol{j} & \boldsymbol{k} \\ {}_yR_x & {}_yR_y & {}_yR_z \\ x_0 & y_0 & z_0 \end{vmatrix}, \tag{8.7}$$

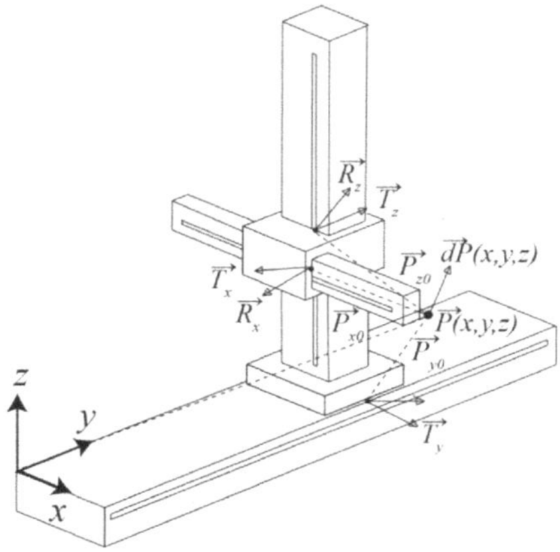

FIGURE 8.14 Error structure of a C-type CMM with *y*-axis errors indicated.

where || stands for the determinant and ***i, j*** and ***k*** are the unit vectors in the *x*-, *y*- and *z*-directions respectively. The total error vector for the y-carriage $d\vec{P}_y(x,y,z)$ becomes:

$$d\vec{P}_y(x,y,z)) = d\vec{P}_y(\vec{R}) + \vec{T}_y = \begin{vmatrix} \boldsymbol{i} & \boldsymbol{j} & \boldsymbol{k} \\ {}_yR_x & {}_yR_y & {}_yR_z \\ x_0 & y_0 & z_0 \end{vmatrix} + \begin{pmatrix} {}_yT_x \\ {}_yT_y \\ {}_yT_z \end{pmatrix}. \tag{8.8}$$

Similarly, the *x*- and *z*-carriage also affect the probe position, contributing $d\vec{P}_x$ and $d\vec{P}_x$. In Figure 8.14 this is indicated for a C-type CMM. The total error vector consists of contributions of all carriages and is given by:

$$\overrightarrow{dP} = \overrightarrow{dP}_x + \overrightarrow{dP}_y + \overrightarrow{dP}_z. \tag{8.9}$$

Additionally between the axes, there will be squareness deviations ${}_jS_i$. These are deviations from the nominal desired angle between two axes, usually 90º. Between the *x* and the *y* axes this is ${}_yS_x$ and acts as a tilting effect through the probe position. Analogously, ${}_zS_x$ and ${}_zS_y$ will also work on the probe position. Normally one speaks of twenty-one error sources of a CMM, counted as follows:

(3 rotation errors + 3 translation errors) times 3 axes, + 3 squarenesses = 21 error sources .

The squareness values are expressed as single numbers (angles), the rotations and translations are principally a function of the axis coordinate from which they originate. This may look complex on one hand, on the other hand, just three coordinates are measured with errors in the micrometer range. A few simplifications are readily available.

A squareness deviation can be described as a constant rotation or as a linearly increasing straightness deviation. This means that, if the squarenesses are taken into account, the straighnesses do not have a linear term and the rotations have no constant term. Sometimes, the straightness errors are estimated by integrating the rotational errors; see also Section 7.5.

With the measuring machine outlined in Figure 8.14, the geometrical deviations are indicated for the deviations in the y-axis. Along the x-axis, measurements take place practically in accordance with the Abbe principle, which means that rotational errors have less or no effect on the finally measured coordinates. The finite stiffness of the machine components cause deformation of the guideways, which in turn causes position deviations. For example, when shifting the x-axis of the measuring machine shown in Figure 8.14, this axis will sag in the z-direction ($dz\,(x)$) and, in addition, the changing moment due to the moving mass of the x-axis will cause a deformation $dx(x)$ of the z-axis.

8.3.4.1.2 Deviations Caused by the Probe System

The deviations in the touch probe occur when touching objects whose contours are to be determined. Each type of touch probe has its own deviation structure and must be analyzed separately. In the case of switching mechanical probing systems, deviations arise from the mechanical support in the switching points. Here, in particular, the deflection of the stylus is dependent on the probing direction. In the case of measuring touch probes there are more sources of deviations, again dependent on the embodiment. This touch probe can be regarded as a small 3D measuring machine. In addition, the stylus-tip deflection again plays a role. Optical probing systems have yet another deviation structure which can be determined in part via calibration, however, the probing of cylindrical or round objects can lead to problems. For example, CCD camera measuring systems are sensitive to the way in which the object to be measured is illuminated and the way in which the information of the CCD elements is processed largely determines the positioning accuracy. Before any probe can be used, its effective diameter and other properties, such as the bending of the shaft at various measurement forces, must be determined. Commonly, this is carried out after any probe change, even when a known probe is used again. For this measurement, a reference sphere is used; the most common is a ceramic ball of 25.4 mm nominal diameter. This can commonly be recognized to be present on any CMM. Basically this procedure is proprietary for any manufacturer. For testing and qualifying the probe, the ISO standard series 10360 (for example ISO 10360-2:2009) prescribe the measurement of a test sphere with a nominal diameter between 10 mm and 50 mm. Twenty-five measurement points are defined that must be used at minimum for qualification of the probe system. The default positioning of these points on the sphere is depicted in Figure 8.15. Through these

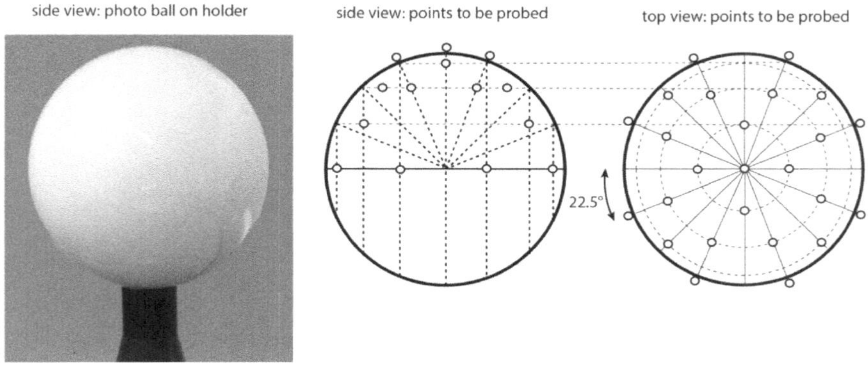

FIGURE 8.15 Default target contact points on a test sphere for probe qualification according to ISO 10360-2:2009.

points, a least-squares sphere is calculated by the CMM software (see also section 10.5.1.3) and for each point, the measured radial distance R_i to the sphere center is taken. The probing error P is the range of the R_i values: $P = \max(R_i)\text{-}\min(Ri)$.

8.3.4.2 Environmental Influences

Different sources of interference from the CMM's environment can influence their measuring accuracy. These include:

- temperature influences;
- vibration;
- moisture;
- magnetic and electric fields.

The temperature influence in particular can be considerable. A temperature distribution $T(x,y,z)$ will always be present in the measuring machine. This depends on the ambient temperature and any heat sources in the machine, such as motors and electronic circuits. The simplest temperature distribution is a constant temperature, for example $T(x, y, z) = 21$ °C. In this case, a correction for the expansion of the measuring systems and the object must be applied to the measurement results; in this case the difference in linear expansion coefficient between the scales and the object is relevant; see Section 2.2. The measuring object can also have a specific temperature distribution, which also needs to be corrected (see also Section 2.2). Additional problems occur when temperature differences occur in the machine. These have two possible consequences:

- linear expansion of machine parts;
- deformation of machine parts.

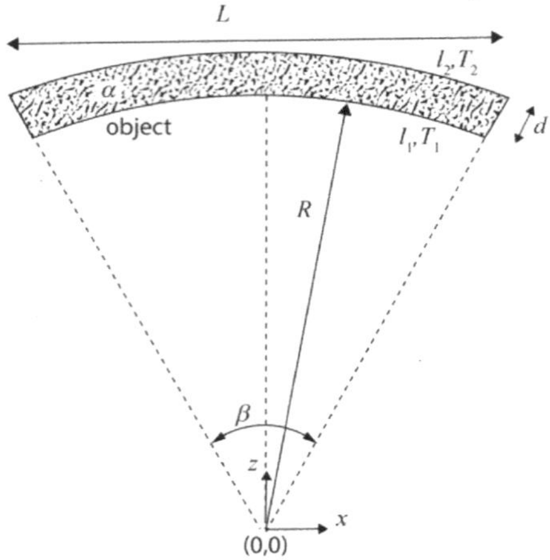

FIGURE 8.16 Warping of a beam due to a temperature difference between upper and lower side.

The linear expansion at constant temperature along an axis has already been discussed in Section 2.2. If an object has a temperature gradient in the x-direction, the correction of a length L measured along the x-axis can be calculated from:

$$\Delta L = \int_0^L \alpha \cdot \left(T(x) - 20\,^\circ\mathrm{C}\right) dx \tag{8.10}$$

where $T(x)$ describes the temperature distribution along the x-axis and α is the linear thermal expansion coefficient of the object.

A temperature difference across the height of a guideway beam can cause the beam to warp. This is caused by the difference in expansion between the top and bottom of the beam. A common case is a block granite plate that has a temperature gradient because of the lightning in a laboratory that is absorbed on the black top surface. This situation is outlined in Figure 8.16. From Figure 8.16, follows, with $R >> d$ and $l_1 \approx l_2 \approx L$ that:

$$l_1 = \beta \cdot R\ ;\ l_2 = \beta(R + d), \tag{8.11}$$

$$l_2 - l_1 = \beta \cdot d = L \cdot \alpha(T_2 - T_1) = \beta \cdot R \cdot \alpha(T_2 - T_1). \tag{8.12}$$

This gives:

$$\frac{1}{R} = \frac{\alpha(T_2 - T_1)}{d} = \alpha \cdot \mathrm{grad}(T_z). \tag{8.13}$$

The warping causes a height difference $dz(\mathrm{x})$; this is a straightness deviation (${}_{x}T_{z}$) and a tilt $\frac{dz(x)}{dz}$, that is a rotational deviation ${}_{x}R_{y}$ when a carriage is moved along the upper side of this beam. With *R>>d and R>>x*, we can write:

$$z(x) = \sqrt{(R+d)^2 - x^2} \approx (R+d)\sqrt{1-\left(\frac{x}{R+d}\right)^2} \approx R - 0.5\frac{x^2}{R}. \quad (8.14)$$

From this it follows that:

$${}_{x}T_{z} = -0.5\frac{x^2}{R} = -0.5 \cdot x^2 \cdot \alpha \cdot \mathrm{grad}\left(T_z\right), \quad (8.15)$$

and

$${}_{x}R_{y} = \frac{dz(x)}{dx} = -\frac{x}{R} = -x \cdot \alpha \cdot \mathrm{grad}\left(T_z\right). \quad (8.16)$$

8.3.4.3 Geometric Error Measurement and Compensation

In order to assess the geometric errors, as described in Section 8.3.4.1.1, several methods are available. An extensive overview can be found in the literature (Schwenke et al., 2008). Most of the instruments and standards used are already described in the previous chapters; here we note some typical aspects. Geometric errors are commonly measured by the manufacturer and are implemented in the CMM software to correct the machine coordinates (often referred to as the A code). This correction can be implemented to be effective already in the CMM positioning or after probing. When assessing a machine that is already software compensated, care must be taken that the errors are assessed including this compensation, in practice this means that the probing system must be involved.

Evaluation of translational errors: ${}_{x}T_{x}$, ${}_{y}T_{y}$, ${}_{z}T_{z}$

These are basically the scale errors. These measurements must take place with a minimum effect of Abbe errors, so scale errors must be evaluated if possible in the line of the axis or parallel to an axis as close as possible to the axis. For this, gage blocks, step gages and displacement measuring laser interferometer systems can be used, and the configuration depends on whether uni- or bi-directional effects are to be included in the measurement or avoided, see section 6.1 and ISO 10360-2:2009. Some possibilities are illustrated in Figure 8.17. For the temperature correction, the scale thermal expansion coefficients must be known, or the machine must be evaluated in the environment in which it is used.

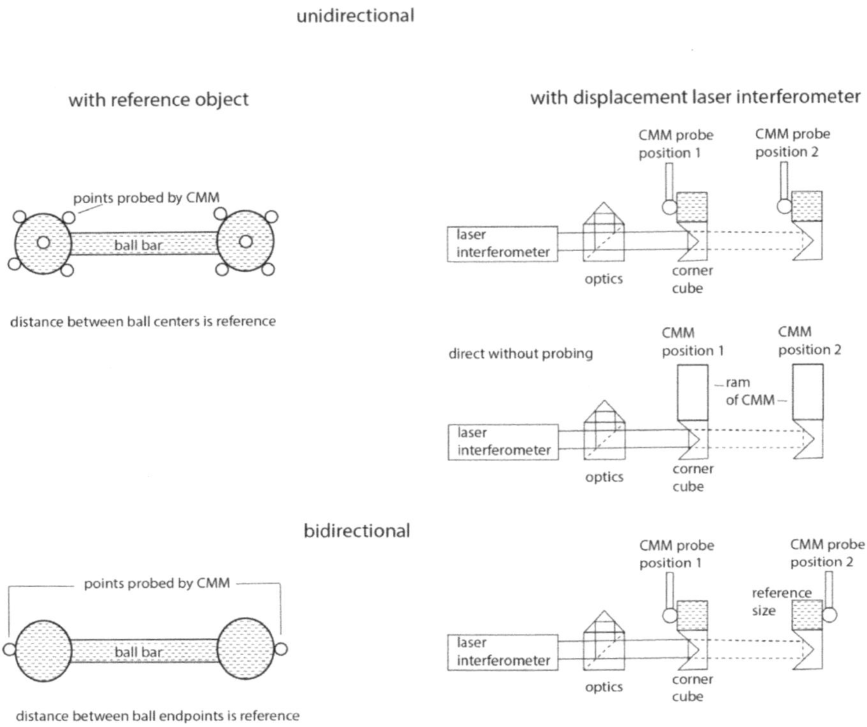

FIGURE 8.17 Possibilities for the calibration of linear distances measured by a CMM.

Evaluation of straightness errors: $_iT_j$ *with* $i,j \in \{x,y,z\}$ *and* $i \neq j$
Straightness errors can be evaluated directly by probing a calibrated straightedge, see Section 6.3.1. Also, the straightness measuring mode and optics of a displacement laser interferometer system can be used, see Section 7.3.3.

Evaluation of rotational errors: $_iR_j$ *with* $i,j \in \{x,y,z\}$
Rotational errors can be evaluated directly (without the probing system being involved) by using electronic levels (see Section 7.4.1) attached to the ram, an autocollimator (see Section 7.4.2) with a flat mirror attached to the ram, or a laser interferometer with rotational optics attached to the ram (see Section 7.4.3). What errors can be assessed by which instrument depends on the CMM configuration. For the common configuration with the z-axis in the vertical direction and the x- and y-axes in the horizontal plane, the possibilities are summarized in Table 8.1. To include the probing and software compensation, indirect methods must be used; for example measuring a step gage at a close and far distance, parallel to the scale involved.

TABLE 8.1
Feasibility of Assessment of Rotational Errors of a Common CMM Configuration

Rotational Error	Autocollimator, Laser Interferometer with Angular Optics	Electronic Level
$_xR_x$	no	yes
$_xR_y$	yes	yes
$_xR_z$	yes	no
$_yR_x$	yes	yes
$_yR_y$	no	yes
$_yR_z$	yes	no
$_zR_x$	yes	yes
$_zR_y$	yes	yes
$_zR_z$	no	no

Evaluation of squareness errors $_iS_j$ *with* $i,j \in \{x,y,z\}$ *and* $i \neq j$
A direct measurement of a squareness error can be carried out by a laser interferometer in the 'straightness' configuration, combined with a pentagon prism, also called an optical square, see Section 7.3.3. The direct calibrations that can be carried out using a laser interferometer system are depicted in Figure 8.18. A ball bar or a gage block can be measured in a diagonal position in a plane of two axes, where most errors except for the squareness cancel if the difference in measured length between two measurements in the opposite diagonal positions is taken. This principle is illustrated in Figure 8.19 that depicts the measurement of a ball bar in position A and in position B in a coordinate system that has x- and y-axes with small squareness deviation α, that gives the y-axis an angle $\beta = \alpha + 90°$ to the x-axis. According to the cosine rule, the length of the bar in position A can be written as:

$$\begin{aligned} L^2 &= \left(x_2 - x_1\right)^2 + \left(y_2 - y_1\right)^2 - 2\left(x_2 - x_1\right)\left(y_2 - y_1\right)\cos\beta \\ &= \left(x_2 - x_1\right)^2 + \left(y_2 - y_1\right)^2 + 2\left(x_2 - x_1\right)\left(y_2 - y_1\right)\sin\alpha. \end{aligned} \quad (8.17)$$

This gives for the measured length in position A:

$$L_{m,A}^2 = \left(x_2 - x_1\right)^2 + \left(y_2 - y_1\right)^2 = L^2 - 2\left(x_2 - x_1\right)\left(y_2 - y_1\right). \quad (8.18)$$

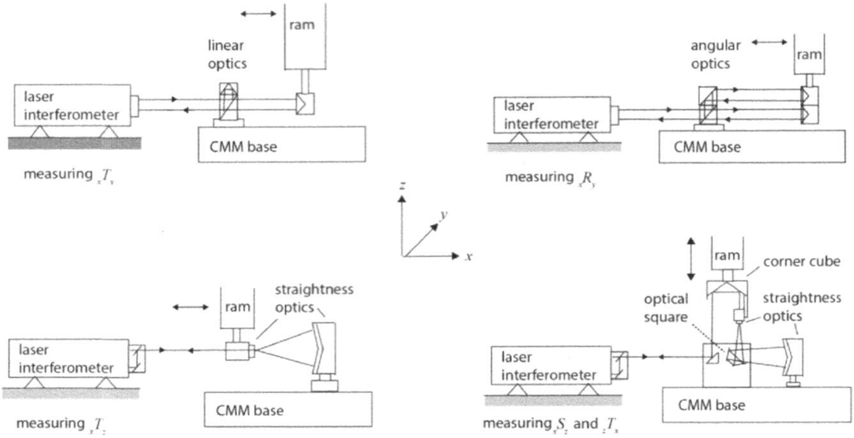

FIGURE 8.18 Direct measurement of geometric deviations of a CMM using a laser interferometer system.

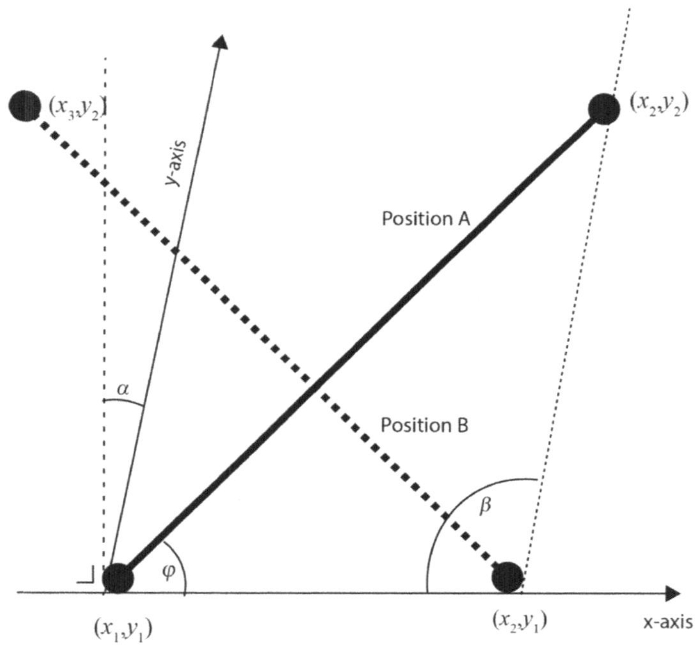

FIGURE 8.19 Configuration for a squareness measurement between two CMM axes using ball bars.

Similarly, this gives for the measured length in position B:

$$L_{m,B}^2 = \left(x_3 - x_1\right)^2 + \left(y_2 - y_1\right)^2 = L^2 + 2\left(x_2 - x_1\right)\left(y_2 - y_1\right)\sin\alpha. \quad (8.19)$$

Defining $\Delta L = L_{m,A} - L_{m,B}$ and using the approximations $\sin(\alpha) \approx \alpha$ and $\Delta L << L$, it can be derived that

$$\alpha = \frac{L \cdot \Delta L}{2\left(x_2 - x_1\right)\left(y_2 - y_1\right)}. \quad (8.20)$$

When the length bar in position A makes an angle φ with the x-axis, we have $x_2 - x_1 = L\cos\varphi$ and $y_2 - y_1 = L\sin\varphi$ and equation (8.20) can be rewritten as:

$$\alpha = \frac{L \cdot \Delta L}{2\left(x_2 - x_1\right)\left(y_2 - y_1\right)} = \frac{\Delta L}{2L\sin\varphi\cos\varphi} = \frac{\Delta L}{L\sin 2\varphi}. \quad (8.21)$$

For a diagonal at $\varphi = 45°$ this simplifies to

$$\alpha_{\varphi = 45°} = \frac{\Delta L}{L}. \quad (8.22)$$

This is an adequate method for determining the squareness deviations between two nominally square axes of a CMM. It can be generalized for room diagonals in three axes (Kruth, 2003).

Overall evaluation

An overall evaluation of a CMM can be carried out by placing reference artefacts, for example a step gage, at random positions and orientations in the measurement volume. A CMM model may be fitted to these measurements so that a direct measurement of straightness, rotations, etc. can be omitted. Also laser interferometer displacement measurements can be carried out along several trajectories in the measurement volume. Further details can be found in the literature (Schwenke et al., 2008).

8.3.5 The Measurement Strategy

The measuring strategy focuses on the entire implementation of the measuring process:

- choice of measuring equipment;
- execution of the measuring process;
- data processing and presentation of data.

When choosing the machine, it is important that the machine can "handle" the measuring task, i.e. the hardware and the software allow the measuring task to be performed with the desired accuracy and precision. The execution of the measuring process concerns, for example, the confirmation of the measuring object, the sequence of the measurements, the selection of the probing positions and the use of auxiliary measuring devices, for example, for temperature measurement. Of particular importance here is the choice of the probing points, since these determine to a large extent whether the measurement outcome can be reliable. For example, three probing points will define a surface, but if they are close to each other, the orientation is poorly defined. Even if three points establish a circle, this circle is badly defined if the points are close to each other. In data processing, it is important that the software is provided with sufficient information to achieve the desired accuracy: thus offer sufficient measuring points for calculating the desired geometric shape. In this way, the accuracy of the determination of a plane increases considerably, if not only three measuring points are taken, but also several well spread across the plane in order to obtain information about the flatness of the plane, which is not obtainable from three points only.

8.4 FRINGE PROJECTION SYSTEMS

Digital fringe projection (DFP) is an extensively used noncontact measurement technique. In this method, fringe patterns are projected onto a surface and observed under an angle by a camera. Often, a sine pattern (far more complex patterns are also used in later models) is projected that is shifted in a few steps, similar to the phase-shifting technique in interferometry and in confocal microscopy; see Sections 4.5.1 and 7.9.3.5. The DFP system comprises at minimum one camera, a projector, and a computer for integration, data acquisition and data processing. The computer-generated fringe patterns (structured light patterns) are projected onto the surface of the object through a projector, and a camera captures the fringe patterns that are deformed depending on the surface coordinates of the object. A basic schematic diagram of a DFP is shown in Figure 8.20. In this figure, the points O_p and O_c denote the points where the optical axes coincide with the apertures. As a reference, first the flat plane is measured, which provides a reference, then the height z corresponding to the object coordinates (x,y) relative to the reference plane is determined. The measurement consists of projecting the sine (or other) patterns that are shifted in a few steps, for example the four steps as described in Section 7.9.3.5. where the fringes are moved in four steps of a quarter of the line pitch p. For every pixel of the CCD camera, the four intensities $I(x,y)$ can be denoted as $I_{A..D}(x,y)$. The effective modulation intensity I_{MOD} can be calculated for every (x,y) coordinate of the object that is illuminated by the projector and can be observed by the camera thus (similar to equation (7.35))

$$I_{MOD}(x,y)=\sqrt{\left(I_D(x,y)-I_B(x,y)\right)^2+\left(I_A(x,y)-I_C(x,y)\right)^2}, \qquad (8.23)$$

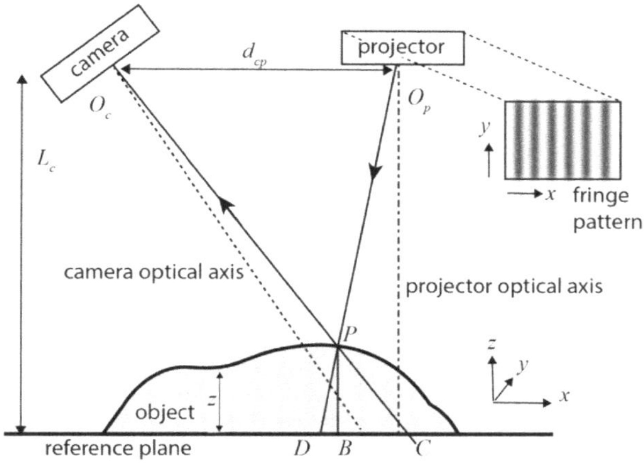

FIGURE 8.20 Simplified schema of a basic fringe projection system.

and the phase $\phi(x, y)$ on the object, as detected by the camera is given similarly to equation (4.27):

$$\phi(x,y) = \tan^{-1}\left(\frac{I_D(x,y) - I_B(x,y)}{I_A(x,y) - I_C(x,y)}\right). \tag{8.24}$$

The projected point P on the object surface has the same phase value as the point D on the reference plane: $\phi_P(x,y) = \phi_D^{ref}(x,y)$. Similarly, as detected by the camera, points D and C have the same phase value: $\phi_C(x,y) = \phi_D^{ref}(x,y)$. By subtracting the phase map as obtained from the reference surface from the measured object phase map, both recorded by the camera, the phase difference at this specific pixel corresponding to point P is acquired:

$$\Delta\phi_{PC} = \phi_P - \phi_C^{ref} = \phi_D^{ref} - \phi_C^{ref} = \Delta\phi_{DC}^{ref}. \tag{8.25}$$

The z-height of point P can also be expressed using the similar triangles O_p-O_c-P and D-C-P:

$$z_P(x,y) = \overline{PB} = \frac{\overline{CD} \cdot L_c}{d_{cp} + \overline{CD}}. \tag{8.26}$$

Assuming $d_{cp} >> \overline{CD}$ and combining equations (8.25) and (8.26) gives:

$$z(x,y) = \frac{L_c}{2\pi \cdot p \cdot d_{cp}} \Delta\phi_{DC}^{ref}(x,y), \tag{8.27}$$

where p is the pitch of the projected fringe pattern at the reference plane and L_c is the distance between the optical center of the camera, as indicated in Figure 8.17. Equation (8.27) gives a basic idea of the method and sensitivities involved, where in the real applications some more issues need to be solved:

- The phase needs to be unwrapped, similar to phase-stepping interferometry, see also section 4.5.1. With the fringe projection method the absolute phase can also be obtained by projecting fringes with different pitch p, which has similarities with the method of exact fractions in gage block interferometry; see section 4.6. A method commonly used is projecting black-white stripes that gives a binary coding of the absolute phase, and combine this with the phase-steps as described above (Posdamer and Altschular, 1982).
- The camera and the projector must be mutually calibrated so that the magnification of both the projector in the line O_p-D and the camera along the camera optical axis are calibrated. Especially the camera pixels must be mapped onto the surface coordinates (x,y), and this mapping will also depend on the measured z-coordinate. This process is commonly called system characterization and is described in detail in the literature (Shaheen, 2021).
- The method requires objects with a diffuse reflectance. For reflecting (steel) objects it is common to spray these with a thin diffusely reflecting coating that can be removed after the measurement. Also, recent embodiments of DFP systems use blue light, which is less sensitive to environmental lighting.

If one camera is used, part of the object will be shaded if features square to the reference plane are present on the object. Therefore, fringe projection systems commonly consist of a fringe projector with two cameras. If an object needs to be scanned completely, it is commonly put on a rotary table and several exposures are made in several angular positions. In order to stitch the geometries together, it is common to attach typical markers to the object and the rotary table that can be recognized by the image processing software, so that the partial geometries can be stitched together. Typical systems may scan and evaluate millions of points within a few seconds of time. This is a major advantage over mechanical CMM measurements. The uncertainty may range from a few micrometers to up tens of micrometers, with outliers and non-measured points frequently encountered.

8.5 X-RAY COMPUTED TOMOGRAPHY (XCT)

XCT systems dedicated to dimensional metrology have appeared on the market since 2005. So far, dimensional XCT metrology is the only technology that is able to measure the inner as well as the outer geometry of a component without the need to cut it and potentially destroy it. As such, it is the only technology for industrial quality control of workpieces having non-accessible internal features (for example, components produced by additive manufacturing) or multi-material components (for example, injection molded plastic parts or plastic parts with metallic inserts). Dimensional XCT metrology also enables quality control of assemblies in assembled states.

As straightforward as the principle and the method may appear, so complicated is a proper data evaluation and interpretation. Extensive review papers and books have appeared describing the principles, detailed features and application (Kruth et al., 2011; Carmignato et al., 2018; De Chiffre et al., 2014). Here a summary of the main aspects is given. Figure 8.21 gives a typical XCT system, consisting of an X-ray source, a rotary table on which the object to be measured must be positioned, and the 2D detector array that resembles a CCD camera. Here the 2000 by 2000 pixels are 0.2 mm by 0.2 mm each and they are sensitive to X-rays. The X-rays should have enough energy to pass through the object, hit the detector and give a significant detector signal. As the X-rays propagate through the workpiece material, the X-rays are attenuated due to absorption or scattering. The attenuation is determined by the length traveled in the absorbing material, by the material composition and its density and by the energy of the X-rays. The intensity measured at the detector pixels defines the total attenuation of the beam going through the material in the line source-detector (pixel). By repeating this for a number of rotational positions of the object, the inside and outside geometry as well as the material composition of the object can be reconstructed from the gray-values of the detector pixels. This mathematical reconstruction results in a voxel-model of the workpiece (a voxel is a 3D equivalent of a pixel) where every voxel can have a value meaning there is material or air, and if there is material a value indicates the absorptivity of the material (i.e. composition, density). The next steps concern the post-processing of the voxel data, including the detection of the workpiece edges (segmentation) and subsequent dimensional measurement and quality control. Some aspects of the operation are treated in the next sections.

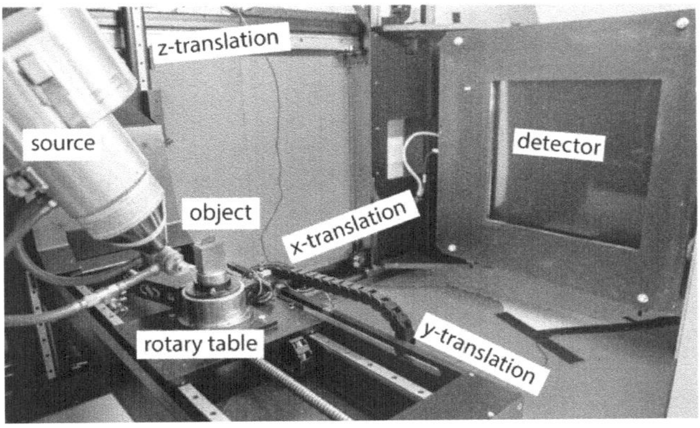

FIGURE 8.21 Inside of an industrial XCT system, showing the source, object and positioning elements.

8.5.1 Hardware Aspects of an XCT System

8.5.1.1 X-Ray Source

In the X-ray source, electrons are accelerated to an energy that characterizes the source, for example, for a source that accelerates electrons using a voltage of 450 kV, the electrons will have an energy of 450 keV (in SI units this is $7.21 \cdot 10^{-14}$ J). These electrons hit a target by which the energy is converted into heat and X-rays. Depending on the target material, the emitted energy of the X-ray photons form a continuous spectrum that ranges up to the electron energy, in this example 450 keV. Ideally, the target acts as a point-source but in practice it will have a finite size. Sources giving spots of about 1 μm cross section are called nano-focus sources; such spots can be achieved with X-ray photon energies up to 250 keV. Sources giving higher energy photons will have larger spots, from 30 μm to 1 mm cross section, and are called micro-focus sources.

8.5.1.2 Rotary Table and Dimensional Positioning

The basic kinematics of industrial metrological XCT systems is shown in Figure 8.21. Basic elements are:

1. A rotary table for stepwise or continuous rotation of the workpiece. The angular positioning and the rotational radial and axial spindle deviations must be accurate enough so as not to give additional error sources.
2. A horizontal translation axis for positioning the rotary table with the workpiece between the source and the detector. This axis is often referred to as the magnification axis: a higher geometrical magnification is achieved by positioning the object closer to the source. The downside of a higher magnification is a larger effect of the finite spot size.
3. A vertical translation axis for moving the turntable up or down, for example, for positioning the center of the workpiece in the measurement volume.
4. Optionally, a horizontal translation axis for moving the rotary table off-axis, for example, for measuring specific parts of a large object.

8.5.1.3 X-Ray Detector Array

Standard 2D detectors have a maximum resolution of 4000 by 4000 pixels at a pixel size of 0.1 mm. They consist of scintillators that convert the X-ray photons into visible light that is detected by photodiodes. The dynamic range is commonly 16 bits and they can record images at about 4 frames/s.

8.5.2 Software Aspects of an XCT System

8.5.2.1 Reconstruction and Edge Detection

The reconstruction of voxel values is commonly carried out through a so-called Radon transform. This model describes the absorption of X-rays when passing through a medium with a varying attenuation coefficient $\mu(r)$ along the line of propagation r:

$$I(r) = I_0 \exp\left(-\int \mu(r)\mathrm{d}r\right), \tag{8.28}$$

where $I(r)$ corresponds to the detected gray value after the beam has passed through the workpiece, taking into account the basic beam divergence of the point-source. This reconstruction is a considerable computational effort; as a basis it takes the workpiece voxels (in fact workpiece coordinates as in CMM evaluation) and basically considers all the angles the workpiece receives the radiation from. One of the complicating factors is the change of the energy spectrum when the beam propagates through the object: lower energy photons can be completely absorbed before they leave the object while the higher energy photons are less affected. This effect is called 'beam hardening'. From all the images taken, the $\mu(r)$ functions are reconstructed for all the source-detector pixel trajectories and from there a $\mu(x,y,z)$ absorptance value for every voxel. The 3D reconstruction is followed by the edge (surface) detection determining the interfaces between material and air or between different materials. An example of such a reconstruction is given in Figure 8.22 for a shell, where in the upper side 4 out of the 3142 images used for the reconstruction are shown, and the lower side gives a visualization in four views of the reconstructed geometry. To obtain such a complete coordinate digitization of an objects surface is impossible by any of the other coordinate or surface measurement methods.

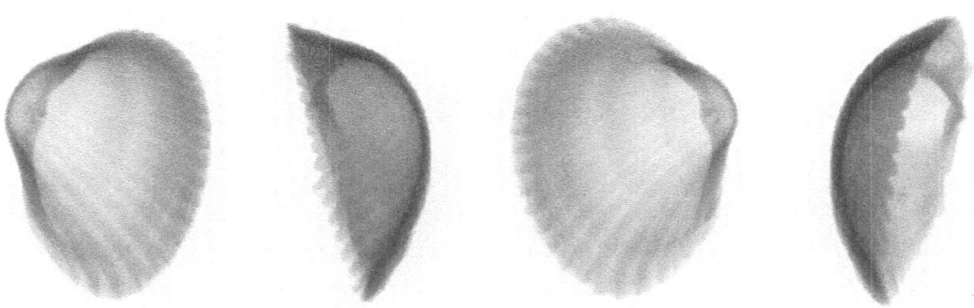

Four radiographic images out of the 3142 exposures taken of a shell

Four quasi 3-D views of the reconstructed 3-D representation of the shell coordinates

FIGURE 8.22 Four radiographs out of the 3142 exposures used to reconstruct a shell in 3D coordinates, of which four images are given at the bottom.

8.5.2.2 Scale Calibration

The conversion from voxels to millimeters is commonly carried out using calibrated reference objects, such as balls and ball bars as described in Section 6.1.6. Also in XCT measurements, balls have the advantage that the ball center calculation is minimally affected by potential error sources such as thresholding and edge detection. If possible, a calibrated ball assembly surrounding the object is measured together with the object. Another possibility is the measurement of well-defined features of the measured object using a CMM and scale the voxels accordingly.

8.5.2.3 Conversion from Voxel to Surface/Geometric Model

The edge detection generally results in a point cloud. This point cloud is then turned into a geometrical surface model by topological association of neighboring points. This association is generally carried out by generating a triangular mesh model that represents an open or closed facetted surface model of the object. A standardized format for such a surface model is the STL (Standard Triangular Language) format that is commonly used in engineering applications.

8.5.2.4 Point, Feature and Coordinate Extraction

Dimensional metrology applications require identification of measuring points and coordinates that represent the feature to be measured. This means that either the points should be converted to geometrical elements that can be evaluated using conventional CMM software (see Figure 8.1), or the point clouds can be used as such within a point-cloud based CMM software, such as the software used in fringe projection systems (see Section 8.4) or laser scanners (see Section 5.9).

8.6 STANDARDIZATION IN COORDINATE MEASURING SYSTEMS

For the coordinate measurement systems discussed in this chapter, the 10360 series of ISO standards with as common nominator *Geometrical Product Specification (GPS) – Acceptance and reverification tests for coordinate measuring systems (CMS)* has been published giving the nomenclature and default evaluation methods for various kind of coordinate measurement systems. Note that the term CMS is used as a catch-all term to include mechanical, optical or XCT systems. However, the standards are not consistent with this and sometimes just the term CMM is used as the catch all. These standards are listed here, highlighting the relevant topics discussed.

ISO 10360-1:2000 Part 1: Vocabulary

This standard gives the definitions of the various components of a CMM such as probes, styli, etc. A relevant definition is the maximum permissible error E_{MPE}: this is defined as the extreme error of indication of a CMM for size measurement, permitted by the specifications of a CMM. This means that a CMM measuring a size should give an error smaller than this specification. A typical specification can be: E_{MPE} =(1.7+3L/1000) μm, where the length L is expressed in millimeters. This way of expressing the E_{MPE} is common in the industry. From a metrological point of view the notation E_{MPE} =1.7 μm +3·10^{-6}·L, like an uncertainty as it is commonly expressed in accreditation scopes, would be preferred, because in that case the unit of L does not need to be specified. A related quantity is the maximum permissible probing error P_{MPE} that specifies the maximum range of values of the radii, measured as described in Section 8.3.4.1.2. This standard also describes the various CMM types.

ISO 10360-2:2009 CMMs used for measuring size
This standard has as major topic: the testing of a CMM to properly determine the E_{MPE} as given in ISO 10360-1; for example, at a minimum of five different sizes per axis, 80% of the total axis length must be covered, etc.

ISO 10360-3:2000 CMMs with the axis of a rotary table as the fourth axis
This standard describes especially the additional tests that can be carried out for specifying the accuracy of a rotary table that is part of the CMM. It uses balls that must be mounted on the table, where the CMM measures the ball centers at several rotational positions.

ISO 10360-4:2000 CMMs used in scanning measuring mode
This standard covers the use of a sphere that must be probed in scanning mode, i.e. continuously, where a number of trajectories must be measured, after which the scanning probe error is determined similarly as in the point-by-point case, described in Section 8.3.4.1.2.

ISO 10360-5:2020 Coordinate measuring machines (CMMs) using single and multiple stylus contacting probe systems using discrete point and/or scanning measuring mode

This standard describes the evaluation of multiple probe systems (such as sketched in Figure 8.10) and the various nomenclatures to specify these systems. Scanning or probing a test sphere as described in Section 8.3.4.1.2 is the basis for the evaluation.

ISO 10360-6:2001 Estimation of errors in computing Gaussian associated features
This standard specifies methods for the CMM-software evaluation of lines (in 2D and 3D), plane, circle, sphere, cylinder, cone and torus. The “Gaussian association” implies that the least-squares method is used (see Appendix). It describes basically the two methods of software validation:

1. Use of reference data sets: these are data sets that must give a defined outcome.
2. Comparison with reference software: a measured or generated data set is evaluated both by – already verified – reference software and the software under test.

ISO 10360-7:2011 CMMs equipped with imaging probing systems
This standard applies to vision systems and contains equivalents to the sphere- and length probing described earlier. Instead of a sphere, a 2D circle must be probed. The length measurements as described in ISO 10360-2:2009 (see above) must be carried out with length standards, where a calibrated length involves a feature that can be measured by the imaging system, for example, a ball bar where small balls are involved that can be captured within the field of view of the vision system.

ISO 10360-8:2013 CMMs with optical distance sensors
A CMM may be equipped with an optical distance sensor such as a confocal sensor, described in section 5.6 or a laser line scanner (LLS) sensor, as described in section 5.9. The tests involve length measurements, for example, on ball bars, measurements on a reference sphere or part of a sphere, and measurement of a flat surface that is positioned at several positions in the measurement volume.

ISO 10360-9:2013 CMMs with multiple probing systems
A CMM may be equipped with both an optical and mechanical probing sensor, or two mechanical probing systems with different measurement principles, for example, point probing and scanning. This test consists of measuring the same reference sphere with the different probing systems and comparing the results.

ISO 10360-10:2021 Laser trackers
The 3D CMM measurement can be enhanced by measuring the probe with a so-called laser tracker; this is an optical distance sensor or a laser interferometer system that measures the distance from a fixed point to the CMM-ram position or another defined point close to the probe; recording both the distance and two angles. The test consists of probing a reference sphere, with the configuration as described in Section 8.3.4.1.2.

ISO/DIS 10360-11:2021 CMSs using the principle of X-ray computed tomography (CT)
In 2021, it was decided that this standard does not really fit in this series, however, because of its connection to Section 8.5, its draft content is mentioned here. A distinction is made between probing error, length measuring error and resolution. For the probing test, a sphere is recommended, for the length measurements ball bars, multi-sphere objects or cylinders at defined distances are mentioned.

ISO 10360-12:2016 Articulated arm coordinate machines (CMM)
An articulated arm CMM is a CMM consisting of a chain of fixed-length segments that are combined by joints that can be rotated and whose rotation is measured, with a probing system at the free end. More informally stated this is a robot-like structure where the tool/gripper is replaced by a probe. The coordinates are calculated from the rotary encoders in the joints. The acceptance/reverification

tests also here consist of the definition of a probing error from the measurement of a sphere, and the length measuring error (uni-directional of bi-directional) by measuring artefacts such as ball bars, step gages or gage blocks.

ISO 10360-13:2021 Optical 3D CMS
This standard basically covers the fringe projection systems as described in Section 8.4. The standard describes the subdivision of the measurement volume in eight sub volumes where the ball bar that is used for the size calibration should extend to at least two of these regions in a large amount of measurements. As with all of the standards in this series, the measurement of a ball is used as the identification of a probing error. Next to these two basic evaluations also a reference flat is recommended that has to be measured in several orientations.

EXERCISES

1. Length measurement on a CMM
   A gage block is measured on a CMM with a touching probe. The gage block is probed exactly in the center of the measurement faces, the probe force is effectively zero.
   The measured coordinates, as indicated by the machine scales, are $\boldsymbol{P}_1 = (x,y,z,)$ = (200.1357, 200.1202, 202.9867) mm, the coordinates on the other side are $\boldsymbol{P}_2$ = (201.1586, 305.0692, 200.0357) mm. The probe ball diameter is 5.0000 mm. The scale and gage block temperature is 20 °C.
   a. Calculate the gage block length.
   b. Indicate which geometrical deviations are relevant – and which are not – if the measurements are carried out with the C-type machine that is sketched in Figure 8.14.
   c. Indicate how a limited probe force would affect the measurements and how this effect could be measured and corrected.
2. Effect of temperature gradient
   A granite surface plate of 600 mm by 800 mm by 200 mm (length by width by thickness) is perfectly flat in absence of a temperature gradient. However, the top of the plate is 0.6 °C warmer than the bottom (with fluorescent lights in a measurement room this is a common value). The linear thermal expansion of granite is $\alpha = 5.5 \cdot 10^{-6}$/K.
   a. Calculate the flatness deviation *FLTt* of the plate
   b. Calculate the angular deviations ${}_xR_y$ and ${}_yR_x$ at the edges of the plate
   c. Some argue that the flatness of a surface plate is defined at 20 °C, so when measuring a surface plate a temperature gradient should be corrected for. Others argue that a surface plate in a measurement room should be calibrated 'as is', meaning including temperature gradient if the lights are switched on. Give arguments for both approaches, and your own conclusion.

3. CMM error evaluation
Consider the CMM as given in Figure 8.8, that depicts a portal-type CMM. Errors in the y-guideway lead to an error vector.

In machine coordinates, the position vector is

$$\vec{P} = \begin{pmatrix} x \\ y \\ x \end{pmatrix} = \begin{pmatrix} 300\text{ mm} \\ 100\text{ mm} \\ 200\text{ mm} \end{pmatrix}.$$

The translation errors vector is:

$$\overrightarrow{{}_yT} = \begin{pmatrix} {}_yT_x \\ {}_yT_y \\ {}_yT_z \end{pmatrix} = \begin{pmatrix} 0.005\text{ mm} \\ 0.001\text{ mm} \\ 0.004\text{ mm} \end{pmatrix}.$$

The rotational errors vector is:

$$\overrightarrow{{}_yR} = \begin{pmatrix} {}_yR_x \\ {}_yR_y \\ {}_yR_z \end{pmatrix} = \begin{pmatrix} 2'' \\ 5'' \\ 3'' \end{pmatrix}.$$

a. Give the deviation vector
b. Suppose all geometric deviations are of the same magnitude; at what location and orientation could you best position a gage block in the x-y plane in order to measure it as accurately as possible?

4. Squareness measurement using reversal.
On a CMM, a granite square is measured at four points at position A. The granite square is 'reversed' and is measured again at position B; see Figure 8.23. The base of the square is aligned with the machine x-axis (y is the same for all coordinates). In machine coordinates, the measured coordinates are given in Table 8.2.
a. Consider a useful workpiece coordinate system and give the measurements in workpiece coordinates.
b. Calculate the squareness deviation of the granite square as well as the CMM squareness ${}_xS_z$ in micrometers.
c. Calculate the straightness deviation of the square and the machine straightness deviation ${}_zT_x$.
d. Give the squareness deviation of the square in arc seconds in order to indicate whether the angle is smaller or larger than 90°.

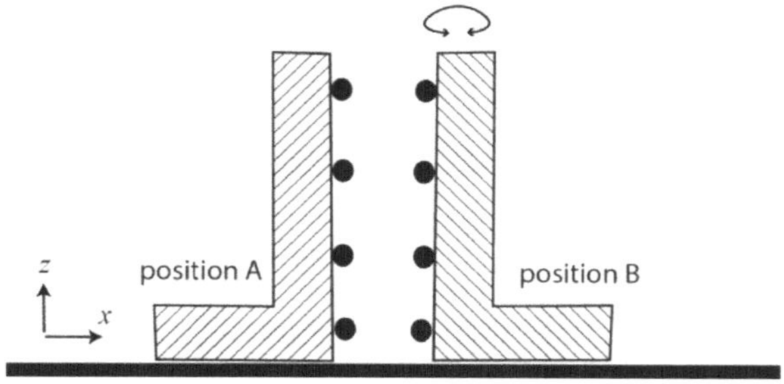

FIGURE 8.23 Two positions of a square measured by a CMM.

TABLE 8.2
Measured Coordinates of a Square in Normal and Reverse Position

z/mm	*x* (Position A)/mm	*x* (Position B)/mm
0	0	50.000
100	-0.002	50.000
200	-0.001	50.003
300	-0.003	50.003

5. Resolution of an XCT measurement
   Consider the XCT-system as depicted in Figure 8.21. The detector system has a resolution of 2000 by 2000 pixels and a size of 400 mm by 400 mm. The distance between the X-ray source and the detector is 1 m. An object of 20 mm by 20 mm by 20 mm has to be measured. The spot size of the source is 20 μm.
   a. Give in general terms the advantages and disadvantages of placing the object close to the source or to the detector.
   b. Estimate the effective resolution when placing the object as close as possible to the source or to the detector.
   c. Estimate the position where the object can be measured with an optimal resolution.
6. Errors in squareness measurements.
   Consider the squareness measurement as depicted in Figure 8.18. The ball bar is measured in position A only.
   a. What additional information is needed to make this a useful measurement?
   b. Give the geometrical error sources that affect this measurement.
   c. Consider which error sources are cancelled when taking the difference in measured length between positions A and B.

7. Correlation between rotational and straightness errors.
   In some CMM models, the straightness errors are derived from the rotational errors or vice versa. Consider how this transformation can be made and give the conditions that must be met to justify this approach.
8. Measurement of $_zR_z$.
   In Table 8.1, it is shown that a direct measurement of $_zR_z$ is not possible using the conventional means for measuring rotations.
   a. Consider for which measurements and/or probe configurations this error is relevant and for which it is not.
   b. Invent a direct measurement method for determining this error.
   c. Invent an object, probe configuration or a combination of measurements and configurations to determine this error.
9. Design of a fringe projection system.
   Figure 8.20 gives a schematic design of a fringe projection system. In this design there are quite some parameters (sizes, angles, resolution, fringe pattern frequencies, etc.) not yet defined.
   Give qualitatively the expected effect of these design parameters on the expected precision and note some other features, advantages and disadvantages for a given object size.
   Specifically, design a system that can measure a nominally flat area of 100 mm by 100 mm with a resolution of 1000 by 1000 pixels, 100 fringes and determine the resolution in x, y and z based on 9° uncertainty in the fringe phase.

REFERENCES

Carmignato, S., Dewulf, W., Leach, R. (Ed.), 2018, *Industrial X-Ray Computed Tomography*, Springer.

Cauchick-Miguel, P.A. and King, Y.G., 1998, Factors which influence CMM touch trigger probe performance, *Int. J. Mach. Tools. Manufact.* **38(4)**363–374.

De Chiffre L de., Carmignato S., Kruth J.-P., Schmitt R., Weckenmann A. 2014 Industrial applications of computed tomography . *Ann of CIRP*, **63(2)**655:677

ISO 10360-1, 2000, Geometrical Product Specification (GPS) – Acceptance and reverification tests for coordinate measuring machines (CMM) – Part 1: Vocabulary, International Organization for Standardization.

ISO 10360-2, 2009, Geometrical Product Specification (GPS) – Acceptance and reverification tests for coordinate measuring machines (CMM) – Part 2: CMMs used for measuring size, International Organization for Standardization.

ISO 10360-3, 2000, Geometrical Product Specification (GPS) – Acceptance and reverification tests for coordinate measuring machines (CMM) – Part 3: CMMs with the axis of a rotary table as the fourth axis, International Organization for Standardization.

ISO 10360-4, 2000, Geometrical Product Specification (GPS) – Acceptance and reverification tests for coordinate measuring machines (CMM) – Part 4: CMMs used in scanning measuring mode. International Organization for Standardization.

ISO 10360-5, 2020, Geometrical Product Specification (GPS) – Acceptance and reverification tests for coordinate measuring systems (CMS) – Part 5: Coordinate measuring machines (CMMs) using single and multiple stylus contacting probe systems using discrete point and/or scanning measuring mode. International Organization for Standardization.

ISO 10360-6, 2001, Geometrical Product Specification (GPS) – Acceptance and reverification tests for coordinate measuring machines (CMM) – Part 6: Estimation of errors in computing Gaussian associated features. International Organization for Standardization.

ISO 10360-7, 2011, Geometrical Product Specification (GPS) – Acceptance and reverification tests for coordinate measuring machines (CMM) – Part 7: CMMs equipped with imaging probing systems. International Organization for Standardization.

ISO 10360-8, 2013, Geometrical Product Specification (GPS) – Acceptance and reverification tests for coordinate measuring machines (CMM) – Part 8: CMMs with optical distance sensors. International Organization for Standardization.

ISO 10360-9, 2013, Geometrical Product Specification (GPS) – Acceptance and reverification tests for coordinate measuring systems (CMS) – Part 9: CMMs with multiple probing systems. International Organization for Standardization.

ISO 10360-10, 2021, Geometrical Product Specification (GPS) – Acceptance and reverification tests for coordinate measuring systems (CMS) – Part 10: Laser trackers. International Organization for Standardization.

ISO 10360-12, 2016, Geometrical Product Specification (GPS) – Acceptance and reverification tests for coordinate measuring systems (CMS) – Part 12: Articulated arm coordinate machines (CMM). International Organization for Standardization.

ISO 10360-13, 2021, Geometrical Product Specification (GPS) – Acceptance and reverification tests for coordinate measuring systems (CMS) – Part 13: Optical 3D CMS. International Organization for Standardization.

ISO/DIS 10360-11, 2021, Geometrical Product Specification (GPS) – Acceptance and reverification tests for coordinate measuring systems (CMS) – Part 11: CMSs usi8ng the principle of X-ray computed tomography. International Organization for Standardization.

Kruth, J.P.; Bartscher, M. Carmignato, S.; Schmitt, R.; De Chiffre, L.; Weckenmann, A. 2011 Computed tomography for dimensional metrology, *Ann CIRP* **60 (2):**821–842.

Kruth, J.P., Zhou, L., Van den Bergh, C. and Vanherck, P., 2003 A Method for Squareness Error Verification on a Coordinate Measuring Machine . *Int. Journ. Adv. Manufac. Tech.* **21:**874–878.

Posdamer, J.L. and Altschuler, M.D., 1982, Surface measurement by space-encoded projected beam systems, *Computer Graphics and Image processing*, **18(1)**:1–17.

Schwenke, H., Knapp. W., Haitjema, H., Weckenmann, A., Schmitt, R., and Delbressine, F., 2008, Geometric Error Measurement and Compensation of Machines, An Update. *Ann. CIRP* **57(2)**:660–675.

Shaheen A., Sims-Waterhouse D., Bointon P., Takushima S., Piano S., Leach R.K. 2021 Characterisation of a multi-view fringe projection system based on the stereo matching of rectified phase maps, *Meas. Sci Techn.* **32 (4)** 045006.

Weckenmann, A., Estler, T., Peggs, G., McMurtry, D. 2004. Probing systems in dimensional metrology. *Ann CIRP* **53(2)**657:684

9 Surface Topography Filtering and Characterization

9.1 INTRODUCTION

In this chapter, the major techniques and aspects are given for the filtering of and characterization of surface topography. This field has extensively developed over the recent decades and has been summarized in textbooks giving the backgrounds (Leach 2024) as well as explicit algorithms (Muralikrishnan and Raja 2009) where the interested reader may find further details, as well as in the relevant ISO specification standards.

9.2 FOURIER ANALYSIS OF ROUNDNESS MEASUREMENTS

The results of roundness measurements as given in a roundness diagram give a signal to which a Fourier analysis can be applied; for that reason the analysis of the roundness diagram of the coin, as given in Figure 7.33, is given here. Note that instead of the roundness diagram, the signal can also be displayed as a linear plot of the (roundness) deviation as a function of the angle. In the linear plot, the arbitrary constant in the roundness plot (see equation (7.23)) is omitted. Figure 9.1 gives both representations of the original measurement that was slightly decentered.

An important aspect of the roundness diagram (and a roundness measurement) is that it is essentially periodic. Describing the measurement as $r(\theta)$, it can be stated that $r(0°) = r(360)°$ and more generally $r(\theta) = r(\theta)+n\cdot 360°$, where n is an integer number. As sine and cosine functions have the same periodic property, the signal can be conveniently decomposed into sine and cosine functions. For N equidistant measurement points, the function $r(\theta_i)$, with i=1..N and $\theta_i = i\cdot 360°/N$, can be written as:

$$r\left(\theta_i\right) = \frac{a_0}{2} + \sum_{k=1}^{N/2} a_k \cos k\theta_i + b_k \sin k\theta_i = \frac{a_0}{2} + \sum_{k=1}^{N/2} c_k \cos k\left(\theta_i + \phi_k\right) = \sum_{k=-N/2}^{N/2} X_k e^{jk\theta_i}, \tag{9.1}$$

DOI: 10.1201/9780367821746-9

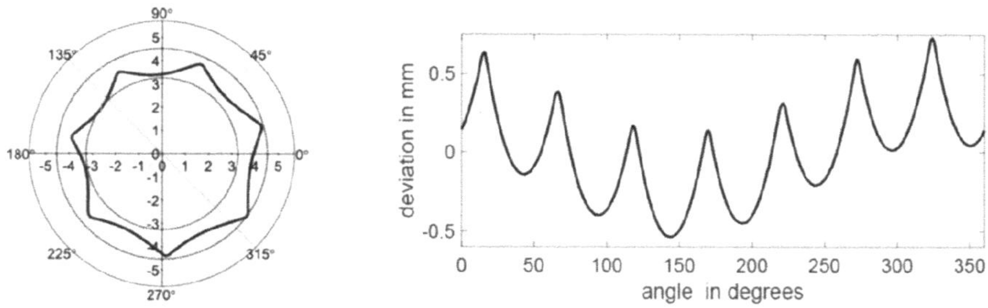

FIGURE 9.1 Roundness measurement of a coin depicted in a roundness diagram (left) as well as in a linear plot (right).

where j is the imaginary unit. Equation (9.1) gives three ways in which the function $r(\theta_i)$ can be written as a series of Fourier components, also called harmonic components. The constants in equation (9.1) are related in the following way:

$$X_k = \frac{1}{2}\sqrt{a^2+b^2}\cdot\exp\left(\mathrm{j}\cdot\tan^{-1}\frac{b_k}{a_k}\right) = \frac{c_k}{2}\exp\left(\mathrm{j}\cdot\phi_k\right), \tag{9.2}$$

where $a_{-k}=a_k$, $b_{-k}=-b_k$, $c_{-k}=c_k$ and $\phi_{-k}=-\phi_k$. The complex coefficients X_k can be described as complex numbers with amplitude $\frac{c_k}{2}$ and phase ϕ_k. A graph of the components c_k as a function of k is called the amplitude spectrum; a graph of the components ϕ_k as a function of k is called the phase spectrum. The number k is called the harmonic component and is the angular frequency, or the number of undulations per revolution (upr). Note that ϕ_k is not defined if c_k is zero. The coefficients can be calculated by taking half of the mean of the product of the function $r(\theta)$ and the harmonic component considered:

$$a_k = \frac{2}{N}\sum_{i=1}^{N} r\left(\theta_i\right)\cos\frac{2\pi ki}{N} \text{ and } b_k = \frac{2}{N}\sum_{i=1}^{N} r\left(\theta_i\right)\sin\frac{2\pi ki}{N}. \tag{9.3}$$

The complex amplitudes X_k can also be calculated directly from

$$X_k = \frac{1}{N}\sum_{i=1}^{N} r\left(\theta_i\right)\exp\left(-\mathrm{j}\frac{2\pi ki}{N}\right). \tag{9.4}$$

The calculation of coefficients a_k and b_k in equation (9.3), and analogously the amplitudes and phases c_k and ϕ_k, and the complex amplitudes X_k is called the Fourier transform. For this, fast algorithms have been developed and software functions performing this operation are mostly called 'fast Fourier transforms' (FFTs). Reconstruction of a function from the Fourier components, such as given in equation (9.1), is the inverse Fourier transform, and fast software functions performing this operation are called 'inverse fast Fourier transforms' (IFFTs or FFT^{-1}). Note that in some applications, such as Matlab or Excel, the $1/N$ factor is omitted in equation (9.4) and applied in equation (9.1) instead. A basic feature of this analysis is that the coefficients and components are independent; adding or removing one Fourier component, for example the decentering or the ellipticity, does not affect the other components. Figure 9.2 illustrates the signal, its transformation in amplitude and phase components and its inverse transform, where the original function is reconstructed as the sum of sine and cosine functions with the amplitudes calculated from the Fourier transform. This example contains the first Fourier component (the decentering), the seventh harmonic and its multiples (higher harmonics). The very small other components that were present in the original signal are omitted. A noteworthy relation between the signal and its Fourier transform can be found when considering the square of the standard deviation; the variance:

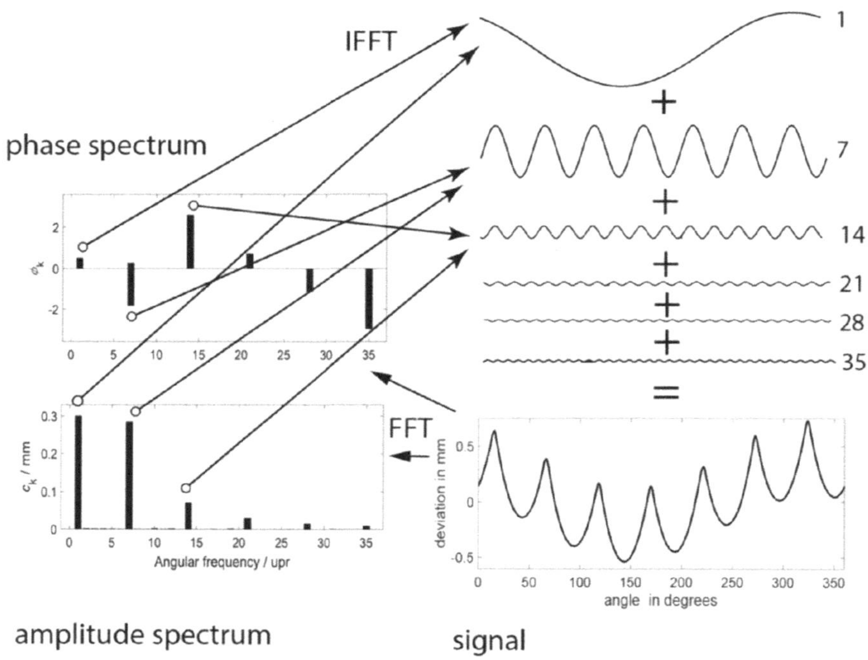

FIGURE 9.2 Signal and its (fast) Fourier transform (FFT) represented as an amplitude and phase spectrum. The inverse (fast) Fourier transform (IFFT) reconstructs the original signal as the sum of its harmonic components.

$$RON_q^2 = \frac{1}{N}\sum_{i=1}^{N} r^2\left(\theta_i\right) = \frac{a_0^2}{4} + \sum_{k=1}^{N/2} \frac{a_k^2 + b_k^2}{2} = \sum_{i=-N/2}^{N/2} \left|X\right|_k^2 . \qquad (9.5)$$

This is called Parseval's theorem. A spectrum of the square of the amplitudes is called the power spectrum. The inverse Fourier transform of the power spectrum is the autocorrelation spectrum. The autocorrelation spectrum is the correlation of a signal with a shifted copy of itself as a function of this shift $\Delta\theta_i$:

$$AC\left(\Delta\theta_i\right) = \frac{1}{N}\sum_{k=1}^{N} r(\theta_k)\cdot r\left(\theta_k + \Delta\theta_i\right) = \sum_{k=1}^{\frac{N}{2}} c_k^2 \cos k(\Delta\theta_i) = \sum_{k=-\frac{N}{2}}^{\frac{N}{2}} \left|X_k\right|^2 \cdot e^{jk''\theta_i} . \qquad (9.6)$$

This equation is called the Wiener-Khinchin theorem. The autocorrelation function may reveal features of a signal that are not directly visible in the signal itself or in the amplitude spectrum. As it contains only cosine terms, it is symmetrical around zero, and the phase information is lost/unused. Figure 9.3 gives the autocorrelation function for the roundness diagram in Figure 9.1. The figure shows the characteristic seven lobes together with the cosine-wave because of the de-centering. Note that the autocorrelation of this coin will remain the same if the coin is rotated over an angle (keeping the same decentering); this gives autocorrelation a more 'general' fingerprint of the signal than the signal itself. For example, when comparing the profile of two coins it may be more useful to compare the autocorrelation functions rather than the profiles themselves, as in that case 'rotating' the diagrams for an optimal fit is not needed. On the other hand, the phase spectrum is sensitive to orientations and locations of features. For example, for the evaluation of the phase in an interference signal, as described in Section 4.5.1, in fact the phase of the first Fourier component is calculated in order to locate a surface height. Rotation of the profile over an angle α causes a shift of $n{\cdot}\alpha$ in the phase of the n-th Fourier

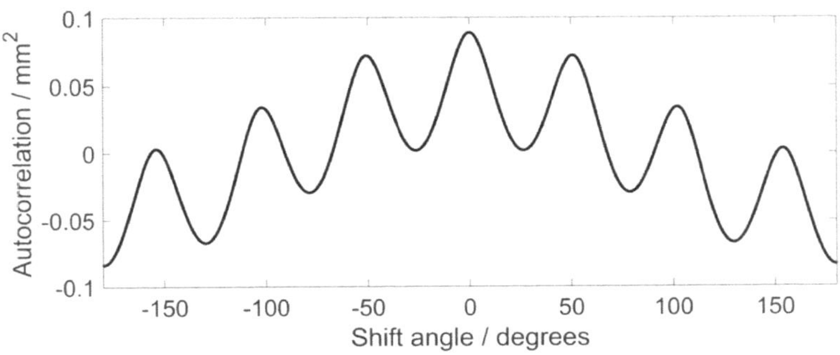

FIGURE 9.3 Autocorrelation function of the roundness diagram given in Figure 9.1.

component. This is called the Fourier shift theorem. This phenomenon can be used to locate a white-light interference fringe, enabling the evaluation of white-light interferometer signals in the spatial frequency domain (De Groot and Deck 2015).

9.3 FORM REMOVAL AND FILTERING FOR ROUNDNESS MEASUREMENTS

The centering of the roundness profile according to the least-squares reference circle can be accomplished by taking the Fourier transform of the signal using equations (9.3) or (9.4), setting the components of the first harmonic (a_1 and b_1 or c_1 or X_1 and X_{-1} in equation (9.1)) equal to zero and taking the inverse transform by applying equation (9.1). An alternative method is to calculate a_1 and b_1 or c_1 or X_1 and X_{-1} by applying equation (9.3) or (9.4) respectively, calculate the decentered circle using equation (9.1) and subtract this from the profile. Explicit expressions for this were already given in equations (7.24) and (7.25). Likewise, the ellipticity can be calculated or removed by considering the second harmonic component (k= 2) in equations (9.1), (9.3) and (9.4). It is important to note that these components are removed completely. This is called form removal and it will be shown that form removal is different from filtering, where the harmonic components are partially removed. As discussed in Section 7.8.3, the default filter for roundness measurements is the Gaussian filter. In general, a filter function is a function that describes how for a certain angle the measured value is replaced by a weighed mean of the measured point and its neighboring points at an angular distance $\Delta\theta$. For a Gaussian filter, this weighing function $s(\Delta\theta)$ has a Gaussian shape as shown in equation (9.7):

$$s_w(\Delta\theta) = \frac{f_c}{2\pi\alpha}\exp\left(-\pi\left(\frac{f_c \cdot \Delta,}{2\grave{\mathrm{A}}\alpha}\right)^2\right), \tag{9.7}$$

where $\alpha = \sqrt{ln(2)/\pi}$ is the constant to provide a 50% transmission characteristic at the cut-off frequency f_c, expressed in undulations per revolution (upr). In numerical evaluations, the function $s_w(\Delta\theta)$ has to be normalized:

$$s_w(\Delta\theta_i) = \frac{\exp\left(-\pi\left(\frac{f_c \cdot \Delta\theta_i}{2\pi\alpha}\right)^2\right)}{\sum_{i=-N/2}^{N/2}\exp\left(-\pi\left(\frac{f_c \cdot \Delta\theta_i}{2\pi\alpha}\right)^2\right)}. \tag{9.8}$$

The filtered profile is given by the convolution of the original profile $r(\theta)$ with the filter function $s_w(\Delta\theta)$, meaning that at every angle θ, the function $r(\theta)$ is replaced by a weighted average over the neighboring points. This convolution gives the filtered function $w(\theta)$:

$$w(\theta)=\int_{-\pi}^{\pi} r(\Delta\theta)s\,(\theta-\Delta\theta)d\Delta\theta \approx \int_{-2\pi/f_c}^{2\pi/f_c} r(\Delta\theta)s_w\,(\theta-\Delta\theta)d\Delta\theta, \quad (9.9)$$

where the limitation of the integral is justified by considering that $s_w\,(\theta-\Delta\theta)\approx 0$ for $|\theta-\Delta\theta|>2\pi/f_c$. The convolution theorem states that the Fourier transform of a convolution of two functions is the product of their Fourier transforms. Denoting the Fourier transform as given in equations (9.3) and (9.4) by *FT,* and the inverse Fourier transform as given in equation (9.1) by FT^{-1}, the convolution integral of equation (9.9) can be rewritten as:

$$w(\theta)=\int_{-\pi}^{\pi} r(\Delta\theta)s\,(\theta-\Delta\theta)d\Delta\theta = r(\theta)\otimes s(\theta) = FT^{-1}\left(FT\left(r(\theta)\right)\cdot FT(s(\theta)\right). \quad (9.10)$$

The $\otimes$ –sign in equation (9.10) denotes the convolution. The right-hand side of equation (9.10) gives the principle of filtering in the frequency domain. As fast algorithms for calculating the Fourier transform are readily available, this may be the fastest way of filtering. The Fourier transform of the Gaussian weighing function $s(\Delta\theta)$ given in equation (9.7) is the angular frequency transmission characteristic $T_w(k)$ given by:

$$T_w\,(k)=\frac{c_k\left(w(\theta)\right)}{c_k\left(r(\theta)\right)}=\exp\left(-\pi\left(\frac{\alpha k}{f_c}\right)^2\right)=2^{-\frac{k}{f_c}}, \quad (9.11)$$

where $\alpha=\sqrt{ln(2)/\pi}$, f_c is the cut-off frequency where the Fourier component (amplitude) $c_k\left(w(\theta)\right)$ of the filtered profile $w(\theta)$ is reduced 50% relative to the Fourier component $c_k\left(r(\theta)\right)$ of the unfiltered profile. The function $T(k)$ is displayed in Figure 7.36 for f_c= 15 upr, 50 upr, 150 upr, 500 upr and 1500 upr. Equation (9.10) denotes the two equivalent routes for Gaussian filtering: the first being filtering in the angular domain: taking the convolution of the roundness $r(\theta)$ and the profile and the filter weighing function $s(\Delta\theta)$. The second route is taking the Fourier transform of the signal, multiply the Fourier components of the roundness profile with the transmission characteristics $T(k)$ and take the inverse Fourier transform of this filtered spectrum. These two routes are graphically depicted in Figure 9.4 for a Gaussian filter with a cut-off frequency of f_c=15 upr, combined with the centering operation: $T(1)$=0. At the left-hand side, the convolution of the filter function $s_w(\Delta\theta)$

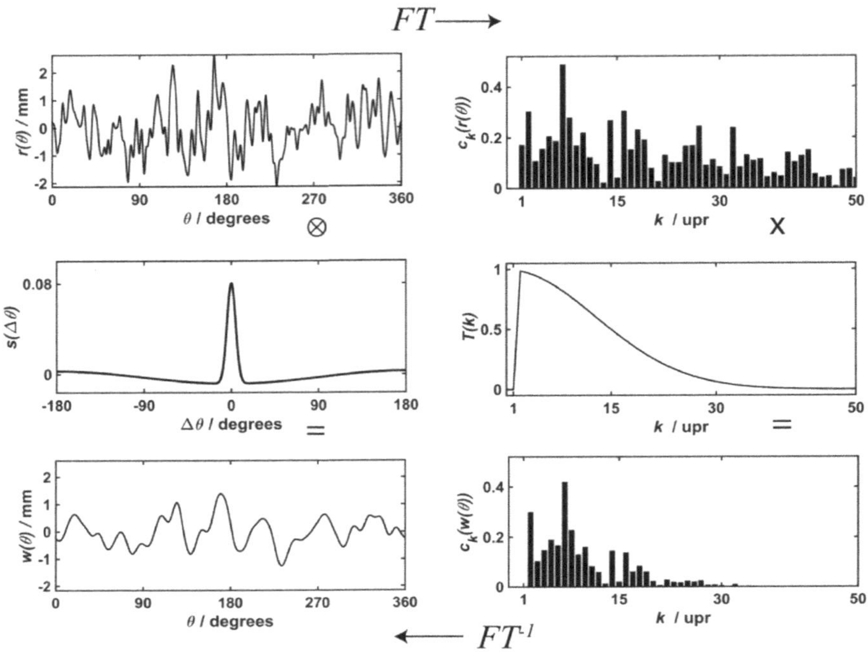

FIGURE 9.4 Graphical impression of the two routes for filtering a roundness measurement, see text.

with the measured profile $r(\theta)$ that gives the filtered profile $w(\theta)$ is depicted. At the right-hand side, the amplitude spectrum obtained by the Fourier transforms of the measured roundness profile that is multiplied by the transmission $T_w(k)$ to obtain the Fourier transform of the filtered profile. The filtered profile $w(\theta)$ is obtained from the inverse Fourier transform of this filtered amplitude spectrum; together with the unchanged phase spectrum. Clearly, the filtered profile is smoother; this is connected to the reduction of the higher frequency components in the amplitude spectrum. This filtering of the higher frequencies is called S-filtering. These higher frequencies may just constitute the profile the user is interested in. The profile containing the higher frequencies $h(\theta)$ may be obtained by subtracting the filtered profile $w(\theta)$ from the original profile $r(\theta)$, however, as in this example $w(\theta)$ was centered, the first harmonic component of $r(\theta)$ must also be removed:

$$h(\theta) = r(\theta) - w(\theta) - c_1 \cos k(\theta_i + \phi_1), \tag{9.12}$$

or:

$$r(\theta) = c_1 \cos k(\theta_i + \phi_1) + w(\theta) + h(\theta). \tag{9.13}$$

Equation (9.12) can be summarized in words as: the profile is the sum of the eccentricity, waviness and roughness. This is illustrated in Figure 9.5 as the roundness diagrams of $r(\theta)$ and its components. The profile $h(\theta_i)$ may also be obtained directly by the convolution of the profile with the filter function $s_h(\Delta\theta_i)$, defined as:

$$s_h(\Delta\theta_i) = \delta(\Delta\theta_i) - \frac{\exp\left(-\pi\left(\frac{f_c \cdot \Delta\theta_i}{2\pi\alpha}\right)^2\right)}{\sum_{i=-N/2}^{N/2} \exp\left(-\pi\left(\frac{f_c \cdot \Delta\theta_i}{2\pi\alpha}\right)^2\right)}, \tag{9.14}$$

where $\delta(\Delta\theta_i)$ is the Dirac delta-function: $\delta(0) = 1,\ \delta(\Delta\theta_i \neq 0) = 0$.

The profile $h(\theta)$ may also be obtained directly by applying the function $T_h(k)$ in the frequency domain, given by:

$$T_h(k) = \frac{c_k(h(\theta))}{c_k(r(\theta))} = 1 - \exp\left(-\pi\left(\frac{\alpha k}{f_c}\right)^2\right) = 1 - 2^{-\frac{k}{f_c}}. \tag{9.15}$$

The Gaussian filter as discussed in this section is the default filter for roundness measurements as specified in ISO 12181-2:2012 and is specified in more detail in ISO 16610-21:2025.

9.4 FORM REMOVAL AND FILTERING FOR PROFILE MEASUREMENTS

Surface topography profiles $z(x)$ are principally different from roundness profiles, but the Fourier analysis and relating filtering is similar to a high degree; for that reason this was treated in Section 9.2 extensively. Similarly to a roundness profile having any eccentricity when measured, a surface topography can have any orientation. Where in roundness measurements the eccentricity can be removed as part of the filtering in the Fourier domain, this is not the case for the orientation of a surface profile. It is common to 'level' the measured profile topography $z'(x)$ before any filtering takes place by subtracting a straight line from the measured profile where the least-squares line is the most common option, giving:

$$z(x) = z'(x) - \alpha - \beta \cdot x, \tag{9.16}$$

where α and β are the constants obtained from the least-squares fit given in Section 7.3.1. Here, it is assumed that $z(x) << L$, where L is the measured length. If this

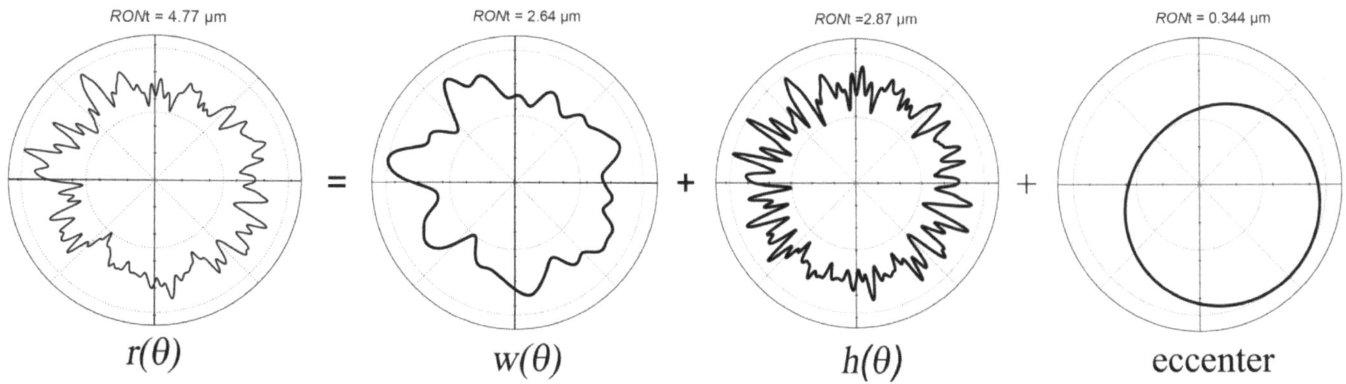

FIGURE 9.5 Example of a roundness diagram $r(\theta)$ and its decomposition in a $w(\theta)$ diagram ('waviness'), a $h(\theta)$diagram ('roughness') and its eccentricity.

condition is not met, the constants α and β must be optimized to the minimum absolute distance to the line; this is called the total least-squares method. After making the profile independent of the basic orientation, further form removal may be necessary, for example, by fitting a polynomial or a (partial) circle, where it must be noted that a partial circle may well be approximated by a second-order polynomial. Removing a form in this way is called the F-operation, where it should be noted that the orientation strictly speaking cannot be regarded as a form component; see also Figure 7.41.

The amplitude spectrum may be obtained from the Fourier transform of the leveled surface similarly as for the roundness measurements, where the abscissa is expressed in undulations per millimeter rather than per revolution, or inversely as a wavelength. The autocorrelation function can be obtained as $AC(\Delta x)$, as in equation (9.6), however, with a modification as the profile $z(x)$ is not by definition periodic:

$$AC(\Delta x_i) = \frac{1}{N-i}\sum_{k=1}^{N} z(x_k)\cdot z(x_k+\Delta x_i) \approx \sum_{k=1}^{\frac{N}{2}} c_k^2 \cos k('' x_i) = \sum_{k=-\frac{N}{2}}^{\frac{N}{2}} \left|X_k\right|^2 \cdot e^{jk'' x_i} \quad (9.17)$$

In ISO 21920-2:2021 the autocorrelation function f_{ACF} is defined as in equation (9.17), however it is normalized to give $f_{ACF}(0)=1$:

$$f_{ACF}(\Delta x_i) = AC(\Delta x_i)/AC(0) = \frac{\frac{1}{N-i}\sum_{k=1}^{N} z(x_k)\cdot z(x_k+\Delta x_i)}{\frac{1}{N}\sum_{k=1}^{N} z^2(x_k)}. \quad (9.18)$$

A typical autocorrelation function may be obtained for a type D surface texture material measure (see section 6.4.2). Figure 9.6 gives the profile z(x) of such a standard together with the autocorrelation function $f_{ACF}(\Delta x)$. The peaks at Δx = + 4 mm and Δx = -4 mm show that the profile repeats every 4 mm, something that is less obvious when looking at the profile itself.

9.4.1 Gaussian Filter for Profile Measurements

The Gaussian filter for profile measurements is similar to the Gaussian filter for roundness measurements, however, there are some differences due to the finite length of the profile and, related to this, effects near the edges, also called 'edge effects'. Where a roundness profile $r(\theta_i)$ has N equidistant measurement points at angles $\theta_i = i\cdot 2\pi/N$ with i=1..N, a surface profile $z(x_i)$ will have N equidistant measurement points at positions $x_i=i\cdot L/N$ with i=1..N and L is the profile length. Where in roundness measurements the Fourier components k have the unit 'upr',

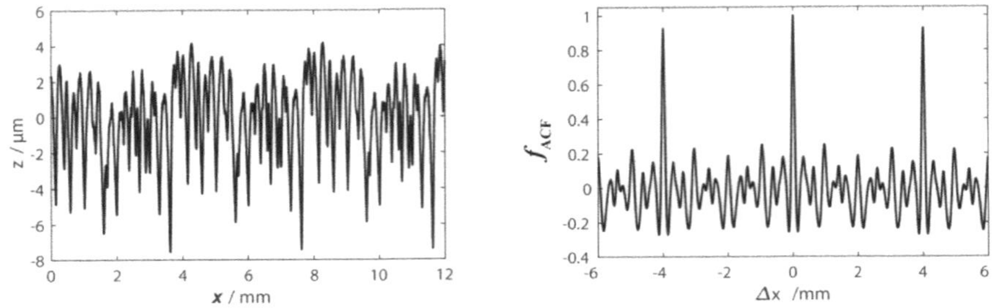

FIGURE 9.6 Profile $z(x)$ of a surface texture type D material measure (left-hand side) and its autocorrelation function $f_{ACF}(\Delta x)$ (right-hand side).

the related Fourier components for profile measurements have the frequency 'per length L' and this needs scaling to 'per millimeter', where the cut-off frequency is given as $f_s = 2\pi L / \lambda_s$. The roughness ('R') profile is obtained by filtering the short-wave components by the short-wavelength filter characterized by λ_s (the S-filter) and the long-wave components by λ_c (the L-filter). For this the weighing function $s(\Delta x)$ becomes:

$$s(\Delta x) = \frac{1}{\alpha\lambda_s}\exp\left(-\pi\left(\frac{\Delta x}{\alpha\lambda_s}\right)^2\right) - \frac{1}{\alpha\lambda_c}\exp\left(-\pi\left(\frac{\Delta x}{\alpha\lambda_c}\right)^2\right), \qquad (9.19)$$

with $\alpha = \sqrt{ln(2)/\pi}$. For a numerical evaluation both terms in the right-hand side of equation (9.19) must be normalized as in equation (9.8). The transmission function $H(\lambda)$ becomes:

$$H(\lambda) = 2^{-\lambda_s/\lambda} \cdot \left(1 - 2^{-\lambda_c/\lambda}\right). \qquad (9.20)$$

The function $H(\lambda)$ was already shown in Figure 7.43 for the default values for λ_s and λ_c given in Table 7.5. For the evaluation, it is noteworthy that the convolution of the profile $z(x)$ and the filter function $s(\Delta x)$ is not defined if $x+\Delta x > L$ or $x-\Delta x<x_1$. For the determination of the spectrum or for filtering in the frequency domain it should be noted that when applying the Fourier transform, the profile is considered to continue periodically ('as if it is a roundness profile'), which may give deviations (end effects) within a distance of $\lambda_c/2$ from the edges. As the second term in equation (9.19) becomes small for $\Delta x > 0.5{\cdot}\lambda_c$, it may be justified to truncate the filter function $s(\Delta x)$ for $|\Delta x| > 0.5{\cdot}\lambda_c$. To be able to apply equation (9.19) in the interval $[x_1; L]$ it is custom to extend the measurement range to the interval $[x_1 - 0.5{\cdot}\lambda_c, L + 0.5{\cdot}\lambda_c]$ or $[x_1 - \lambda_c, L + \lambda_c]$ so equation (9.19) can be applied. For filtering in the frequency domain this means that the measured profile in the interval $[x_1 - 0.5{\cdot}\lambda_c; L + 0.5{\cdot}\lambda_c]$ or $[x_1 - \lambda_c, L + \lambda_c]$ is filtered similar to the procedure depicted in Figure 9.4, then the interval for further evaluation is restricted to $x_i \in [x_1; L]$. In most software of texture measuring instruments, the user may choose to have either a pre-and post-travel of $0.5 \cdot \lambda_c$ or λ_c for this reason. This may be no problem if the object to be measured has plenty of space to take a profile, but it is awkward that quite some profile information is not used, or in fact it is only used to properly define the filter. Therefore, several methods have been designed to enable a reasonable filtering and evaluation over the full measured profile. Some of these methods are also summarized in ISO 16610-21:2025 and ISO 16610-28:2015. Two relatively easy to implement methods are the following:

1. The function $s(\Delta x)$ is (further) truncated – and re-normalized – in the regions $[x_1, 0.5 \cdot \lambda_c]$ and $[L\text{-}0.5 \cdot \lambda_c, L]$, so that the profile boundaries are not crossed when taking the weighed mean.
2. The profile is extended symmetrically over half a length L on both sides: for $x \in [-L/2, x_1]$: $z(-x)=z(x)$ and for $x \in [L, L+0.5L]$ $z(x+L)=z(L-x)$. This extended profile can be filtered in the frequency domain, and afterwards the filtered profile of the original interval $x \in [x_1, L]$ is taken for further evaluation. This operation makes the function continuous, so no high frequencies are introduced if there is a sharp jump between $z(x_1)$ and $z(L)$. This is equivalent to using the cosine-transform instead of the Fourier transform.

The Gaussian filter is the default filter for nearly all surface texture profile parameters defined in ISO 21920-2:2021, and for straightness measurements according to ISO 12780-2:2011; see also Table 7.1.

9.4.2 Robust Profile Filters: Gaussian Regression Filters

The Gaussian filter can be sensitive to peaks in the profile: the mean line will somehow be moved towards this peak and the filtered profile will be distorted. Filters that are adapted to counter this effect are called robust filters. A simplified version is the so-called '3σ – filter'; points that deviate more than three standard deviations from the mean line are disregarded or put equal to the mean line value and the filter is applied again to calculate the mean line. In the filter standardized in ISO 16610-31:2025, the robustness is implemented by giving points a reduced weight depending to their distance from the mean line for a second iteration of the Gaussian filter. The measures against end-effects given in Section 9.4.1 may not be sufficient if the profile is strongly curved near the edges. Therefore, the regression filter was introduced: the Gaussian weighing function is applied combined with a local fit to a regression line of a certain order, which causes a modification of the Gaussian weighing function (Brinkman and Bodschwinna 2003). Such filters are called Gaussian regression filters. In the ISO 16610-31:2025 standard, these two concepts are combined. In ISO 21920-2:2022 it is specified that the robust Gaussian, second order regression filter should be used for the surface texture profile parameters *Rk, Rpk, Rvk, Rpkx, Rvkx, Rmrk1, Rmrk2, Rak1, Rak2, Rpq, Rmq* and *Rvq*. These parameters are derived from curves based on the distribution of the height coordinates, called the material ratio curve.

9.4.3 Morphological Profile Filter; the Mechanical Surface

If a surface topography is measured using a mechanical probe with a spherical tip, the position of the center of this sphere is naturally taken as the z-coordinate. Depending on the probe radius, the probe will not follow the finer structures when it is moving over the surface; the probe will not measure dales if it touches the tops of two adjacent hills simultaneously. In its function, the surface may be used for a mechanical bearing where a ball moves over the surface and it may be relevant to know what part of the surface will not be in contact with the ball. Therefore, the morphological profile filter is designed with a ball – a disk in 2D – as the default structuring element and the profile is calculated which is when the surface is probed from the top (dilation) or from the bottom (erosion), and when a subsequent correction is made, knowing the ball/disk radius (closing: the erosion of the dilation profile with the same diameter disk, and opening: the dilation of the erosion profile with the same diameter disk respectively). Zero measurement force/indentation is assumed. These basic concepts are illustrated in Figure 9.7 and are defined in ISO 16610-41:2015. In the calculation, the *z*-coordinates of both the structuring element (the disk) and the profile in the region $x = x \pm r$, where r

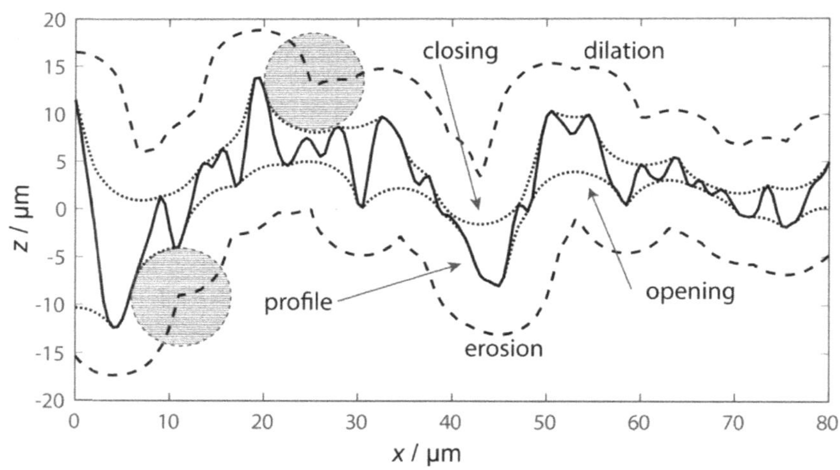

FIGURE 9.7 Surface profile and the morphological filter operations: dilation and closing when the profile is probed with a circular disk of 5 µm radius from the upper side, and erosion and opening when the profile is (hypothetically) probed from downside.

is the disk radius, are considered, so that $z(\text{disk}) > z(\text{profile})$ in that region. The coordinates of the disk with radius r in the vicinity of $z(x)$ are given by:

$$z_{\text{disk}}(x+\Delta x) = z(x) + r\left(1-\sqrt{1-\left(\frac{\Delta x}{r}\right)^2}\right) \text{ for } |\Delta x| < r. \tag{9.21}$$

The dilation profile z_d is found from

$$z_d(x) = z(x) + r + \max\left(z_{disk}(x+\Delta x) - z(x+\Delta x)\right) \text{ for } |\Delta x| < r. \tag{9.22}$$

As the calculation of equation (9.22) must be repeated for every x-coordinate, it has some similarity with the convolution given in equation (9.9), therefore, this is sometimes called the convolution of the profile with the probe; and the erosion of the measured values is called the deconvolution, but these denotations are not strictly mathematically correct (convolution is a linear process, and in this case, it is nonlinear). It can be calculated whether a probe/disk of radius r can touch the bottom of a sinusoidal profile with amplitude A and wavelength λ, denoting the surface profile near the bottom for $x << \lambda/\pi$ as:

$$z(x) = -A\cos\frac{2\pi x}{\lambda} \approx -A\left(1+\frac{1}{2}\left(\frac{2\pi x}{\lambda}\right)^2\right), \tag{9.23}$$

and the disk for $x<<r$ as:

$$z_{disk}(x) \approx r\left(1-\left(1-\frac{1}{2}\left(\frac{x}{r}\right)^2\right)\right) \approx \frac{x^2}{2r}. \tag{9.24}$$

The probe will just fit if the curvatures are the same. This means that the second derivative of both equations (9.23) and (9.24) near x=0 must be the same, giving as a criterion for the radius:

$$r \leq \frac{\lambda^2}{4\pi^2 A}. \tag{9.25}$$

The closing profile can be taken as the reference line, and the roughness profile as the difference of the measured profile with the closing profile. As an example, Figure 9.8 gives the profile of specimen with its dilation with a 5 mm disk, and the filtered profile, taking the dilation as the reference line.

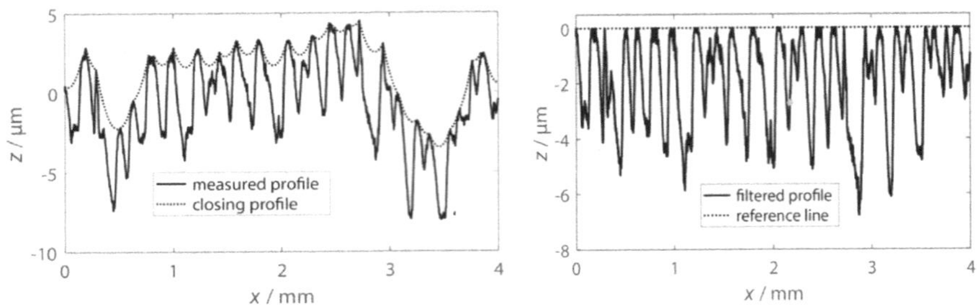

FIGURE 9.8 Surface profile with the closing profile using a 5 mm disk (left-hand side) and the morphologically filtered profile taking the closing profile as the reference line (right-hand side).

The mechanical surface

The mechanical surface is defined as the erosion of a measured surface with a given probe radius, with that same radius. In Figure 9.7 this means that the 'real' profile is probed with a given radius, giving the dilation profile (assuming no measuring force or friction); and the mechanical profile is obtained from the erosion of this profile, giving the closing profile in Figure 9.7. This is the best estimate that can be made of the 'real' profile, based on the information in the measured dilation profile and knowing the probe radius, and this is called the mechanical surface that is the default surface for measurement, filtering and parameter calculations, see also ISO 21920-2:2021 and ISO 14406:2010.

9.5 FORM REMOVAL AND FILTERING FOR AREAL SURFACE TOPOGRAPHY MEASUREMENTS

The form removal and filtering for areal surface topography measurements is similar to the corresponding profile case as described in the previous sections of this chapter. The basic orientation of a planar surface topography can be removed by 'leveling' the profile topography by subtracting a plane from the measured areal topography $z'(x,y)$ where the least-squares plane is the most common option, giving:

$$z(x,y) = z'(x,y) - a - b \cdot x - c \cdot y, \tag{9.26}$$

where a, b and c are the constants obtained from the least-squares fit, given in Section 7.6.1.

More complicated forms can be removed using higher-order polynomials; these can be written as orthogonal (independent) forms as Legendre polynomials. Surface topographies defined inside of a circle-shaped surface can be written in polar coordinates that can be decomposed into Zernike polynomials, that consist of modified Legendre polynomials for the radial parts and sine and cosine functions for the azimuthal parts (Noll 1976). A special case is aspheric surfaces that can be described by a spherical form with a conical deviation, and further deviations as polynomials for the radial coordinate r from the center, the optical axis:

$$z(r) = \frac{r^2 / R}{1 + \sqrt{1 - (1 + K)(r / R)^2}} + a_2 r^2 + a_4 r^4 + \ldots, \tag{9.27}$$

where R is the radius of the surface, K is the conic constant and a_n are the coefficients of the higher-order polynomials. This expression and some more variants are described in ISO 10110-12: 2015.

The Gaussian filter can be applied similarly as the profile filter to an areal topography measurement $z(x,y)$ with weighing function for the waviness topography (ISO 16610-61:2015):

$$s(x,y) = \frac{1}{\alpha^2 \lambda_c^2} \exp\left(-\frac{\pi}{\alpha^2} \left(\frac{x^2 + y^2}{\lambda_c^2} \right) \right), \tag{9.28}$$

with $\alpha = \sqrt{ln(2) / \pi}$ and λ_c the cut-off wavelength. Because of the linear character of the Gaussian filter, the filter operation can also be achieved by filtering all profiles $z(x,y_i)$ and $z(x_i,y)$ subsequently. The filtering can also be done in the 2-D frequency domain using the 2-D Fourier transform. Computing the 2D Fourier transform of a matrix of z(x,y) values is equivalent to first computing the 1D transform of each column of z(x,y) and then taking the 1D transform of each row of the result. The 2D Fourier transform and its inverse are available in Matlab as 'fft2' and 'ifft2'. Next to filtering it can also be used to compute the 2D autocorrelation function. According to ISO 25178-3:2012, the Gaussian filter is the default filter for the assessment of areal surface texture. If applied to areal topography data, the robust regression filters and the morphological filter need an areal assessment. A cylindrical surface can be regarded as a combination of roundness and profile topographies, and the extracted roundness and straightness profiles can be filtered subsequently, see also section 7.8.5. As an illustration, Figure 9.9 gives a measurement of a 18.5 mm by 19.6 mm area of a potato chip, taken with a mechanical surface texture measuring instrument. 200 profiles each containing 2000 points were taken. The measured surface texture was first leveled, then the saddle-shape form was removed by fitting a hyperbolic paraboloid to the measurements. The remaining surface texture was decomposed in the higher and lower spatial frequency components by using a 2D Gaussian filter with cut-off wavelength $\lambda_s = 0.24$ mm and $\lambda_c = 0.4$ mm. The surface containing the wavelengths between 0.24 mm and 0.4 mm is called the S-L surface; the surface containing the wavelengths above 0.4 mm but with the form removed is called the S-F surface. See also Figure 7.41.

9.6 THE ISO 16610-FRAMEWORK

The filters treated in this chapter are part of a wider endeavor for classifying and standardizing the comprehensive possibilities of filtering. Within the ISO 16610-framework the classification is made as summarized in Table 9.1. Examples of

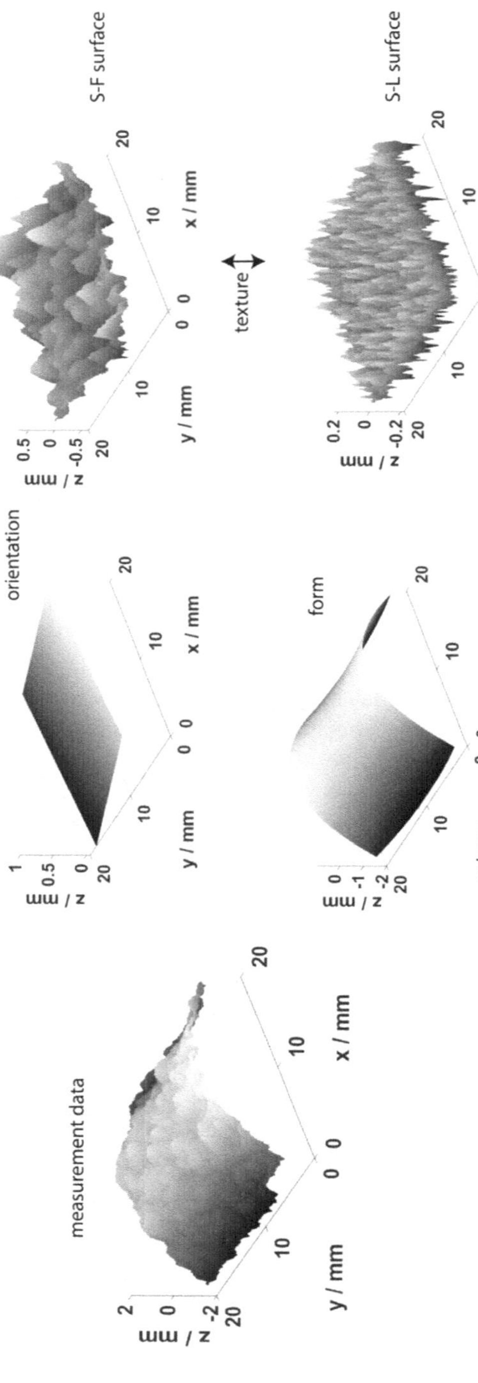

FIGURE 9.9 Surface topography measurement of a potato chip and its decomposition into the orientation, form and texture, the texture being separated into the longer wavelengths (S–F surface) and the shorter wavelengths (S–L surface).

TABLE 9.1
Matrix of the Present and Foreseen Standards in the ISO 16610 Series

	General: ISO 16610-1					
	Profile Filters			Areal Filters		
	ISO 16610-11			ISO 16610-12		
Fundamental	**Linear**	**Robust**	**Morphological**	**Linear**	**Robust**	**Morphological**
Basic concepts	Part 20	Part 30	Part 40	Part 60	Part 70	Part 80
Particular filters	Parts 21–25	Parts 31–35	Parts 41-45	Parts 61–66	Parts 71–75	Parts 81–85
How to filter	Parts 26–28	Parts 36–38	Parts 46–48	Parts 66–68	Parts 76–78	Parts 86–88
Multiresolution	Part 29	Part 39	Part 49	Part 69	Part 79	Part 89

filters in this framework that not treated in this chapter are the spline filters and the spline wavelet filters. Time will learn which filters will find application in surface metrology.

9.7 AREAL SURFACE TEXTURE PARAMETERS

Areal texture parameters are defined in ISO 25178-2:2012. The most commonly used amplitude parameters were already mentioned in Section 7.9.2. Areal parameters are generally denoted by S (surface) or V (volume). The notation S_{xxx} or V_{xxx} is used in product documentation, drawings and data sheets, where the xxx denotes the specific parameter as defined in the ISO standard. When used as a symbol in an equation the notation is S_{xxx} or V_{xxx}. Next to the extensions from the profile parameter, there are specific parameters that depend in the directional character of the surface. An example is the spatial parameter 'autocorrelation length'. The roughness profile parameter Ral(s) is defined as the horizontal distance where the autocorrelation function $f_{ACF}(\Delta x)$ decays to the value *s;* the default value for *s* is 0.2. The areal parameter Sal(s) 'autocorrelation length' is defined as the horizontal distance of the 2D autocorrelation function $f_{ACF}(\Delta x,\Delta y)$ which has the fastest decay to a specified value *s*. Figure 9.10 gives the autocorrelation function $f_{ACF}(\Delta x)$ of the central profile in the *x*-direction of the S-F surface depicted in Figure 9.9, as well as the 2D autocorrelation profile $f_{ACF}(\Delta x,\Delta y)$ of this S-F surface, and this 2D autocorrelation with the $s=$ 0.2 threshold indicated. The figure shows that both 'autocorrelation length' definitions are principally different. Another areal parameter that can be derived is the texture aspect ratio Str(s), that is the ratio of the horizontal distance of the $f_{ACF}(\Delta x,\Delta y)$ that has the fastest decay to a specified value *s* to the horizontal distance of the $f_{ACF}(\Delta x,\Delta y)$, which has the slowest decay (indicated as Sml in Figure 9.10). This parameter can be derived from Figure 9.10 (with the default s=0.2) as $S_{tr}=S_{al}/S_{ml}$.

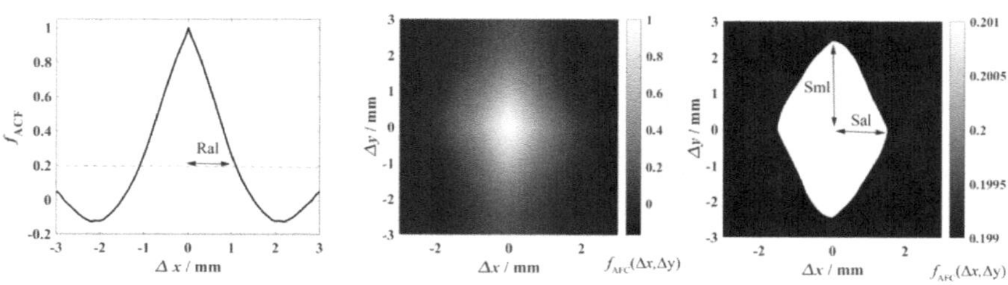

FIGURE 9.10 Autocorrelation from a profile from the S–F topography given in Figure 9.9 (left-hand side), the 2D autocorrelation of this topography in the center, and the same with a threshold $s = 0.2$ (right-hand side).

EXERCISES

1. Roughness measurement
   A workpiece is measured using a mechanical texture measuring instrument having a 2 μm probe diameter. The profile can be described by the function given in equation (9.29):

$$z(x) = \cos(\pi \cdot x/10) + 5 \cdot \sin(\pi \cdot x/400) \tag{9.29}$$

   with x and z both expressed in micrometers.

   The profile length is 4 mm. The profile consists of a larger amplitude with a wavelength $\lambda = 800$ μm and a smaller amplitude with a wavelength $\lambda = 20$ μm.
   a. Calculate the parameter Rt of this profile.
   b. Estimate the value of parameter Ra. A graphical method is sufficient.
   c. Determine whether the probe can measure this profile without affecting the Rt parameter.
   d. Give the expression of the autocorrelation function f_{ACF} (Δx) of this profile.

   This profile is filtered using a Gaussian filter having a cut-off wavelength $\lambda_c = 0.8$ mm.
   e. Give the expression of the amplitude transmission of this filter.
   f. Give the expression of the filtered profile.
   g. Calculate Pq and Rq.
2. Discrete roundness measurement
   In a roundness measurement, the radius values R are measured as a function of the angle φ in degrees and R in micrometers; they are given in Table 9.2.
   a. Determine the concentricity deviation of the workpiece relative to the rotary axis of the roundness tester.
   b. Calculate the roundness deviation relative to the least-squares center (*RON*t(LSC)).
   c. Determine the first four Fourier coefficients of this signal.
   d. Determine the 2-point and 3-point out of roundness.

TABLE 9.2
Radius *R* as a Function of the Angle φ of a Roundness Measurement

φ/°	0	45	90	135	180	225	270	315
R / μm	11	10.7	10	10.7	11	9.3	8	9.3

TABLE 9.3
Height z as a Function of the Position x of a Profile Measurement

x/μm	0	1	2	3	4	5
z/μm	0	1	1	−1	−1	0

3. Discrete roughness measurement
We consider a periodic digitized surface profile. The length is a few millimeters, the period is 6 μm. This profile is fully described by the following z-coordinates as a function of x as indicated in Table 9.3.
   a. Determine the parameters Pa and Pq.
   b. Estimate the maximum probe radius that can measure this profile undistorted.
   c. What is the maximum frequency that can be determined in the power spectrum?
   d. Determine a_0, a_k and b_k so that the signal can be described by the Fourier series:

$$z(x) = \frac{a_0}{2} + \sum_{k=1}^{3} a_k \cos\frac{2\pi kx}{6} + b_k \sin\frac{2\pi kx}{6}. \tag{9.30}$$

How could you have determined the coefficients of the third harmonic directly?
   e. Determine the power spectrum.
   f. Give the autocorrelation spectrum.
   g. Calculate the dilation of this profile with a spherical probe with 2 μm radius.
   This profile is filtered with a Gaussian filter with a cut-off length $\lambda_c = 3$ μm.
   h. Give the amplitude transmission of this filter.
   i. Calculate the filtered profile.
4. Roundness reversal
Consider the roundness reversal measurement in section 7.8.4. The roundness deviations of the spindle are denoted by $S(\theta)$, the roundness deviations of the roundness standard are denoted by $R(\theta)$. The measurements in the normal position $r_A(\theta)$ and reverse position $r_B(\theta)$ are given by:

$$\begin{aligned} r_A(\theta)/\mu\text{m} &= \sin\theta + \cos 3\theta - 3\sin 4\theta, \text{ and } r_B(\theta)/\mu\text{m} \\ &= -\cos\theta - 3\cos 3\theta + \sin 4\theta, \end{aligned} \tag{9.31}$$

Give $R(\theta)$ and $S(\theta)$ as the deviation from the least-squares circle from the measurement values in equation (9.31).

5. Roundness multi-step method
Because not every roundness tester allows a measurement in the reverse position as indicated in Figure 9.37, an alternative method for separating the deviations from the roundness standard and the spindle of the roundness tester has been designed that is called the 'multi-step' method. It consists of rotating the specimen to be calibrated in N steps of 360°/N each and a taking a measurement in every angular position of the specimen. The evaluation is a combination of rotations and averaging of the obtained roundness measurements $r_i(\theta)$ with i = 1..N. Here as an example the four-step method is taken: the specimen measured in the normal position as indicated in Figure 7.37, and three more measurements in which the roundness specimen is rotated 90°, 180° and 270° respectively relative to the initial position.
   a. Give the expected measurements $r_i(\theta)$ where the roundness deviation of the specimen is given by $R(\theta) = -\cos(3\theta) - \sin(4\theta)$ and the spindle deviation is given by $S(\theta) = 2\cos(3\theta) - 2\,sin(4\theta)$.
   b. Design an optimum evaluation method to reconstruct $S(\theta)$ from the measurements.
   c. Design an optimum evaluation method to reconstruct $R(\theta)$ from the measurements.
   d. Which harmonic components can be separated and which remain undefined for an n-step method and what would be the best way to handle these components?

6. White-light correlogram analysis by filtering in the frequency domain
In evaluation of white-light correlograms as given in section 4.5.3 an amplitude $E_i = E(d_i)$ and a phase are attributed to every height coordinate d_i based on the intensity measurements $I_{i-2}..I_{i+2}$. These can be summarized in a complex amplitude $K(d_i)=E(d_i)\exp(j\phi(d_i))$.
   a. Check that the complex amplitude can be obtained by the convolution of the intensities $I(d_i)$ with the complex filter function given by $f(\Delta i)$ =(-0.5,-j,1,j,-0.5) for Δi=(-2..2).
   b. According to the convolution theorem the function $K(d_i)$ can be obtained by taking the inverse Fourier transform of the multiplication of the Fourier transforms of both the intensities E_i and the filter function $f(\Delta i)$. Do this in Matlab, taking the measurements E_i as given in exercise 7 of Chapter 4 and the filter function as given above. Note that the negative indices are reflected in the upper half of the filter function, so in Matlab indices, with a profile length of twenty-four elements: $f(1)$ =1; $f(2)$=j; $f(3)$=-0.5; $f(23)$ =-0.5; $f(24)$=-j. Other elements are zero. Check that the function $K(i)$ has the amplitude and phase similar to those indicated in Figure 4.12. and as calculated in exercise 7. Try 'plot(unwrap(angle(K)))' in Matlab and interpret what you see.

You may note that the FFT of the filter function is (almost) zero for the upper half of the spectrum, that contains the transmission of the 'negative' frequencies. Applying such a filter to an interferogram; transmitting the relevant positive frequencies and omitting the negative frequencies is called the 'carrier frequency' method and can be used – with some modification – to evaluate 1D and 2D interferograms (Takeda 1982).

REFERENCES

Brinkman, S. and Bodschwinna, H., 2003, Advanced Gaussian Filters. In: Blunt, L. and Jiang, X. (ed), *Advanced Techniques for Assessment of Surface Topography. Development of a Basis for 3D Surface Texture Standards 'Surfstand'*, pp. 63–75. Amsterdam, Elsevier.

De Groot, P., and Deck, L., 2015, Surface profiling by analysis of white-light interferograms in the spatial frequency domain. *Journal of Modern Optics* **42(2)**:389–401.

ISO 10110-12, 2019, Optics and Photonics – Preparation of Drawings for Optical Elements and Systems – Part 12: Aspheric Surfaces. Geneva, International Organization for Standardization.

ISO 12181-2, 2012, Geometrical Product Specifications (GPS) – Roundness, Part 2: Specification Operators. Geneva, International Organization for Standardization.

ISO 12780-2, 2011, Geometrical Product Specifications (GPS) – Straightness, Part 2: Specification Operators. Geneva, International Organization for Standardization.

ISO 14406, 2010, Geometrical Product Specifications (GPS) – Extraction. Geneva, International Organization for Standardization.

ISO 16610-21, 2025, Geometrical Product Specifications (GPS) – Filtration – Part 21: Linear Profile Filters: Gaussian Filters. Geneva, International Organization for Standardization.

ISO 16610-28, 2016, Geometrical Product Specifications (GPS) – Filtration – Part 28: Profile Filters: End Effects. Geneva, International Organization for Standardization.

ISO 16610-31, 2025, Geometrical Product Specifications (GPS) – Filtration – Part 31: Robust Profile Filters: Gaussian Regression Filters. Geneva, International Organization for Standardization.

ISO 16610-41, 2016, Geometrical Product Specifications (GPS) – Filtration – Part 41: Morphological Profile Filters: Disk and Horizontal Line-Segment Filters. Geneva, International Organization for Standardization.

ISO 16610-61, 2015, Geometrical Product Specifications (GPS) – Filtration – Part 21: Linear Areal Filters: Gaussian Filters. Geneva, International Organization for Standardization.

ISO 21920-2, 2021, Geometrical Product Specifications (GPS), Surface Texture: Profile – Part 2: Terms, Definitions and Surface Texture Parameters. Geneva, International Organization for Standardization.

ISO 25178-2, 2021, Geometrical Product Specifications (GPS), Surface Texture: Areal – Part 2: Terms, Definitions and Surface Texture Parameters. Geneva, International Organization for Standardization.

ISO 25178-3, 2012, Geometrical Product Specifications (GPS), Surface Texture: Areal – Part 3: Specification Operators. Geneva, International Organization for Standardization.

Leach, R.L, 2024, *Characterization of Areal Surface Texture*. Berlin, Springer.

Muralikrishnan, B. and Raja, J., 2009, *Computational Surface and Roundness Metrology*. Berlin, Springer.

Noll, R.J., 1976, Zernike Polynomials and Atmospheric Turbulence. *J. Opt. Soc. Am.* **66**:207–211.

Takeda, M., Ina, H. and Kobayashi, S., 1982, Fourier-Transform Method of Fringe-Pattern Analysis for Computer-Based Topography and Interferometry. *J. Opt. Soc. Am.* **72:**156–160.

Appendix

The Least-Squares Method

A.1 THE MEAN, WEIGHTED MEAN AND CHI-SQUARED DISTRIBUTION

A.1.1 THE MEAN AND THE Q^2 FUNCTION

For a Gaussian distribution of measurement values, the probability density function for any measurement value x_i with standard deviation σ_i is given by

$$F(x) = \frac{1}{\sigma_i \sqrt{2\pi}} e^{-(x_i - X)^2 / 2\sigma_i^2}, \tag{A.1}$$

where X is the mean. For a number n of measurements x_i, the overall probability is given by the product of the probability densities, which can be written as:

$$F_{\text{tot}}(q) = \prod_{i=1}^{n} \left(\frac{1}{\sigma_i \sqrt{2\pi}} \right) \prod_{i=1}^{n} e^{-(x_i - X)^2 / 2\sigma_i^2} = e^{-\sum_{i=1}^{n} (x_i - X)^2 / 2\sigma_i^2} \prod_{i=1}^{n} \left(\frac{1}{\sigma_i \sqrt{2\pi}} \right). \tag{A.2}$$

The overall probability is maximal when the function

$$Q^2(X) = \sum_{i=1}^{n} \left(\frac{x_i - X}{\sigma_i} \right)^2 \tag{A.3}$$

is minimal. This is the least-squares principle. Minimizing $Q^2(X)$ to X gives the best estimate for X:

$$\frac{\partial Q^2}{\partial X} = 0: \; X = x_w = \frac{\sum_{i=1}^{n} \frac{x_i}{\sigma_i^2}}{\sum_{i=1}^{n} \frac{1}{\sigma_i^2}}. \tag{A.4}$$

This is the common expression for the weighted mean. If the standard deviation is the same for every measurement x_i, equation (A.4) reduces to:

$$X = \overline{x} = \frac{\sum_{i=1}^{n} x_i}{N}. \tag{A.5}$$

This is the arithmetic mean, as might be anticipated. It is an important principle that determining an average basically consists of minimizing the Q^2 function and maximizing the probability that the value X fits the given measurement values x_i. The minimum value of Q^2 is commonly referred to as the Chi-squared value:

$$\chi^2 = \sum_{i=1}^{n} \left(\frac{x_i - \frac{\sum_{i=1}^{n} x_i}{N}}{\sigma} \right)^2 = \frac{s^2 (n-1)}{\sigma^2} = n - 1 = v, \tag{A.6}$$

with $s^2 = \dfrac{\sum_{i=1}^{n} \left(x_i - \frac{\sum_{i=1}^{n} x_i}{n} \right)^2}{n-1}$, the experimentally determined standard deviation,

see also equation (3.3). v is the degrees of freedom, that is, a more general parameter describing the number of measurements (n) minus the number of parameters calculated from these measurements; in this case, the mean (1). From the propagation of uncertainties, equation (3.26), it can be derived that the standard deviation in the weighted mean is given by:

$$\sigma_w^2 = \Sigma \left(\frac{\partial x_w}{\partial x_i} \right)^2 \sigma_i^2 = \Sigma \left(\frac{\frac{1}{\sigma_i^2}}{\Sigma \left(\frac{1}{\sigma_i} \right)^2} \right)^2 \sigma_i^2 = \Sigma \frac{1/\sigma_i^2}{\left(\Sigma \left(1/\sigma_i^2 \right) \right)^2} = \left(\sum_{i=1}^{n} \frac{1}{\sigma_i^2} \right)^{-1}. \tag{A.7}$$

This gives $\sigma(\overline{x}) = \sigma(x_i) / \sqrt{n}$ if the standard deviation is the same for every measurement x_i.

The second derivative, the 'curvature' of the function in the minimum is:

$$\frac{\partial^2 Q^2}{\partial X^2} = -\frac{2}{\sigma_w^2} = -2\frac{n}{\sigma^2}. \tag{A.8}$$

The function Q^2 around the minimum as function of a chosen value for X can be described by:

$$Q^2\left(X\right)=\chi^2+\frac{n}{\sigma^2}\left(X-\overline{x}\right)^2. \tag{A.9}$$

This function reaches the value $Q^2\left(X\right)=\chi^2+1$ if $X=\overline{x}\pm\frac{\sigma}{\sqrt{n}}$. The standard uncertainty in X can thus be obtained from the intersection of the function $Q^2\left(X\right)$ with the line $Q^2=\chi^2+1$.

The function $Q^2(X)$ reaches the value $Q^2\left(X\right)=2\cdot\chi^2$ if $X=\overline{x}\pm\sigma\sqrt{\frac{n-1}{n}}$. This can be used to estimate the effective value of σ. As an illustration, the thread measurements shown in Table 3.1 are taken, with mean value $\overline{x}$ = 18.3360 mm; standard deviation $\sigma = s$ = 4.2 µm and $\sigma_w = s_m$ = 1.9 µm. From the Q^2 function, the best value for the mean, the standard deviation and the standard deviation in the mean can be derived. These values can also be calculated directly from the measured values but in more general cases a plot of Q^2 can also be applied.

A.1.2 The Weighted Mean and the X^2 Function

In the diameter measurement, an equal value for the standard deviation σ of all the individual measurements was assumed and this was estimated by the experimental standard deviation s. It may happen that a mean has to be calculated from a number of measurement data that each have their own standard deviation, that is, standard uncertainty type A. As an example, we again take the diameter measurements from Table 3.1; however, now the first three measurements have a type-A uncertainty characterized by a standard deviation of 0.003 mm and the last two have a standard deviation of 0.002 mm. This is shown in Table A.1. From the table, we find for the weighted mean value, using equation (A.4): $\overline{M}=18.3375\,\text{mm}$, and using equation (A.7) for the standard uncertainty in the mean value: $s_{\text{internal}}(M)=0.0011\,\text{mm}$. The χ^2 value calculated according to equation (A.6) is 11.5 while the expected value is equal to the degrees of freedom: v = 4. In general, the following situations occur:

- $\chi^2 \gg v$ indicates that one or more standard deviations have been underestimated.
- $\chi^2 \ll v$ indicates that presumably a common uncertainty, for example, a systematic bias, has been included in all the standard uncertainty values, leading to an overestimation of the combined uncertainty.

For a statistical evaluation, the reduced χ^2, which is χ^2 divided by the number of degrees of freedom, should be considered:

$$\chi_r^2=\frac{\chi^2}{\upsilon}. \tag{A.10}$$

TABLE A.1
Measurement Values with Standard Uncertainty Values

Measurement Number	Measured Value	Standard Uncertainty	Weighting Factor w	Contribution Chi-Squared
	M_i/mm	s_i/mm	$w_i = 1/s_i^2$	$\left(\frac{M_i - \bar{M}}{s_i}\right)^2$
1	18.3314	0,003	111111	4.1
2	18.3332	0,003	111111	2.0
3	18.3347	0,003	111111	0.84
4	18.3412	0,002	250000	3.51
5	18.3395	0,002	250000	1.05
Sum	$\Sigma M_i \cdot w_i$=15.2812·10^6	$s_{\bar{M},\text{ innteral}} = \sqrt{\frac{1}{\Sigma \frac{1}{s_i^2}}} = 0.0011$	833.333·10^3	11.5

For an 'ideal' measurement, $\chi_r^2 \approx 1$. A measure of the probability of 'something being wrong' can be calculated from the integral over the chi-squared distribution, analogous to the t-distribution in Section 3.4.1. The chi-squared function has a probability density function that can be used to calculate probabilities. Table A.2 gives an example for 95% probability limits. These can be obtained from published statistical tables or calculated, for example, using the MATLAB function *chi2inv*; >>p=chi2inv(0.95,v)/v with v being the number of degrees of freedom. If the χ_r^2 -value exceeds the limit values shown in Table A.2, the uncertainties may have been underestimated, and the model used may be incorrect or inadequate. In the example of Table A.2, we find $\chi_r^2 = 11.5/4 = 2.9$, so the standard deviations appear to be underestimated. The uncertainty as calculated from equation (A.7) is called 'internal error' and is meaningful if $\chi_r^2 \approx 1$. If $\chi_r^2 \gg 1$, one can consider the dispersion in the measurement points and assume that the standard deviations have been taken a common factor too small (or too large for $\chi_r^2 \ll 1$). The best estimate for this common factor is $\sqrt{\chi_r^2}$. Multiplying by this factor, one obtains the 'external error':

$$s_{\text{external}}(\overline{M}) = \frac{1}{\Sigma \frac{1}{s_i^2}} \cdot \chi_r^2 = \frac{1}{\Sigma \frac{1}{s_i^2}} \cdot \sum_{i=1}^{n} \left(\frac{M_i - \bar{M}}{s_i}\right)^2 \cdot \frac{1}{n-1}. \qquad \text{(A.11)}$$

TABLE A.2
Limit Values for 95% Confidence of the Reduced Chi-Squared Function χ_r^2

Degrees of Freedom ν	Limit Value of χ_r^2 for 95% Confidence
1	3.84
2	2.99
3	2.60
4	2.37
5	2.21
7	2.00
10	1.83
30	1.46
100	1.24

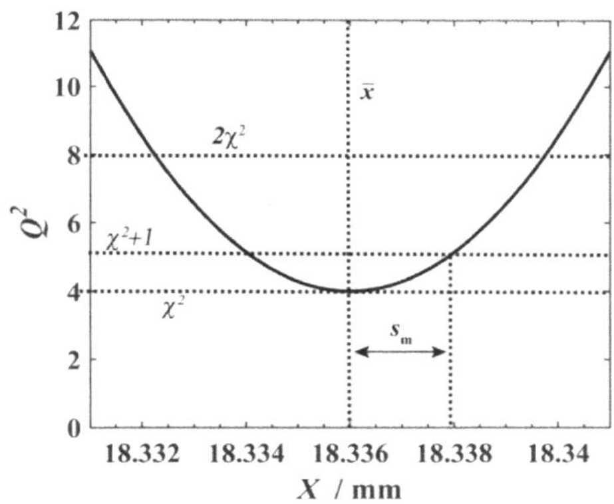

FIGURE A.1 The Q^2 function for the diameter measurement according to equation (A.9).

In the example in Table 6.2, it is given by: $s_{\text{external}}(\overline{M}) = s_{\text{internal}}(\overline{M})\sqrt{\chi_r^2} = 1.9$ µm. This is similar to $s_m = 1.9$ µm found for the unweighted mean in Section 3.4.1 and in Figure A.1. It is usual practice to take the largest of the 'internal' and 'external' error/uncertainty values. Disregarding the internal errors, and calculate a χ^2-value from the unweighted measurements, is equivalent to attributing the same uncertainty to every individual measurement (for example, '1') and calculate the external error. This is effectively what is carried out when calculating the standard deviation in the mean from a series of measurements.

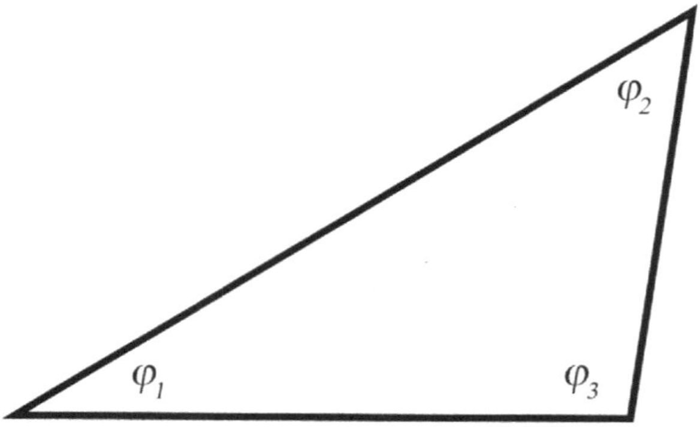

FIGURE A.2 Triangle with the angles indicated.

A.2 GENERAL LINEAR LEAST-SQUARES EVALUATION

In general, more measurements are taken of a quantity than strictly necessary, for example, to determine a measurement value, a number of measurements are taken, or a distance is measured from different directions. Often, there is a linear relationship between the measured quantity and the quantity to be finally determined. If this relationship is not linear, it can often be linearized. As an example, three angles of a triangle are measured. This can be an accurate measurement, as in surveying or polygon measurements, or a relatively inaccurate one as, for example, with an angle gage or a geo-triangle. Considering the triangle in Figure A.2, three angles are measured: φ_1, φ_2 and φ_3 with the measured values $y_1 = 30.3°$; $y_2 = 59.6°$ and $y'_3 = 91.8°$, each with a standard uncertainty of 0.5°. Since the sum of the three angles must equal to 180°, we can interpret the last measurement as a measurement of the sum of the first two angles: $y_3 = 180° - y'_3 = 88.2°$. The measurement can be represented as $\boldsymbol{a}_{ij} \cdot \boldsymbol{a}_j = \boldsymbol{y}_i$ with $\boldsymbol{a}_j$ the quantities sought, $\boldsymbol{y}_i$ the measurements and $\boldsymbol{a}_{ij}$ the matrix capturing the linear relationship, thus

$$\pm\cdot\begin{pmatrix}\varphi_1\\ \varphi_2\end{pmatrix} = \begin{pmatrix}1 & 0\\ 0 & 1\\ 1 & 1\end{pmatrix}\begin{pmatrix}\varphi_1\\ \varphi_2\end{pmatrix} = \begin{pmatrix}y_1\\ y_2\\ y_3\end{pmatrix}. \tag{A.12}$$

The squared sum Q^2 can be written as:

$$Q^2 = \sum_{i=1}^{3}\left(\frac{\sum_{i=1}^{2}\begin{pmatrix}1 & 0\\ 0 & 1\\ 1 & 1\end{pmatrix}\begin{pmatrix}\varphi_1\\ \varphi_2\end{pmatrix} - \begin{pmatrix}y_1\\ y_2\\ y_3\end{pmatrix}}{s_i}\right)^2. \tag{A.13}$$

Taking the minimum of Q^2: $\frac{\partial Q^2}{\partial \varphi_k} = 0$ gives:

$$\boldsymbol{\beta}'\boldsymbol{\beta}\boldsymbol{a} = \boldsymbol{\beta}'\boldsymbol{z} \tag{A.14}$$

where $\beta_{ij} = \alpha_{ij}/s_i$ and $z_i = y_i/s_i$ and the '-sign denotes the transposed matrix. This gives for the coefficients $\boldsymbol{a}$:

$$\boldsymbol{a} = \left(\boldsymbol{\beta}'\boldsymbol{\beta}\right)^{-1}\boldsymbol{\beta}'\boldsymbol{z}. \tag{A.15}$$

The matrix $\left(\boldsymbol{\beta}'\boldsymbol{\beta}\right)^{-1}\boldsymbol{\beta}'$ in equation (A.15) is also called the pseudo-inverse. MATLAB has the function *pinv* for calculating the pseudo-inverse, and even applies this method automatically with the backslash ('\') function, so $\boldsymbol{a} = \boldsymbol{\beta}\backslash\boldsymbol{y}$. The solution in this case, with $s_i = 0.5$, becomes:

$$\begin{pmatrix}\varphi_1\\ \varphi_2\end{pmatrix} = \left(\boldsymbol{\beta}'\boldsymbol{\beta}\right)^{-1}\boldsymbol{\beta}' \cdot \begin{pmatrix}z_1\\ z_2\\ z_3\end{pmatrix} = \begin{pmatrix}1/3 & -1/6 & 1/6\\ -1/6 & 1/3 & 1/6\end{pmatrix}\begin{pmatrix}60.6\\ 118.2\\ 182.8\end{pmatrix} = \begin{pmatrix}29.7\\ 59.1\end{pmatrix}. \tag{A.16}$$

For the solution φ_1 =29.7°, φ_2 =59.0° and, implicitly, φ_3 =180° – 29.7° – 59.1° = 91.2°, we can see that the discrepancy between the measurements and the solution is more or less evenly distributed over the measurements. This is a general property of the least-squares method: just as taking an average is in fact the solution of a least-squares problem, a more general solution often comes down to taking some kind of average and distributing the 'residual' part of the nonperfect fitting evenly over the measurements. For the sum of squares, χ^2, we find:

$$\chi^2 = \sum_{i=1}^{3}\left(\frac{y_i - \varphi_i}{s_i}\right)^2 = \left(\frac{30.3 - 29.7}{0.5}\right)^2 + \left(\frac{59.6 - 59.0}{0.5}\right)^2 + \left(\frac{88.1 - 88.7}{0.5}\right)^2 = 4.32. \tag{A.17}$$

The internal variances and covariances are given by the so-called internal error matrix $\boldsymbol{E}_{\text{int}}$:

$$\boldsymbol{E}_{\text{int}=}\left(\boldsymbol{\beta}'\cdot\,\boldsymbol{\beta}\right)^{-1}, \tag{A.18}$$

and the external error matrix $\boldsymbol{E}_{\text{ext}}$ is obtained by multiplying the internal error matrix by χ_r^2:

$$E_{\text{ext}} = \chi_r^2 \cdot (\boldsymbol{\beta}' \cdot \boldsymbol{\beta})^{-1} = 4.32 \cdot \begin{pmatrix} 1/6 & -1/12 \\ -1/12 & 1/6 \end{pmatrix} = \begin{pmatrix} 0.72 & -0.36 \\ -0.36 & 0.72 \end{pmatrix}. \tag{A.19}$$

This gives as standard uncertainty $u(\varphi_1) = u(\varphi_2) = \sqrt{0.72} = 0.85°$. The correlation coefficient between φ_1 and φ_2 is calculated as:

$$r_{\varphi_1,\varphi_2} = \frac{u(\varphi_1,\varphi_2)}{u(\varphi_1)\cdot u(\varphi_2)} = \frac{-0.36}{\sqrt{0.72}\cdot\sqrt{0.72}} = -0.5. \tag{A.20}$$

The correlation coefficient must be used when calculating uncertainties in quantities that are a combination of the two given angles. As an example, we calculate the uncertainty in $\varphi_3 = 180° - \varphi_1 - \varphi_2$ using equations (3.26) and (3.27):

$$\begin{aligned} u_c^2(\varphi_3) &= \sum_{i=1}^{N}\left(\frac{\partial f}{\partial \varphi_i}\right)^2 u^2(\varphi_i) + 2\sum_{i=1}^{N-1}\sum_{j=i+1}^{N}\frac{\partial f}{\partial \varphi_i}\frac{\partial f}{\partial \varphi_j}u(\varphi_i)\cdot u(\varphi_j)\cdot r(\varphi_i,\varphi_j) \\ &= u^2(\varphi_1) + u^2(\varphi_2) + 2\cdot u(\varphi_1)\cdot u(\varphi_2)\cdot -0.5 = 0.72 + 0.72 - 0.72. \end{aligned} \tag{A.21}$$

This gives $u(\varphi_3) = \sqrt{0.72} = 0.85°$, as expected, the same uncertainty as $u(\varphi_1)$ and $u(\varphi_2)$.

A.3 NUMERICAL METHODS FOR LEAST-SQUARES EVALUATION

The examples in Sections A.1 and A.2 were solvable with analytical methods and some matrix manipulations. In cases where this fails, for example, if the minimization cannot be performed analytically as in equation (A.4), and the measurements and parameters do not have a linear relationship as in equation (A.12), numerical methods can still be used. Numerical methods can also be used if one fails to find one of the other formulations. Again, the sum of squares is considered:

$$Q^2 = \sum_{i=1}^{n}\left(\frac{y_i - f_i(x_1, x_2..x_n)}{s_i}\right)^2. \tag{A.22}$$

Here, y_i are the measured values and the function f is a known, or assumed, function with parameters x_i that must be determined from measurements y_i. Thus, in the case of angular measurement, the function f can be written as: $f_1 = \varphi_1$, $f_2 = \varphi_2$, $f_3 = 180 - \varphi_1 - \varphi_2$.

The function Q^2 then becomes:

$$Q^2 = \left(\frac{30.3 - \varphi_1}{0.5}\right)^2 + \left(\frac{59.6 - \varphi_2}{0.5}\right)^2 + \left(\frac{88.1 - \varphi_1 - \varphi_2}{0.5}\right)^2. \tag{A.23}$$

The function Q^2 must be minimized for two parameters: φ_1 and φ_2. This is a general problem from mathematics: minimizing a function with multiple variables. This is carried out by varying the variables and then finding a minimum. This can best be done with a software package where this feature is already implemented.

As an example, the function Q^2 is programmed in MATLAB as:

```
function Qsq=triangles(y)
Qsq =4*((30.3-y(1)).^2+(59.6-y(2)).^2+(88.1-y(1)-y(2)).^2);
end
```

with initial values:

```
v = [20, 40];
```

Next, the general function that minimizes a function: '*fminsearch*' is used:

```
a = fminsearch(@triangles,v)
```

and this gives in $\boldsymbol{a}$ as variables $a(1)$ and $a(2)$ the values for φ_1 and φ_2, respectively, which were found earlier. The minimization can be considered as descending into a valley where the landscape is given by the function Q^2 (φ_1,φ_2). The variance–covariance matrix can be computed by numerically calculating the second derivative, and the mixed partial derivative, in the minimum by varying the variables by a small amount, in this case 1°:

$$\begin{aligned}\frac{\partial^2\left(Q(\varphi_1,\varphi_2)\right)}{\partial\varphi_1^2} &\approx \frac{Q(\varphi_1+h,\varphi_2)+Q(\varphi_1-h,\varphi_2)-2Q(\varphi_1,\varphi_2)}{h^2} \\ &= \frac{Q(29.7+1,59)+Q(29.7-1,59)-2Q(29.7,59)}{1}=8.\end{aligned} \tag{A.24}$$

Analogously, the mixed derivative can be calculated:

$$\begin{aligned}&\frac{\partial^2\left(Q(\varphi_1,\varphi_2)\right)}{\partial\varphi_1\partial\varphi_2} \\ &\approx \frac{Q\left(\varphi_1+\frac{h}{2},\varphi_2+\frac{k}{2}\right)+Q\left(\varphi_1-\frac{h}{2},\varphi_2-\frac{k}{2}\right)-Q\left(\varphi_1-\frac{h}{2},\varphi_2+\frac{k}{2}\right)-Q\left(\varphi_1+\frac{h}{2},\varphi_2-\frac{k}{2}\right)}{h\cdot k} \\ &= \frac{Q(29.7+0.5,59+0.5)+Q(29.7-0.5,59-0.5)-Q(29.7-0.5,59+0.5)-Q(29.7+0.6,59-0.5)}{1} \\ &= 4.\end{aligned} \tag{A.25}$$

The matrix $\boldsymbol{H}$ with elements $H_{ij} = \dfrac{\partial^2 Q}{\partial \varphi_i \partial \varphi_j}$ is also known as the Hessian matrix. This gives for the internal error matrix $\boldsymbol{E}_{\text{int}}$:

$$\boldsymbol{E}_{ij} = \left(\frac{1}{2}\boldsymbol{H}\right)_{ij}^{-1} = \left(0.5\begin{pmatrix} 8 & 4 \\ 4 & 8 \end{pmatrix}\right)^{-1} = \begin{pmatrix} 1/6 & -1/12 \\ -1/12 & 1/6 \end{pmatrix}, \quad \text{(A.26)}$$

and for the external error matrix:

$$\boldsymbol{E}_{\text{ext}} = \chi_r^2 \cdot \left(\frac{1}{2}\boldsymbol{H}\right)_{ij}^{-1} = 4.32 \cdot \begin{pmatrix} 1/6 & -1/12 \\ -1/12 & 1/6 \end{pmatrix} = \begin{pmatrix} 0.72 & -0.36 \\ -0.36 & 0.72 \end{pmatrix}. \quad \text{(A.27)}$$

This is identical to equation (A.19). In the literature (Bevington and Robinson 2003), more examples and considerations for this process are given.

A.4 DATA FITTING TO LINES AND POLYNOMIALS

A.4.1 Linear Least-Squares Line

Table A.3 gives the measurements of a gage block length, measured in a gage block interferometer, as a function of temperature. In general, the length L as a function of temperature T, can be expressed in first order approximation (see Section 2.2):

$$L_T = L_{20} + L_{20} \cdot \alpha_E \left(T - 20\,°C\right) \quad \text{(A.28)}$$

with α_E the linear thermal expansion coefficient. We want to find the values for L_{20} and α_E and their type A standard uncertainties from the data in Table A.3. A linear function can generally be written as:

TABLE A.3
Gage Block Measurement as a Function of Temperature

Measurement #	Temperature /°C	Length /mm	Temperature – 20 °C/°C	Length – 100 mm/μm
1	19	99.998 80	-1	–1,20
2	19.5	99.999 63	-0.5	–0.57
3	20	100.000 02	0	+0.02
4	20.5	100.000 61	0.5	+0.61
5	21	100.001 26	1.0	+1.26
6	21.5	100.001 84	1.5	+1.84

$$y = a_0 + a_1 \cdot x. \qquad \text{(A.29)}$$

For n number of measurements (x_i, y_i), omitting the uncertainties in y_i the Q^2 function can be written as:

$$Q^2 = \sum_{i=1}^{n} \left(y_i - a_0 - a_1 \cdot x_i\right)^2. \qquad \text{(A.30)}$$

In the minimum, the derivatives of Q^2 with respect to both a_o and a_I must equal zero. This gives:

$$\sum_{i=1}^{n} y_i = na_0 + a_1 \cdot \sum_{i=1}^{n} x_i \text{ and} \sum_{i=1}^{n} x \cdot y_i = a_0 \sum_{i=1}^{n} x_i + a_1 \sum_{i=1}^{n} x_i^2. \qquad \text{(A.31)}$$

This can be rewritten as:

$$\begin{pmatrix} \Sigma y_i \\ \Sigma x_i y_i \end{pmatrix} = \begin{pmatrix} n & \Sigma x_i \\ \Sigma x_i & \Sigma x_i^2 \end{pmatrix} \cdot \begin{pmatrix} a_0 \\ a_1 \end{pmatrix} \text{so} \begin{pmatrix} a_0 \\ a_1 \end{pmatrix} = \begin{pmatrix} n & \Sigma x_i \\ \Sigma x_i & \Sigma x_i^2 \end{pmatrix}^{-1} \cdot \begin{pmatrix} \Sigma y_i \\ \Sigma x_i y_i \end{pmatrix},$$

with the solution:

$$\begin{pmatrix} a_0 \\ a_1 \end{pmatrix} = \frac{1}{n \Sigma x_i^2 - \left(\Sigma x_i\right)^2} \begin{pmatrix} \Sigma x_i^2 & -\Sigma x_i \\ -\Sigma x_i & n \end{pmatrix} \begin{pmatrix} \Sigma y_i \\ \Sigma x_i y_i \end{pmatrix}. \qquad \text{(A.32)}$$

Equation (A.32) was already given explicitly in Section 7.3.1. It will be available in most software packages and even pocket calculators. Calculations concerning the error matrix can be carried out as shown in the next section. For the linear least-squares method, including the weighing with the standard deviation or standard uncertainty at every point, direct calculations can be carried out by defining: the weights $w_i = s_i^{-2}$ and the total weight $W = \sum_{i=1}^{n} w_i$. The means are defined as $\overline{x} = \Sigma x_i w_i / W$; $\overline{x^2} = \Sigma x_i^2 w_i / W$ and similar for $\overline{y}, \overline{y^2}$ and $\overline{xy}$. With this, the variances can be defined as:

$\Delta = \overline{x^2} - \overline{x}^2$; $\varepsilon = \overline{y^2} - \overline{y}^2$ and $\mu = \overline{x \cdot y} - \overline{x} \cdot \overline{y}$. The resulting parameters are:

$$a_0 = \overline{y} - \frac{\mu}{\Delta}\overline{x} \; ; a_1 = \frac{\mu}{\Delta} \; ; \chi_r^2 = \left(\varepsilon - \frac{\mu^2}{\Delta}\right) \cdot \frac{W}{n-2}. \qquad \text{(A.33).}$$

Standard uncertainties and correlations are given by:

$$s(a_0)_{int} = \sqrt{\overline{x^2}/(W\Delta)};$$
$$s(a_0)_{ext} = \sqrt{\frac{\overline{x^2}}{n-2}\left(\frac{\varepsilon}{\Delta}-\left(\frac{\mu}{\Delta}\right)^2\right)};\quad s(a_1)_{int} = \sqrt{1/(W\Delta)};$$
$$s(a_1)_{ext} = \sqrt{\frac{1}{n-2}\left(\frac{\varepsilon}{\Delta}-\left(\frac{\mu}{\Delta}\right)^2\right)} \tag{A.34}$$
$$\text{and } r(a_0,a_1) = -\frac{\overline{x}}{\sqrt{\overline{x^2}}}.$$

The calculations are simplified considerably by shifting the x-axis to the (weighted) mean of the x-values; in particular, then there is no correlation between a_0 and a_1. The y-axis can also be shifted to the mean y-value. In general, the least-squares line always passes through the (weighted) mean of the x and y values. The external standard uncertainty (standard deviation) in y' for a given value of x' is given by:

$$\begin{aligned}
u_c^2(y') &= \sum_{i=1}^{N}\left(\frac{\partial f}{\partial a_i}\right)^2 u^2(a_i) + 2\sum_{i=1}^{N-1}\sum_{j=i+1}^{N}\frac{\partial f}{\partial a_i}\frac{\partial f}{\partial a_j}u(a_i)\cdot u(a_j)\cdot r(a_i,a_j)\\
&= u^2(a_0) + x'^2\cdot u^2(a_1) + 2\cdot x'\cdot u(a_0)\cdot u(a_1)\cdot r(a_0,a_1)\\
&= \frac{\overline{x^2}}{n-2}\left(\frac{\varepsilon}{\Delta}-\left(\frac{\mu}{\Delta}\right)^2\right) + x'^2\cdot\frac{1}{n-2}\left(\frac{\varepsilon}{\Delta}-\left(\frac{\mu}{\Delta}\right)^2\right) - 2x'\cdot\frac{1}{n-2}\left(\frac{\varepsilon}{\Delta}-\left(\frac{\mu}{\Delta}\right)^2\right)\cdot\overline{x}\\
&= \frac{1}{n-2}\left(\frac{\varepsilon}{\Delta}-\left(\frac{\mu}{\Delta}\right)^2\right)\left(\overline{x^2}+x'^2-2\cdot x'\cdot\overline{x}\right).
\end{aligned} \tag{A.35}$$

As an example, the data from Table A.3 are taken. In order to increase the numerical stability and to reduce the correlation, the data for T-20 °C and L_T-100 mm are considered. The data are taken unweighted, giving $W=n$ with $n=6$. In this approach the internal error is only used in the calculations. This gives the values and external uncertainties as shown in Table A.4. Figure A.3 shows the distance of the measured points to the calculated least-squares line and the uncertainty limits of the calculated y-values using equation (A.35). The uncertainty is lowest at the mean x-value of 20.25 °C. When extrapolating outside the measurement range from 19 °C to 21.5 °C, the uncertainty increases significantly. The equations (A.33) can be rewritten for a line through the origin $(x,y)= (0,0)$; defining a_0 as zero and giving for a_1 and χ_r^2:

$$a_1 = \overline{xy}/\overline{x^2};\quad \chi_r^2 = \left(\frac{\overline{x^2}\cdot\overline{y^2}-\overline{xy}^2}{\overline{x^2}}\right)\cdot\frac{W}{n-1} \tag{A.36}$$

TABLE A.4
Values Found for the Linear Least-Squares Fitting to the Data in Table A.3

Quantity	Value	External Standard Uncertainty
a_0	0.0227 µm	0.0083 µm
a_1	1.2160 µm/°C	0.0074 mm/°C
$r(a_0, a_1)$	–0.281	
$\sqrt{\chi_r^2}$	0.0174 µm	
L_{20}	100 mm + 0.0227 µm	0.0083 µm
α_E	$12.160 \cdot 10^{-6}$ //°C	$0.074 \cdot 10^{-6}$ /°C
Degrees of freedom ν	6–2=4	

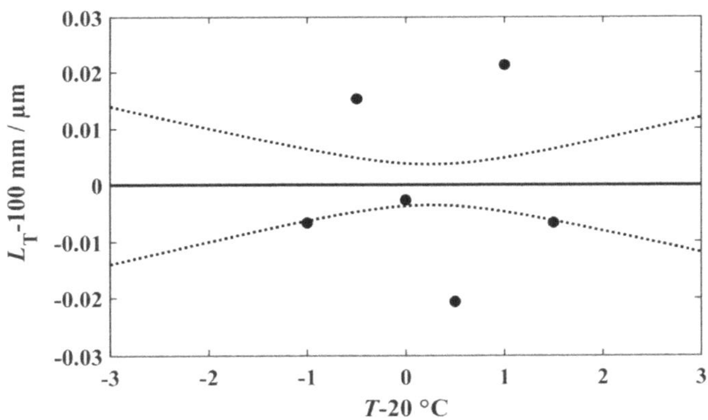

FIGURE A.3 Deviations of the length measurements from the fitted straight line with indicated standard uncertainty limits of the predicted values.

and uncertainties:

$$s(a_1)_{int} = \sqrt{1/\left(W\overline{x^2}\right)}\ ;\ s(a_1)_{ext} = \sqrt{\frac{1}{n-1}\left(\frac{\overline{y^2}}{\overline{x^2}} - a_1^2\right)}. \tag{A.37}$$

Equations (A.36) and (A.37) can be useful when calibrating a measuring instrument using reference standards at a number of points within the measuring range. The value of a_1 will give the 'calibration factor' that enables a correction for the instrument's output factor, and a significant χ_r^2-value may indicate a high degree of non-linearity, see also Section 1.6.3.

A.4.2 Least-Squares Fit to Higher-Order Polynomials

In general, a function can be written as a series of polynomials:

$$y = a_0 + a_1 \cdot x + a_2 \cdot x^2 + .. + a_m \cdot x^m. \tag{A.38}$$

For n number of measurements (x_i, y_i), the Q^2 function can be written as:

$$Q^2 = \sum_{i=1}^{n} \left(\frac{y_i - a_0 - a_1 \cdot x_i - a_2 \cdot x^2 - .. - a_m \cdot x^m}{s_i} \right)^2, \tag{A.39}$$

where s_i is the standard deviation, or the type A uncertainty, in the measured values y_i. For the optimum values $a_1..a_n$ the derivatives of Q^2 to each coefficient a_j must equal 0. This gives:

$$\begin{aligned}
\sum_{i=1}^{n} \frac{y_i}{s_i^2} &= \frac{na_0}{s_i^2} + a_1 \cdot \frac{\sum_{i=1}^{n} x_i}{s_i^2} + a_2 \cdot \frac{\sum_{i=1}^{n} x_i^2}{s_i^2} + \ldots \\
\sum_{i=1}^{n} \frac{x_i y_i}{s_i^2} &= a_0 \frac{\sum_{i=1}^{n} x_i}{s_i^2} + a_1 \frac{\sum_{i=1}^{n} x_i^2}{s_i^2} + a_2 \frac{\sum_{i=1}^{n} x_i^3}{s_i^2} \ldots \\
\sum_{i=1}^{n} \frac{x_i^2 \cdot y_i}{s_i^2} &= a_0 \frac{\sum_{i=1}^{n} x_i^2}{s_i^2} + a_1 \frac{\sum_{i=1}^{n} x_i^3}{s_i^2} + \ldots.
\end{aligned} \tag{A.40}$$

....

Equation (A.40) can be rewritten as:

$$\begin{pmatrix} \Sigma \frac{1}{s_i^2} & \cdots & \Sigma \frac{x_i^m}{s_i^2} \\ \vdots & \ddots & \vdots \\ \Sigma \frac{x_i^m}{s_i^2} & \cdots & \Sigma \frac{x_i^{2m}}{s_i^2} \end{pmatrix} \cdot \begin{pmatrix} a_0 \\ \vdots \\ a_m \end{pmatrix} = \begin{pmatrix} \Sigma \frac{y_i}{s_i^2} \\ \vdots \\ \Sigma \frac{x_i^m y_i}{s_i^2} \end{pmatrix}, \text{or}$$

$$\begin{pmatrix} \Sigma\dfrac{y_i}{s_i^2} \\ \vdots \\ \Sigma\dfrac{x_i^m y_i}{s_i^2} \end{pmatrix} = \begin{pmatrix} \Sigma\dfrac{1}{s_i^2} & \cdots & \Sigma\dfrac{x_i^m}{s_i^2} \\ \vdots & \ddots & \vdots \\ \Sigma\dfrac{x_i^m}{s_i^2} & \cdots & \Sigma\dfrac{x_i^{2m}}{s_i^2} \end{pmatrix}^{-1} \cdot \begin{pmatrix} a_0 \\ \vdots \\ a_m \end{pmatrix} = \Gamma^{-1} \cdot \begin{pmatrix} a_0 \\ \vdots \\ a_m \end{pmatrix}. \tag{A.41}$$

The inverse matrix Γ^{-1} gives the internal errors and the covariances of the coefficients $a_0..a_m$. The value of χ^2 is found by substituting the values of $a_0..a_m$ from equation (A.39).

Because of the generality of this problem, several calculation packages offer it by default. MATLAB has the function *polyfit* to calculate the coefficients $a_0..a_m$, and the function *polyval* to calculate the function values of the polynomial found. The indication whether the adjustment is good or better is the value of the reduced χ^2, which is the value of Q^2 in the minimum found, divided by the number of degrees of freedom. In the case of unweighted measurements this gives:

$$\chi_r^2 = \frac{\sum_{i=1}^{n} \left(y_i\left(measured\right) - y_i\left(modeled\right) \right)^2}{n-m-1}. \tag{A.42}$$

Here, *(n-m-1)* is the number of degrees of freedom. For a zero-order fit, i.e. y is constant, we find the mean for a_0, the variance is given by χ_r^2 and the standard deviation is $\sqrt{\chi_r^2}$. A higher-order fit always leads to the same or better fit than a lower-order one, but the usefulness of higher order fits must be considered in relation to the number of degrees of freedom. As an example, the $\sqrt{\chi_r^2}$ –value for a fit with an increasing polynomial order of the data from Table A.3 is given in Table A.5. The table shows that the major improvement is found when going from the zero- to the first order polynomial fit. Higher order fits give reducing improvement, indicating that with these measurements it cannot be shown whether the gage block has a non-linear dependence with the temperature. When involving the uncertainties in the measurement data and the internal error, a check whether the χ_r^2 -value is significantly larger than unity, and whether higher order coefficients are significantly different from zero, are other methods to consider to what polynomial degree a fit is useful. When applying the second order Gaussian regression filter to a measured profile (see Section 9.4.2), a second order fit is made over the length considered, commonly for point $z(x)$ this is in the range $(x-\lambda_c;x+\lambda_c)$, where the measured points $z(x)$ are weighed by $s_h(\Delta x)$ as given in equation (9.14), instead of s_i^{-2} in equation (A.39), and the weighting function is truncated near the edges so as not to cross them.

TABLE A.5
Quality of Fit, Shown as a Standard Deviation, with Increasing Polynomial Order of the Measurements from Table A.3

Polynomial Order	Degrees of Freedom	$\sqrt{\chi_r^2}$
0	5	1.13 µm
1	4	0.017 µm
2	3	0.020 µm
3	2	0.024 µm
4	1	0.013 µm
5	0	Meaningless, perfect fit

A.5 APPLICATIONS IN DIMENSIONAL METROLOGY

A.5.1 Geometric Elements

The main application of coordinate measurements is to determine geometric elements, such as lines, planes, spheres, cylinders, etc. These are commonly calculated following the least-squares principle. Comprehensive examples and programs are given in the literature (Muralikrishnan and Raja 2009; Forbes 1991).

A.5.1.1 Line

The least-squares fit to a straight line has already been covered in Section A.4.1 which is applicable in this context. However, it must be kept in mind that this approach assumes that there is a theoretical ideal straight line and the deviations are measuring errors. This does not hold when the least-squares line is used as the reference line to determine a straightness deviation; then the deviations are 'real' and not errors. The uncertainties in the coefficients make little sense in that case, however, the root of the unweighted χ_r^2 -value can still be used as the *STR*q(LSLI)-value; see Section 7.3.1. If the weighted χ_r^2 -value is calculated and is not significant, the straightness is likely better than what the measuring method is able to determine. A line in three dimensions is more complex.

A.5.1.2 Plane

The least-squares fit to a plane is an extension of the fit to a line, and in fact a variation of the fitting to a polynomial. A plane $z(x,y)$ can be defined with the following:

$$z(x,y) = a_0 + a_1 \cdot x + a_2 \cdot y. \quad \text{(A.43)}$$

For n number of measurements $z_i(x_i,y_i)$, omitting the uncertainties in $z_{i,}$ the Q^2 function can be written as:

$$Q^2 = \sum_{i=1}^{n}\left(z_i - a_0 - a_1 \cdot x_i - a_2 \cdot y_1\right)^2 . \tag{A.44}$$

In the minimum, the derivatives of Q^2 to both a_o, a_1 and a_2 must be equal to zero. This gives:

$$\sum_{i=1}^{n} z_i = na_0 + a_1 \cdot \sum_{i=1}^{n} x_i + a_2 \cdot \sum_{i=1}^{n} x_i, \sum_{i=1}^{n} x_i \cdot z_i = a_0 \sum_{i=1}^{n} x_i + a_1 \sum_{i=1}^{n} x_i^2 + a_2 \sum_{i=1}^{n} x_i \cdot y_i \text{ and}$$
$$\sum_{i=1}^{n} y_i \cdot z_i = a_0 \sum_{i=1}^{n} x_i + a_1 \sum_{i=1}^{n} x_i \cdot y_i + a_2 \sum_{i=1}^{n} y_i^2 . \tag{A.45}$$

This can be written as

$$\begin{pmatrix} \Sigma z_i \\ \Sigma x_i \cdot z_i \\ \Sigma y_i \cdot z_i \end{pmatrix} = \begin{pmatrix} \Sigma 1 & \Sigma x_i & \Sigma y_i \\ \Sigma x_i & \Sigma x_i^2 & \Sigma y_i \cdot x_i \\ \Sigma y_i & \Sigma x_i \cdot y_i & \Sigma y_i^2 \end{pmatrix} \begin{pmatrix} a_0 \\ a_1 \\ a_2 \end{pmatrix} . \tag{A.46}$$

The coefficients then follow from:

$$\begin{pmatrix} a_0 \\ a_1 \\ a_2 \end{pmatrix} = \begin{pmatrix} \Sigma 1 & \Sigma x_i & \Sigma y_i \\ \Sigma x_i & \Sigma x_i^2 & \Sigma y_i \cdot x_i \\ \Sigma y_i & \Sigma x_i \cdot y_i & \Sigma y_i^2 \end{pmatrix}^{-1} \cdot \begin{pmatrix} \Sigma z_i \\ \Sigma x_i \cdot z_i \\ \Sigma y_i \cdot z_i \end{pmatrix} . \tag{A.47}$$

This was already stated in Equation (7.16). The matrix to be inverted depends on the x-y configuration, but not on the measured z-values. This approach is valid for nominally levelled configurations, $z << x,y$. A more general approach, the 'total least-squares' takes the perpendicular distance to the plane to be minimized. This may be approximated by taking the calculation in equation (A.47) as a first approximation, and then transforming the $(x,y,z,)$ coordinates by a rotation by an angle of -arctan(a_1) around the y-axis and a rotation of -arctan(a_2) around the x-axis and repeating equation (A.47) for the transformed coordinates; see also Section 8.3.1. Instead of a (flat) plane, higher order (quadratic and more) forms can be fitted by extending the fit to $z(x,y) = a_0 + a_1 \cdot x + a_2 \cdot y + a_3 \cdot x^2 + a_4 \cdot y^2 + a_5 \cdot x \cdot y +$ ..., rewriting equation (A.44) and following the procedure similar to equations (A.45) to (A.47).

A.5.1.3 Sphere and Circle

The equations and procedure for a circle in the x-z plane are obtained by omitting the y-coordinate in the calculations for a ball that is given here. A sphere can be described by:

$$(x-x_0)^2+(y-y_0)^2+(z-z_0)^2=R^2, \tag{A.48}$$

with $x_{0,}$ $y_{0,}$ z_0 the coordinates of the center and R the radius. This gives as a logical choice for the error function Q^2:

$$Q^2(x_0,y_0,z_0,R)=\sum_{i=1}^{n}\left(R-\sqrt{(x_i-x_0)^2+(y_i-y_0)^2+(z_i-z_0)^2}\right)^2. \tag{A.49}$$

For a chosen center point (x_0,y_0,z_0) and measured x, y and z coordinates (x_i,y_i,z_i), a radius R_i can be calculated:

$$(x_i-x_0)^2+(y_i-y_0)^2+(z_i-z_0)^2=R_i^2. \tag{A.50}$$

For a number of points (x_i,y_i,$z_{i,}$) the mean value of R_i will be the optimum choice for R. This means that in the minimizing of equation (A.49), problems with correlation may be avoided by defining R as the mean of the radii defined in equation (A.50) rather than making it one of the parameters to be optimized; for example, by defining the function in MATLAB, similar to the function defined in Section (A.3):

```
function Qsq=FunSphere(x,data);
xc=x(1); % x-coordinate of sphere center
yc=x(2); % y-coordinate of sphere center
yc=x(3); % z-coordinate of sphere center
xd=data(:,1)-xc;
yd=data(:,2)-yc;
zd=data(:,3)-zc;
radii=sqrt(xd.^2+yd.^2+zd.^2);
rm=mean(radii);
sqradii=(radii-rm).^2;
Qsq=sum(sqradii);
```

This minimization can be carried by numerical methods as discussed in Section A.3; then the value and uncertainty in R must be evaluated separately. A direct least-squares method can be formulated by rewriting equation (A.48) as:

$$x_i^2+y_i^2+z_i^2=2x_i\cdot x_0+2y_i\cdot y_0+2z_i\cdot z_0+R^2-x_0^2-y_0^2-z_0^2. \tag{A.51}$$

In matrix notation, equation (A.51) can be written as follows:

$$\begin{pmatrix} x_1^2 + y_1^2 + z_1^2 \\ \cdots \\ \cdots \\ x_n^2 + y_n^2 + z_n^2 \end{pmatrix} = \begin{pmatrix} 2x_1\ 2y_1\ 2z_1\ 1 \\ \cdots \\ \cdots \\ 2x_n\ 2y_n\ 2z_n\ 1 \end{pmatrix} \cdot \begin{pmatrix} x_0 \\ y_0 \\ z_0 \\ R^2 - x_0^2 - y_0^2 - z_0^2 \end{pmatrix}. \tag{A.52}$$

This is a general least-squares problem that can be solved with the pseudo-inverse, see equation (A.15) and Section A.5.

A MATLAB function might look like this:

```
function [x0,y0,z0,r]=spherefit(x,y,z);
w = [2*x(:) 2*y(:) 2*z(:) ones(size(x(:)))]\[x(:).^2+y(:).^2+z(:).^2];
x0=w(1);y0=w(2);z0=w(3);r=sqrt((w(4)+x0^2+y0^2+z0^2));
```

Here, the matrix to be inverted does depend on the measured values. Especially for small circle/sphere segments, the matrix 'w' may become difficult to invert, and one may get a more stable solution by minimizing the Q^2 function, equation (A.49), from an appropriate starting position. Several approaches for this can be found in the literature (Muralikrishnan and Raja 2009).

A.5.1.4 Minimum-Zone Evaluations

In the minimum-zone evaluation, the range of values (maximum to minimum) is minimized, rather than the squared sum of differences with the mean. In a numerical evaluation, the squared sum may be replaced by the range in the function to be minimized. The least-squares solution may be used as a starting point. For example, for obtaining the minimum-zone sphere center the function can be analogous to the least-squares sum in the previous section:

```
function Qmz=FunmzSphere(x,data);
xc=x(1); % x-coordinate of sphere center
yc=x(2); % y-coordinate of sphere center
yc=x(3); % z-coordinate of sphere center
xd=data(:,1)-xc;
yd=data(:,2)-yc;
zd=data(:,3)-zc;
radii=sqrt(xd.^2+yd.^2+zd.^2);
Qmz=range(radii);
```

Or for a minimum-zone plane:

```
function Qmz=Funmzp(a,x,y,z)
values=z-a(1)*x-a(2)*y;
Qmz=range(values);
```

Here the further evaluation can be carried out analogous to the procedure given in Section A.3. This approach may appear over-simplified and disregards potential major mathematical complications, such as local minima and discontinuities, but may work well in practical situations. More fundamental approaches can be found in the literature (Muralikrishnan and Raja 2009).

A.5.2 Surface Plate

The flatness of a surface plate is usually measured using electronic levels. A common configuration is a square grid with a pitch of 100 mm. Electronic levels measure the height difference between the nodes relative to the gravity vector, with an unknown bias, in both the x- and y direction. As an example, here the flatness measurement on a grid of 2 × 3 points is illustrated. Figure A.4 shows the measurement situation. An electronic level is used to measure the height differences between points on the grid $z(x,y)$. For example, $m_1 = z(1,0) - z(0,0)$. The vertex is defined as $z(0,0) = 0$. The remaining points $z(x,y)$ then depend on the measurements as follows:

$$\begin{pmatrix} 1 & 0 & 0 & 0 & 0 \\ -1 & 1 & 0 & 0 & 0 \\ 0 & 0 & -1 & 1 & 0 \\ 0 & 0 & 0 & -1 & 1 \\ 0 & 0 & 1 & 0 & 0 \\ -1 & 0 & 0 & 1 & 0 \\ 0 & -1 & 0 & 0 & 1 \end{pmatrix} \cdot \begin{pmatrix} z(1,0) \\ z(2,0) \\ z(0,1) \\ z(1,1) \\ z(2,1) \end{pmatrix} = \begin{pmatrix} m_1 \\ m_2 \\ m_3 \\ m_4 \\ m_5 \\ m_6 \\ m_7 \end{pmatrix}. \tag{A.53}$$

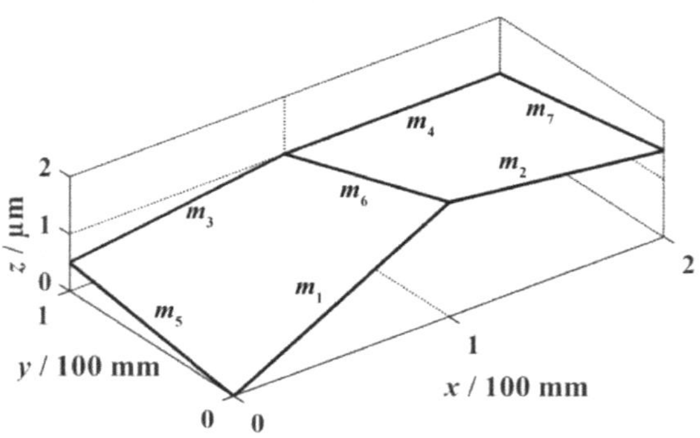

FIGURE A.4 Image of a surface plate showing angular measurements (height differences) m_i between points.

By taking the pseudo-inverse, see equation (A.15), the heights can be derived from the measurements:

$$\begin{pmatrix} z(1,0) \\ z(2,0) \\ z(0,1) \\ z(1,1) \\ z(2,1) \end{pmatrix} = \frac{1}{15}\begin{pmatrix} 11 & -1 & 4 & 1 & 4 & -3 & -1 \\ 10 & 10 & 5 & 5 & 5 & 0 & -5 \\ 4 & 1 & -4 & -1 & 11 & 3 & 1 \\ 8 & 2 & 7 & -2 & 7 & 6 & 2 \\ 9 & 6 & 6 & 9 & 6 & 3 & 6 \end{pmatrix} \cdot \begin{pmatrix} m_1 \\ m_2 \\ m_3 \\ m_4 \\ m_5 \\ m_6 \\ m_7 \end{pmatrix}. \quad \text{(A.54)}$$

This method can be found in the literature (Birch and Cox 1973). After calculating the height values, the heights relative to the least-squares plane can be calculated using equation (A.47). For larger surface plates, setting up the matrix as in equation (A.53) may be cumbersome, especially if there are points missing or if the configuration is not rectangular. In that case, an iterative approach to data evaluation may be considered based on the least-squares principle that the mean of values is the least-squares solution. For example, the point (1,1) in Figure A.4 is determined relative to its neighboring points in three ways for which the mean value can be taken:

$$z(1,1) = \frac{\left(z(0,1)+m_3\right)+\left(z(2,1)-m_4\right)+\left(z(1,0)+m_6\right)}{3}. \quad \text{(A.55)}$$

This contributes to the least-squares sum:

$$Q^2(1,1) = \left(z(0,1)+m_3-z(1,1)\right)^2+\left(z(2,1)-m_4-z(1,1)\right)^2+\left(z(1,0)+m_6-z(1,1)\right)^2. \quad \text{(A.56)}$$

This substitution will affect the other heights if these are defined similarly. Equation (A.55) and the similar equation for the other points can be repeated for all the points until the sum of the Q^2 contribution of all the points as in equation (A.56) stabilizes to a minimum value (Haitjema 1996). In case straightness measurements are taken that do not have a common reference direction, such as with autocollimators or laser interferometer angular optics on a base, diagonal measurements must be added, complicating the calculations. This is shown in the literature for surface plates (Meijer 1981) and for optical flats, calibrated with a set-up involving an autocollimator and penta-prism, see Figure 7.25 (Geckeler 2002). Note that this principle has a respectable history; see Figure 1.17.

A.5.3 Gage Block Measurement by Multiple Wavelength Interferometry

In Section 4.5, it was mentioned that the $Q^2(L)$ function for an interferometric gage block measurement of length L is given by the function:

$$Q^2(L) = \sum_{i=1}^{K}\left(L - \frac{\lambda_i}{2n_i}(N_i + f_i)\right)^2, \qquad \text{(A.57)}$$

where λ_i is the used vacuum wavelength, n_i is the the air refractive index, f_i is the measured fraction, and N is the integer number of wavelengths fitting in the length L:

$$N_i = round\left(\frac{2n_iL}{\lambda_i} - f_i\right), \qquad \text{(A.58)}$$

where *round* is a function that rounds a number to the nearest integer. As an illustration the Q^2 function is given Figure A.5 in the range 5 mm ± 1.5 µm for a measurement of a gage block with a nominal length of 5 mm, using the wavelengths and fraction measurements given in Table 4.2. The figure shows the minimum at the solution of L= 5 mm + 0.353 µm that was found in Section 4.5 with an arrow. It can be observed that, within the range of 5 mm ± 1 µm, this is the minimum with the lowest Q^2-value, however, the adjacent minima are not that strictly worse, and outside the interval there appears to be a better minimum at a length of L = 5 mm + 1.34 µm. This illustrates the importance for accurate measurements; in this case not only for an accurate determination at the minimum Q^2-value, but even more for determining the proper interference

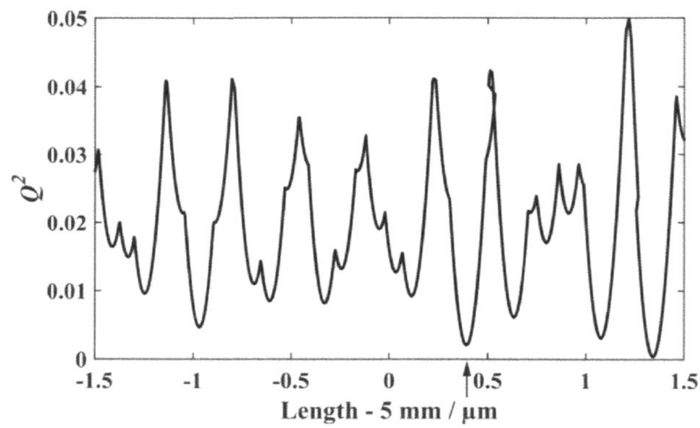

FIGURE A.5 The squared sum function Q^2 of an interferometric gage block measurement as a function of the length near the nominal length.

order. For a single wavelength, the Q^2-value reaches the value zero that is repeated every half wavelength. For more wavelengths, the solution becomes more unique, provided these measurements are taken with sufficient accuracy. The Q^2 function illustrates this.

A.5.4 Polygon Measurement Using Autocollimators at Multiple Angles

Assume that two autocollimators are at a fixed position and directed to two faces of a polygon. The polygon can be rotated around its center using a rotary table. Every polygon face is rotated such that Autocollimator 1 has a zero reading, then four readings ($M_1..M_4$) are taken from Autocollimator 2 that is first located at nominally 90°, where the polygon is rotated in four steps of 90° then another four readings ($M_5..M_8$) are taken from Autocollimator 2 that is located at nominally 180°, where the polygon is again rotated in four steps of 90°. This is illustrated in Figure A.6. With measurements in arcseconds, with a standard uncertainty of 1", the polygon angles α_i are related to the measurements M_i, similarly to equation (A.12), thus:

$$\begin{pmatrix} 0.75 & -0.25 & -0.25 & -0.25 \\ -0.25 & 0.75 & -0.25 & -0.25 \\ -0.25 & -0.25 & 0.75 & -0.25 \\ -0.25 & -0.25 & -0.25 & 0.75 \\ 0.5 & 0.5 & -0.5 & -0.5 \\ -0.5 & 0.5 & 0.5 & -0.5 \\ -0.5 & -0.5 & 0.5 & 0.5 \\ 0.5 & -0.5 & -0.5 & 0.5 \end{pmatrix} \begin{pmatrix} \alpha_1 - 90^\circ \\ \alpha_2 - 90^\circ \\ \alpha_3 - 90^\circ \\ \alpha_4 - 90^\circ \end{pmatrix} = \begin{pmatrix} M_1 \\ M_2 \\ M_3 \\ M_4 \\ M_5 \\ M_6 \\ M_7 \\ M_8 \end{pmatrix}. \tag{A.59}$$

Note that in equation (A.56), the unknown angle between the autocollimators is already eliminated, see also Chapter 7, Exercise 1. The least-squares solution for equation (A.59), see equation (A.15), is:

$$\begin{pmatrix} \alpha_1 - 90^\circ \\ \alpha_2 - 90^\circ \\ \alpha_3 - 90^\circ \\ \alpha_4 - 90^\circ \end{pmatrix} = \frac{1}{12} \begin{pmatrix} 5 & -3 & 1 & -3 & 2 & -2 & -2 & 2 \\ -3 & 5 & -3 & 1 & 2 & 2 & -2 & -2 \\ 1 & -3 & 5 & -3 & -2 & 2 & 2 & -2 \\ -3 & 1 & -3 & 5 & -2 & -2 & 2 & 2 \end{pmatrix} \begin{pmatrix} M_1 \\ M_2 \\ M_3 \\ M_4 \\ M_5 \\ M_6 \\ M_7 \\ M_8 \end{pmatrix} \tag{A.60}$$

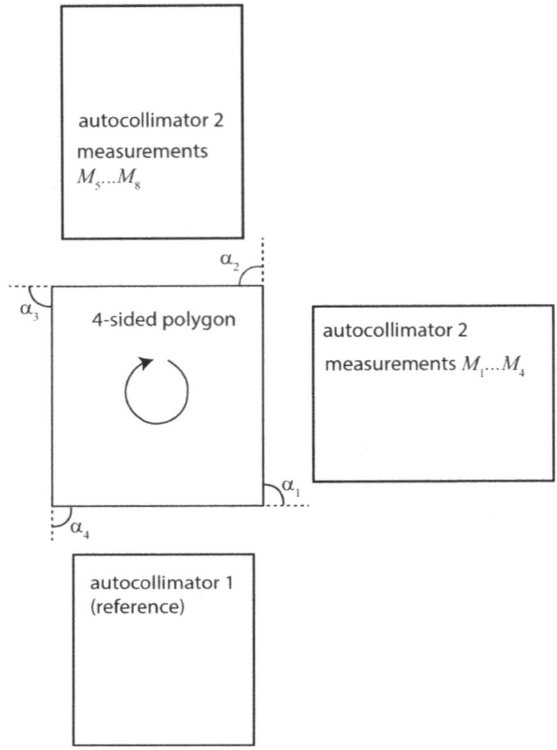

FIGURE A.6 Schematic of a set-up for the calibration of a four-sided polygon using two autocollimators.

with a variance-covariance matrix:

$$E_{int} = \frac{1}{12}\begin{pmatrix} 5 & -3 & 1 & -3 \\ -3 & 5 & -3 & 1 \\ 1 & -3 & 5 & -3 \\ -3 & 1 & -3 & 5 \end{pmatrix}. \tag{A.61}$$

The internal uncertainty in the angles α_i, based on $u(M_i) = 1''$, is $u(\alpha_i) = \sqrt{5/12}$=0.65". The uncertainties in combinations of angles, for example $\alpha_1+\alpha_2$, can be obtained from equation (A.61), taking the co-variances into account using equations (3.26) and (3.27), or by expressing $\alpha_1+\alpha_2$ in terms of the measurements M_i using equation (A.60). A thorough least-squares analysis for polygons with more sides and more autocollimator positions is given in the literature (Cook 1954). When all possible autocollimator positions are taken, and all possible measurements are taken, as was also carried out here for a four-sided polygon, this is called the "Rosette Method".

A.5.5 Multilateration

Multilateration is the principle that an absolute coordinate of an object can be obtained by measuring its distance to fixed points of which the coordinates are defined. The most prominent use of this principle is in the Global Positioning System, where from the distance of a number of satellites, a user can determine its position on earth using a mobile phone. However, in geodesy this principle is often used (Widdison and Long 2024), as well as in coordinate metrology (Linares et al. 2014). The principle of multilateration is shown in Figure A.7. Given n points $P_i(x_i,y_i,z_i)$ and the measured distances L_i to the point $N(x_0,y_0,z_0)$, the coordinates x_0,y_0,z_0 must be determined. In general, this method gives an overdetermined set of the equations where the least-squares solution gives the optimum coordinates x_0, y_0 and z_0. Omitting weighing factors and uncertainties, the Q^2-function can be written as:

$$Q^2(x,y,z) = \sum_{i=i}^{n} (x_0 - x_i)^2 + (y_0 - y_i)^2 + (z_0 - z_i)^2 - L_i^2. \tag{A.62}$$

Similar to the fitting of a sphere, the terms in equation (A.59) can be written as:

$$x_i^2 + y_i^2 + z_i^2 - L_i^2 = 2x_i \cdot x_0 + 2y_i \cdot y_0 + 2z_i \cdot z_0 - x_0^2 - y_0^2 - z_0^2. \tag{A.63}$$

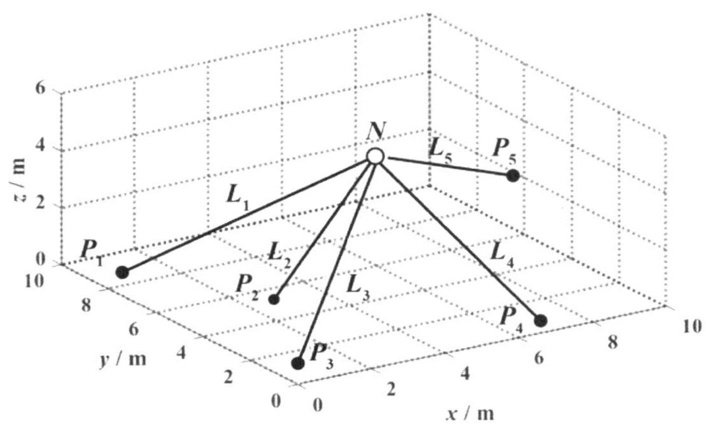

FIGURE A.7 Illustration of the principle of multilateration: from a number of defined points P_i, five in this example, the distances L_i to a point N are measured, defining the coordinates of point N.

This means that, similar to equation (A.52), a set of equations can be composed:

$$\begin{pmatrix} 1-2x_1 & -2y_1 & -2z_1 \\ & \cdots & \\ & \cdots & \\ 1-2x_n & -2y_n & -2z_n \end{pmatrix} \cdot \begin{pmatrix} x_0^2+y_0^2+z_0^2 \\ x_0 \\ y_0 \\ z_0 \end{pmatrix} = \begin{pmatrix} L_1^2 - x_1^2 - y_1^2 - z_1^2 \\ \cdots \\ \cdots \\ L_n^2 - x_n^2 - y_n^2 - z_n^2 \end{pmatrix}. \quad \text{(A.64)}$$

This set of equations can be solved by the pseudo-inverse, see equation (A.15) in Section A.5. However, the solution is not always well-defined and unique. Therefore, many strategies have been developed to minimize equation (A.62) in a stable and fast way. One strategy is the following: a reasonable first estimate of the point $N(x_0,y_0,z_0)$ is chosen. Then from every reference point P_i a new estimate (x_{0i},y_{0i},z_{0i}) is calculated by defining a point at a distance L_i from point P_i in the direction of $N(x_0,y_0,z_0)$. The mean coordinate $\left(\overline{x_{0i}},\overline{y_{0i}},\overline{z_{0i}}\right)$ of these n points (x_{0i},y_{0i},z_{0i}) is taken as the value of $N(x_0,y_0,z_0)$ in the next iteration and this procedure is repeated until the coordinates stabilize within some criterion. This principle was applied to obtain absolute 2D coordinates of a 2D grid from distance measurements in four directions (Haitjema 1996). Other strategies can be found in the literature (Widdison and Long 2024). If all distances L_i are given the same value, this results in a fit to a sphere of known diameter. Especially when fitting a partial sphere this may omit problems with the correlation between the radius and the sphere center, giving large uncertainties when using the method given in Section A.5.1.3.

A.5.6 Least-Squares Fit to an Ellipse: The Heydemann Correction

In interferometry, errors occur if the signals deviates from the ideal sine- and cosine forms given in Section 4.4.1. In a classical paper, Heydemann proposed a method for measuring and correcting for these deviations (Heydemann 1981) that is commonly called the Heydemann correction. This involves a least-squares fit to an ellipse that is described in this section.

As shown in Section 4.4.1, the interferometer signals for an ideal homodyne or heterodyne interferometer can be written as: $u_1 = R\cos\phi$ and $u_2 = R\sin\phi$; and the displacement ΔL for a wavelength λ, interpolated in-between the counts can be obtained from $\Delta L = \frac{4\pi}{\lambda}\tan^{-1}\frac{u_2}{u_1}$. However, in real systems the signals may have an offset, a mutually different amplification and the phase difference between the signals may be not exactly 90º. In that case, when plotting u_1 against u_2, no pure circle is obtained, but a shifted and tilted ellipse, described by the distorted signals: u_1^d and u_2^d given by:

$$u_1^d = u_1 + p \text{ and } u_2^d = \frac{1}{r}\left(u_2 \cos\alpha - u_1 \sin\alpha\right) + q. \tag{A.65}$$

The original circle $u_1^2 + u_2^2 = R^2$ is distorted to an ellipse given by:

$$\left(u_1^d - p\right)^2 + \left(\frac{\left(u_2^d - q\right)r + \left(u_1^d - p\right)\sin\alpha}{\cos\alpha}\right)^2 = R^2. \tag{A.66}$$

Here r is the amplification ratio between the two signals, p and q are the offsets of both signals and α is the phase shift from 90º between the channels. This can be rewritten in a general expression for an ellipse, where u_1^d and u_2^d satisfy the implicit equation:

$$A\left(u_1^d\right)^2 + B\left(u_2^d\right)^2 + Cu_1^d u_2^d + Du_1^d + Eu_2^d = 1, \tag{A.67}$$

where

$$\begin{aligned} &A = \left(R^2\cos^2\alpha - p^2 - r^2q^2 - 2rpq\sin\alpha\right)^{-1}, \\ &B = Ar^2,\ C = 2Ar\sin\alpha,\ D = -2A\left(p + rq\sin\alpha\right), \\ &E = -2A\left(rq + p\sin\alpha\right). \end{aligned} \tag{A.68}$$

A least-squares fit may be derived from n ($n \geq 5$) measured intensities $u_{11}^d \ldots u_{1n}^d$ and $u_{21}^d .. u_{2n}^d$ written as a matrix equation, thus:

$$\begin{pmatrix} \left(u_1^d\right)_1^2 & \left(u_2^d\right)_1^2 & \left(u_1^d\right)_1\left(u_2^d\right)_1 & \left(u_1^d\right)_1 & \left(u_2^d\right)_1 \\ \cdots & \cdots & \cdots\cdots & \cdots\cdots & \cdots\cdots \\ \left(u_1^d\right)_n^2 & \left(u_2^d\right)_n^2 & \left(u_1^d\right)_n\left(u_2^d\right)_n & \left(u_1^d\right)_n & \left(u_2^d\right)_n \end{pmatrix} \begin{pmatrix} A \\ B \\ C \\ D \\ E \end{pmatrix} = \begin{pmatrix} 1 \\ .. \\ 1 \end{pmatrix}. \tag{A.69}$$

The coefficients $A..E$ can be found from the pseudo-inverse, see equation (A.15) in Section A.5. The non-linearity parameters can be derived from the determined coefficients:

$$\alpha = \sin^{-1}\frac{C}{\sqrt{4AB}};\ r = \sqrt{\frac{B}{A}};\ p = \frac{2BD - EC}{C^2 - 4AB};\ q = \frac{2AE - DC}{C^2 - 4AB}. \tag{A.70}$$

This enables the correction towards 'ideal' signals u_1 and u_2 without bias, equal magnifications and with 90⁰ phase difference:

$$u_1 = u_1^d - p \text{ and } u_2 = \frac{1}{\cos\alpha}\left[\left(u_1^d - p\right)\sin\alpha + r\left(u_2^d - q\right)\right]. \quad \text{(A.71)}$$

From these corrected signals, the counting increases or decreases may be applied at the zero crossings and $\Delta L = \frac{4\pi}{\lambda}\tan^{-1}\frac{u_2}{u_1}$ gives the interpolated additional displacement, as described in Section 4.1.1. If the offsets are considerable compared to the signals u_1^d and u_1^d, the accuracy of the fit can be improved by deriving p and q from equations (A.69) and (A.70), in a first iteration, and applying equations (A.69) and (A.70) in a second iteration to both signals, corrected for the offsets only: $u_1^c = u_1^d - p$ and $u_2^c = u_2^d - q$ in order to obtain accurate values for r and α. Considering the plot of u_1^d against u_2^d as a roundness diagram, the offsets p and q are related to the first harmonic deviation ('decenter') and r and α are related to the second harmonic deviation: the ellipticity. In the displacement measurement, these deviations are reflected in deviations that have a sinusoidal character fitting once and twice respectively in the period of half a wavelength of displacement.

EXERCISES

1. Measurement of a square
   Of a square, the four corners are measured with a standard uncertainty of 0.5°. Result:
   $\phi_1 = 91°$, $\phi_2 = 89°$, $\phi_3 = 91°$, $\phi_4 = 90°$,
   a. Give the angles and their internal and external standard uncertainties.
   b. What is the reduced Chi-squared value and what can be concluded about this?
2. Fitting a circle segment
   Of a circle segment, five measured points (x,z) are given in Table A.6. The standard uncertainty in every z-value is 0.01 mm.
   a. What is the number of degrees of freedom?
   b. Calculate the radius R and the function Q^2 for three cases of the coordinates of the centre in millimeters: (x_c,z_c) = (0,200); (0,-2500); (0,1250);
   b. Calculate the centre (x_m,z_m) of the circle and its radius R by trying different values of (x_m,y_m) and minimizing Q^2.
   c. Determine the variance-covariance matrix (numerically).

TABLE A.6
Measured Coordinates on a Circle Segment

x/mm	−100	−50	0	50	100
z/mm	4.06	1.01	0.03	1.05	4.10

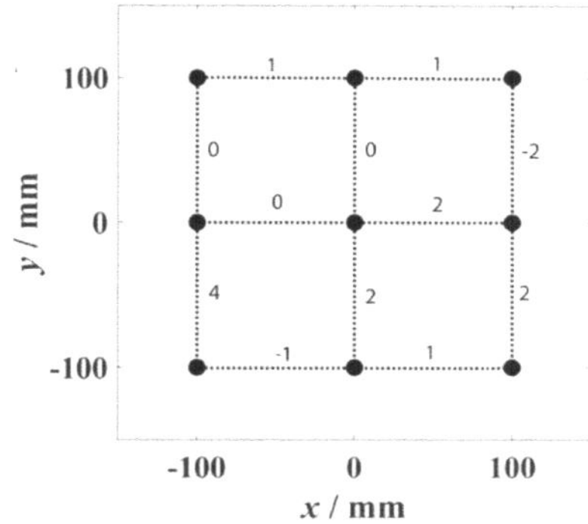

FIGURE A.8 Lay-out of inclination measurements in μrad and measured points on a 300 mm by 300 mm surface plate.

d. Evaluate the internal and external standard uncertainties in x_m, y_m and R
e. Fix the radius to a rounded nominal value close to the value of R found in d. Evaluate x_m and y_m and their standard uncertainty for this case.

3. Flatness measurement
A 300 mm square surface plate is measured at nine points over the central 200 mm by 200 mm using electronic levels with a pitch of 100 mm. The readings are in microradians, also expressed as μm/m. Figure A.8 depicts the measurement as seen from above with the measurement values indicated along the lines.
For example, the (positive) angle between points (-100,-100) and (-100,0) is 4 μrad.
a. Take the solution found in exercise 6, chapter 7 as a starting point.
b. Calculate the Q^2 value of this solution.
c. Adjust the points intuitively and calculate again the Q^2 value.
d. Calculate the z-height of the points from the least-squares solution, defining the point (-100,-100) as $z = 0$.
e. Estimate the standard uncertainty in the points from the Q^2 value considering the degrees of freedom.
f. Consider how much this uncertainty improved from the first estimation calculated in b.
4. Least-squares line
Make a least-squares fit to the data in the second and third column of Table A.3 without subtracting mean values. Note the correlation coefficient. Calculate the length and the uncertainty in the length at 0 °C and 20 °C.
5. Polygon measurement
Consider the polygon measurement discussed in Section A.5.4.
a. Calculate the standard uncertainty in the quantity $P=\alpha_1+\alpha_2$ using the variance-covariance matrix in equation (A.61).

b. Calculate the standard uncertainty in the quantity $P=\alpha_1+\alpha_2$ from the expression of $P(M_1..M_8)$ derived from equation (A.60).
c. Explain why P does not depend on M_6 and M_8.

6. Multilateration in 2D
In the x–y plane, the point to be calculated has positive (x_0,y_0) coordinates and has a distance of 6 mm from point (0,0), 9 mm from point (12,0) and 4 mm from point (4,0).
   a. Take the point (x_0,y_0)=(5,5) as a first estimate and calculate the Q^2 value.
   b. Minimize Q^2 iteratively/graphically.
   c. Make a 3D plot of $Q^2(x,y)$ in a useful range around its minimum.
   d. Calculate the minimum point, calculate the variance-covariance matrix, the standard uncertainty in the x- and y-coordinates and the correlation coefficient.
7. Nonlinearity in interferometric measurements
In a homodyne interferometer, the detector detecting the cosine-component has a bias of +5% of the mean value, the detector detecting the sine-component has a bias of –3% of the detected mean value, the difference in amplification is 10% and the phase error (of the second detector) is 10°.
The wavelength λ is 640 nm. Calculate the error when measuring a displacement of 0 nm, 40 nm, 80 nm and 160 nm, taking the point where the phase is zero as the reference.

REFERENCES

Bevington, P.R. and Robinson, D.K, 2003, *Data Reduction and Error Analysis for the Physical Sciences.* Boston, McGraw-Hill.

Birch, K.G. and Cox, M.G., 1973, *Calculation of the Flatness of Surfaces: A Least Squares Approach.* London, NPL MOM report 5.

Cook, A.H., 1954, The Calibration of Circular Scales and Precision Polygons. *Br. J. Appl. Phys.* **5**:367–371.

Forbes, A.B., 1991, Least-Squares Best-Fit Geometric Elements, NPL Report DITC 140/89, NPL, London, https://eprintspublications.npl.co.uk/5050/ accessed 2024-12-26.

Geckeler, R.D., 2002, Sub-nm Topography Measurements of Large Flats: An Ultra-Precise Flatness Standard for Semiconductor Industry. *Proc. SPIE* **4779**:1–12.

Haitjema, H., 1996, Iterative Solution of Least-Squares Problems Applied to Flatness and Grid Measurements. In: Ciarlini, et al. (ed.), *Advanced Mathematical Tools in Metrology II*, p. 171. London, World Scientific.

Heydemann, P.L.M., 1981, Determination and Correction of Quadrature Fringe Measurement Errors in Interferometers. *Appl. Opt.* **20(19):**3382–3384.

Linares, J.-M., Chaves-Jacob, J., Schwenke, H., Longstaff, A., Fletcher, S., Flore, J., Uhlmann, E. and Wintering J., 2014, Impact of Measurement Procedure When Error Mapping and Compensating a Small CNC Machine Using a Multilateration Laser Interferometer. *Precis. Eng.* **38(3)**:578–588.

Meijer, J. and de Bruijn, W., 1981, Determination of Flatness from Straightness Measurements and Characterization of the Surface by Four Parameters. *Precis. Eng*. **3**:17–22.

Muralikrishnan, B. and Raja, J., 2009, *Computational Surface and Roundness metrology*. Berlin, Springer.

Widdison, E. and Long, D.G., 2024, A Review of Linear Multilateration Techniques and Applications. *IEEE Access* **12**:26251–26266.

Index

Note: Figures are indicated by *italics*. Tables are indicated by **bold**.

I

K

L

M

N

O

P